Lecture Notes in Computer Science 3218

Commenced Publication in 1973
Founding and Former Series Editors:
Gerhard Goos, Juris Hartmanis, and Jan van Leeuwen

Jürgen Gerhard

Modular Algorithms in Symbolic Summation and Symbolic Integration

 Springer

Author

Jürgen Gerhard
Maplesoft
615 Kumpf Drive, Waterloo, ON, N2V 1K8, Canada
E-mail: Gerhard.Juergen@web.de

This work was accepted as PhD thesis on July 13, 2001, at

Fachbereich Mathematik und Informatik
Universität Paderborn
33095 Paderborn, Germany

Library of Congress Control Number: 2004115730

CR Subject Classification (1998): F.2.1, G.1, I.1

ISSN 0302-9743
ISBN 3-540-24061-6 Springer Berlin Heidelberg New York

Springer is a part of Springer Science+Business Media

springeronline.com

© Springer-Verlag Berlin Heidelberg 2004
Printed in Germany

Typesetting: Camera-ready by author, data conversion by Boller Mediendesign
Printed on acid-free paper   SPIN: 11362159   06/3142   5 4 3 2 1 0

To Herbert Gerhard (1921–1999)

Foreword

This work brings together two streams in computer algebra: symbolic integration and summation on the one hand, and fast algorithmics on the other hand.

In many algorithmically oriented areas of computer science, the *analysis of algorithms* – placed into the limelight by Don Knuth's talk at the 1970 ICM – provides a crystal-clear criterion for success. The researcher who designs an algorithm that is faster (asymptotically, in the worst case) than any previous method receives instant gratification: her result will be recognized as valuable. Alas, the downside is that such results come along quite infrequently, despite our best efforts.

An alternative evaluation method is to run a new algorithm on examples; this has its obvious problems, but is sometimes the best we can do. George Collins, one of the fathers of computer algebra and a great experimenter, wrote in 1969: "I think this demonstrates again that a simple analysis is often more revealing than a ream of empirical data (although both are important)."

Within computer algebra, some areas have traditionally followed the former methodology, notably some parts of polynomial algebra and linear algebra. Other areas, such as polynomial system solving, have not yet been amenable to this approach. The usual "input size" parameters of computer science seem inadequate, and although some natural "geometric" parameters have been identified (solution dimension, regularity), not all (potential) major progress can be expressed in this framework.

Symbolic integration and summation have been in a similar state. There are some algorithms with analyzed run time, but basically the mathematically oriented world of integration and summation and the computer science world of algorithm analysis did not have much to say to each other.

Gerhard's progress, presented in this work, is threefold:

- a clear framework for algorithm analysis with the appropriate parameters,
- the introduction of modular techniques into this area,
- almost optimal algorithms for the basic problems.

One might say that the first two steps are not new. Indeed, the basic algorithms and their parameters – in particular, the one called dispersion in Gerhard's work – have been around for a while, and modular algorithms are a staple of computer algebra. But their combination is novel and leads to new perspectives, the almost optimal methods among them.

A fundamental requirement in modular algorithms is that the (solution modulo p) of the problem equal the solution of the (problem modulo p). This is generally not valid for all p, and a first task is to find a nonzero integer "resultant" r so that the requirement is satisfied for all primes p not dividing r. Furthermore, r has to be "small", and one needs a bound on potential solutions, in order to limit the size and number of the primes p required. These tasks tend to be the major technical obstacles; the development of a modular algorithm is then usually straightforward. However, in order to achieve the truly efficient results of this work, one needs a thorough understanding of the relevant algorithmics, plus a lot of tricks and shortcuts.

The integration task is naturally defined via a limiting process, but the Old Masters like Leibniz, Bernoulli, Hermite, and Liouville already knew when to treat it as a symbolic problem. However, its formalization – mainly by Risch – in a purely algebraic setting successfully opened up perspectives for further progress. Now, modular differential calculus is useful in some contexts, and computer algebra researchers are aware of modular algorithms. But maybe the systematic approach as developed by Gerhard will also result in a paradigm shift in this field. If at all, this effect will not be visible at the "high end", where new problem areas are being tamed by algorithmic approaches, but rather at the "low end" of reasonably domesticated questions, where new efficient methods will bring larger and larger problems to their knees.

It was a pleasure to supervise Jürgen's Ph.D. thesis, presented here, and I am looking forward to the influence it may have on our science.

Paderborn, 9th June 2004   Joachim von zur Gathen

Preface

What fascinated me most about my research in symbolic integration and symbolic summation were not only the strong parallels between the two areas, but also the differences. The most notable non-analogy is the existence of a polynomial-time algorithm for rational integration, but not for rational summation, manifested by such simple examples as $1/(x^2 + mx)$, whose indefinite sum with respect to x has the denominator $x(x + 1)(x + 2) \cdots (x + m - 1)$ of exponential degree m, for all positive integers m. The fact that Moenck's (1977) straightforward adaption of Hermite's integration algorithm to rational summation is flawed, as discussed by Paule (1995), illustrates that the differences are intricate.

The idea for this research was born when Joachim von zur Gathen and I started the work on our textbook *Modern Computer Algebra* in 1997. Our goal was to give rigorous proofs and cost analyses for the fundamental algorithms in computer algebra. When we came to Chaps. 22 and 23, about symbolic integration and symbolic summation, we realized that although there is no shortage of algorithms, only few authors had given cost analyses for their methods or tried to tune them using standard techniques such as modular computation or asymptotically fast arithmetic. The pioneers in this respect are Horowitz (1971), who analyzed a modular Hermite integration algorithm in terms of word operations, and Yun (1977a), who gave an asymtotically fast algorithm in terms of arithmetic operations for the same problem. Chap. 6 in this book unites Horowitz's and Yun's approaches, resulting in two asymptotically fast and optimal modular Hermite integration algorithms. For modular hyperexponential integration and modular hypergeometric summation, this work gives the first complete cost analysis in terms of word operations.

Acknowledgements. I would like to thank:

My thesis advisor, Joachim von zur Gathen.

Katja Daubert, Michaela Huhn, Volker Strehl, and Luise Unger for their encouragement, without which this work probably would not have been finished.

My parents Johanna and Herbert, my sister Gisela, my brother Thomas, and their families for their love and their support.

My colleagues at Paderborn: the Research Group Algorithmic Mathematics, in particular Marianne Wehry, the MuPAD group, in particular Benno Fuchssteiner, and SciFace Software, in particular Oliver Kluge.

My scientific colleagues all over the world for advice and inspiring discussions: Peter Bürgisser, Frédéric Chyzak, Winfried Fakler, Mark Giesbrecht, Karl-Heinz Kiyek, Dirk Kussin, Uwe Nagel, Christian Nelius, Michael Nüsken, Walter Oevel, Peter Paule, Arne Storjohann, and Eugene Zima.

Waterloo, 23rd July 2004   Jürgen Gerhard

Table of Contents

List of Figures and Tables

List of Algorithms

1. Introduction

Modular algorithms are successfully employed in various areas of computer algebra, e.g., for factoring or computing greatest common divisors of polynomials or for solving systems of linear equations. They control the well-known phenomenon of intermediate expression swell, delay rational arithmetic until the very last step of the computation, and are often asymptotically faster than their non-modular counterparts.

At the time of writing, modular algorithms only rarely occur in computational differential and difference algebra. The goal of this work is to bring these worlds together. The main results are four modular algorithms for symbolic integration of rational and hyperexponential functions and symbolic summation of hypergeometric terms, together with a complete cost analysis. To our knowledge, this is the first time that a cost analysis is given at all in the case of hyperexponential integration and hypergeometric summation.

In the remainder of this introduction, we illustrate the main ideas with a rational integration example. The algorithm for integrating rational functions given in most undergraduate calculus textbooks decomposes the denominator into a product of linear – or at most quadratic – factors, performs a partial fraction decomposition, and then integrates the latter term by term. Consider the following rational function:

$$\frac{f}{g} = \frac{x^3 + 4x^2 + x - 1}{x^4 + x^3 - 4x^2 + x + 1} \in \mathbb{Q}(x) \,.$$

The irreducible factorization of the denominator polynomial g over the real numbers is

$$g = (x - 1)^2 \left(x - \frac{3 + \sqrt{5}}{2}\right)\left(x - \frac{3 - \sqrt{5}}{2}\right) \,,$$

and the partial fraction decomposition is

$$\frac{f}{g} = \frac{1}{(x-1)^2} + \frac{7}{5} \cdot \frac{1}{x-1} - \frac{1}{5} \cdot \frac{1}{x - \frac{3+\sqrt{5}}{2}} - \frac{1}{5} \cdot \frac{1}{x - \frac{3-\sqrt{5}}{2}} \,.$$

Thus the integral is

$$\int \frac{f}{g} = \left(-\frac{1}{x-1}\right) + \frac{7}{5} \ln(x - 1) - \frac{1}{5} \ln\left(x - \frac{3 + \sqrt{5}}{2}\right) - \frac{1}{5} \ln\left(x - \frac{3 - \sqrt{5}}{2}\right) \,.$$

J. Gerhard: Modular Algorithms, LNCS 3218, pp. 1-5, 2004.
© Springer-Verlag Berlin Heidelberg 2004

This method has some computational drawbacks: it requires the complete factorization of the denominator into irreducible factors, and it may involve computation with algebraic numbers. Therefore the symbolic integration algorithms that are implemented in modern computer algebra systems pursue a different approach, which goes back to Hermite (1872). The idea is to stick to rational arithmetic as long as possible. Hermite's algorithm computes a decomposition

$$\frac{f}{g} = \left(\frac{c}{d}\right)' + \frac{a}{b} , \tag{1.1}$$

where $a, b, c, d \in \mathbb{Q}[x]$ are such that the rational function a/b has only simple poles, i.e., the polynomial b has only simple roots if a and b are coprime. The remaining task of integrating a/b is handled by methods due to Rothstein (1976, 1977); Trager (1976); and Lazard & Rioboo (1990). In general, the latter algorithms cannot completely avoid computations with algebraic numbers, but they reduce them to a minimum.

Hermite's algorithm can be executed using arithmetic operations on polynomials in $\mathbb{Q}[x]$ only. Moreover, it does not require the complete factorization of the denominator, but only its *squarefree decomposition*

$$g = g_n^n \cdots g_2^2 g_1 , \tag{1.2}$$

with squarefree and pairwise coprime polynomials $g_n, \ldots, g_1 \in \mathbb{Q}[x]$. The squarefree decomposition splits the irreducible factors according to their multiplicities, but factors with the same multiplicity are not separated: g_i is the product of all distinct irreducible factors of multiplicity i in g. The g_i's can be computed by using essentially gcd computations in $\mathbb{Q}[x]$. In our example, the squarefree decomposition is

$$g = (x - 1)^2 (x^2 + 3x + 1),$$

so that $g_2 = x - 1$ and $g_1 = x^2 + 3x + 1$. Hermite's method first computes the partial fraction decomposition of f/g along this partial factorization of the denominator and then integrates term by term:

$$\begin{aligned}
\frac{f}{g} &= \frac{1}{(x-1)^2} + \frac{7}{5} \cdot \frac{1}{x-1} - \frac{1}{5} \cdot \frac{2x+3}{x^2+3x+1} \\
&= \left(-\frac{1}{x-1}\right)' + \frac{7}{5} \cdot \frac{1}{x-1} - \frac{1}{5} \cdot \frac{2x+3}{x^2+3x+1}
\end{aligned} \tag{1.3}$$

or equivalently,

$$\int \frac{f}{g} = -\frac{1}{x-1} + \int \frac{7}{5} \cdot \frac{1}{x-1} + \int -\frac{1}{5} \cdot \frac{2x+3}{x^2+3x+1} .$$

Computing the second remaining integral by the method of Lazard, Rioboo and Trager leads to the closed form

$$\int \frac{f}{g} = -\frac{1}{x-1} + \frac{7}{5} \ln(x-1) - \frac{1}{5} \ln(x^2+3x+1) ,$$

which does not involve any algebraic numbers.

In general, both the partial fraction decomposition and the term by term integration require arithmetic with rational numbers. With a modular approach, rational numbers show up only in the very last step of the computation. The idea of the small primes modular algorithm is to choose "sufficiently many" "lucky" prime numbers $p_1, p_2, \ldots \in \mathbb{N}$, to perform Hermite's algorithm modulo each p_i independently, and to reconstruct a, b, c, d from their modular images by the Chinese Remainder Algorithm and rational number reconstruction.

During the term by term integration, we need to divide by integers of absolute value at most $\max\{\deg f, \deg g\}$, so we require our primes to exceed this lower bound. In the example, the bound is 4, and we choose $p_1 = 19$ as our first prime. When we use symmetric representatives between -9 and 9, the image of f/g modulo 19 is just f/g. The squarefree decomposition of g modulo 19 is

$$g \equiv (x - 1)^2 \cdot (x^2 + 3x + 1) \bmod 19 .$$

(Note that $x^2 + 3x + 1 \equiv (x - 3)(x + 6) \bmod 19$ is reducible modulo 19, but the algorithm does not need this information.) We compute the partial fraction decomposition and integrate the first term:

$$
\begin{aligned}
\frac{f}{g} &\equiv \frac{1}{(x - 1)^2} + \frac{9}{x - 1} + \frac{-8x + 7}{x^2 + 3x + 1} \\
&= \left(-\frac{1}{x - 1} \right)' + \frac{9}{x - 1} + \frac{-8x + 7}{x^2 + 3x + 1} \bmod 19 . \quad (1.4)
\end{aligned}
$$

This, in fact, is the image modulo 19 of the decomposition (1.3).

Now let us take $p_2 = 5$. Again, we take symmetric representatives, this time between -2 and 2, and the image of f/g modulo 5 is

$$\frac{x^3 - x^2 + x - 1}{x^4 + x^3 + x^2 + x + 1} .$$

The squarefree decomposition of g modulo 5 is

$$g \equiv (x - 1)^4 \bmod 5 ,$$

and partial fraction decomposition and term by term integration yield

$$\frac{f}{g} \equiv \frac{2}{(x - 1)^3} + \frac{2}{(x - 1)^2} + \frac{1}{x - 1} = \left(-\frac{1}{(x - 1)^2} - \frac{2}{x - 1} \right)' + \frac{1}{x - 1} \bmod 5 .$$

This is *not* the image modulo 5 of the decomposition (1.3), so what went wrong? The reason is that the squarefree decomposition of g mod 5 is not the image modulo 5 of the squarefree decomposition of g: 1 is only a double root of g, but it is a quadruple root of g mod 5. We say that 5 is an "unlucky" prime with respect to Hermite integration.

Our next try is $p_3 = 7$. Then f/g modulo 7 is

$$\frac{x^3 - 3x^2 + x - 1}{x^4 + x^3 + 3x^2 + x + 1} .$$

The squarefree decomposition of g mod 7 is

$$g \equiv (x - 1)^2 \cdot (x^2 + 3x + 1) \bmod 7 ,$$

so 7 is a "lucky" prime. Again, we compute the partial fraction decomposition and do a term by term integration:

$$\frac{f}{g} \equiv \frac{1}{(x-1)^2} + \frac{x-2}{x^2+3x+1} = \left(-\frac{1}{x-1}\right)' + \frac{x-2}{x^2+3x+1} . \qquad (1.5)$$

Although there seems to be a term missing, this decomposition is the image modulo 7 of the decomposition (1.3); the missing term is to be interpreted as $0/(x-1)$.

When we have sufficiently many lucky primes, we reconstruct the result. We assume that we have computed the squarefree decomposition (1.2) of the denominator in advance; this can also be used to detect unlucky primes. (In fact, the squarefree decomposition can also be computed by a modular algorithm). If $\deg f < \deg g$, then the decomposition (1.1) is unique if we take $b = g_1 \cdots g_n$ and $d = g/b$ as denominators and stipulate that $\deg a < \deg b$ and $\deg c < \deg d$. Using the partial fraction decomposition of a/b, we therefore know that the decomposition (1.3) in our example, which we want to reconstruct, has the form

$$\frac{f}{g} = \left(\frac{c}{x-1}\right)' + \frac{a_1}{x-1} + \frac{a_2 x + a_3}{x^2 + 3x + 1} ,$$

with rational numbers c, a_1, a_2, a_3. From our two successful modular computations (1.4) and (1.5), we obtain the congruences

$$c \equiv -1 \bmod 19 , \quad a_1 \equiv 9 \bmod 19 , \quad a_2 \equiv -8 \bmod 19 , \quad a_3 \equiv 7 \bmod 19 ,$$
$$c \equiv -1 \bmod 7 , \quad a_1 \equiv 0 \bmod 7 , \quad a_2 \equiv 1 \bmod 7 , \quad a_3 \equiv -2 \bmod 7 .$$

With the Chinese Remainder Algorithm, we find

$$c \equiv -1 \bmod 133, \quad a_1 \equiv 28 \bmod 133, \quad a_2 \equiv -27 \bmod 133, \quad a_3 \equiv 26 \bmod 133 .$$

Finally, we apply rational number reconstruction with the bound $8 = \lfloor\sqrt{133/2}\rfloor$ on the absolute values of all numerators and denominators, and obtain the unique solution

$$c = -1, \quad a_1 = \frac{7}{5}, \quad a_2 = -\frac{2}{5}, \quad a_3 = -\frac{3}{5} .$$

Thus we have found the decomposition (1.3) by our modular algorithm.

The example shows two main tasks that usually have to be addressed when designing a modular algorithm:

- Determine an *a priori* bound on the size of the coefficients of the final result. This is necessary to determine the maximal number of required prime moduli in order to reconstruct the result correctly from its modular images.
- Find a criterion to recognize "unlucky" primes, and determine an *a priori* bound on their number. When we choose our moduli independently at random, then this provides a lower bound on the success probability of the resulting probabilistic algorithm.

We address these two tasks for all the modular algorithms presented in this book, and give rigorous correctness proofs and running time estimates, for both classical and asymptotically fast arithmetic. Often the latter estimates are – up to logarithmic factors – asymptotically optimal.

2. Overview

Differential Algebra and Symbolic Integration. The main objects of investigation in differential algebra are symbolic ordinary differential equations of the form $f(y^{(n)}, \ldots, y', y, x) = 0$, where n is the *order* of the differential equation, x is the independent variable, f is some "nice" function that can be described in algebraic terms, and y is a unary function in the independent variable x. Of particular interest are *linear* differential equations, where f is a linear function of its first $n + 1$ arguments. The general form of an nth order linear differential equation is

$$a_n y^{(n)} + \cdots + a_1 y' + a_0 y = g , \qquad (2.1)$$

where the *coefficients* $a_n, \ldots, a_0$, the *perturbation function* g, and the unknown function y are all unary functions of the independent variable x. It is convenient to rewrite (2.1) in terms of an nth order *linear differential operator* $L = a_n D^n + \cdots + a_1 D + a_0$, where D denotes the usual differential operator mapping a function y to its derivative y':

$$Ly = g .$$

The letter L is both an abbreviation for "linear" and a homage to Joseph Liouville (1809–1882; see Lützen 1990 for a mathematical biography), who may be regarded as the founder of the algebraic theory of differential equations.

Differential algebra studies the algebraic structure of such linear differential operators and their solution manifolds. In contrast to numerical analysis, the focus is on *exact* solutions that can be represented in a symbolic way. Usually, one restricts the coefficients, the perturbation function, and the unknown function to a specific subclass of functions, such as polynomials, rational functions, hyperexponential functions, elementary functions, or Liouvillian functions.

The algebraic theory of differential equations uses the notion of a *differential field*. This is a field F with a *derivation*, i.e., an additive function $' : F \longrightarrow F$ satisfying the Leibniz rule

$$(fg)' = fg' + f'g$$

for all $f, g \in F$. A *differential extension field* K of F is an extension field in the algebraic sense such that the restriction of the derivation on K to the subfield F coincides with $'$. Usually the derivation on K is denoted by $'$ as well. An element $f \in K$ is *hyperexponential over* F if $f' = a \cdot f$ for some $a \in F$. An element is *hyperexponential* if it is hyperexponential over the field $\mathbb{Q}(x)$ of univariate rational functions.

J. Gerhard: Modular Algorithms, LNCS 3218, pp. 7–25, 2004.
© Springer-Verlag Berlin Heidelberg 2004

For example, if $'$ is the usual derivation, then $f = \exp(x^2)$ and $f = \sqrt{x}$ are both hyperexponential, as shown by $\exp(x^2)' = 2x \cdot \exp(x^2)$ and $(\sqrt{x})' = \sqrt{x}/2x$. On the other hand, $f = \ln(x)$ is not hyperexponential, since $f'/f = 1/(x\ln(x))$ is not a rational function. Trivially, every element of F is hyperexponential over F.

In this work, we discuss the special case of *symbolic integration*, where we are given an element g in a differential field F and look for a closed form representation of $\int g$. More precisely, this is an *antiderivative*, or *indefinite integral*, f in F or in some differential extension field of F, such that $f' = g$, i.e., $y = f$ is a solution of the very special linear first order differential equation

$$Dy = g \ .$$

(In general, f belongs to a proper extension field of F.) We study algorithms for *rational integration*, where $F = \mathbb{Q}(x)$ and D is the usual derivative with $Dx = 1$ and the integral is always a rational function plus a sum of logarithms, and for *hyperexponential integration*, where we are looking for a hyperexponential antiderivative of a hyperexponential element over $\mathbb{Q}(x)$. Solving the latter problem also involves finding polynomial solutions of linear first order differential equations with polynomial coefficients, i.e., computing a $y \in \mathbb{Q}[x]$ that satisfies

$$(aD + b)y = c$$

for given $a, b, c \in \mathbb{Z}[x]$.

We also discuss algorithms for the related problem of *squarefree factorization* in $\mathbb{Z}[x]$. A nonzero polynomial $f \in \mathbb{Z}[x]$ is *squarefree* if it has no multiple complex roots. Given a polynomial $f \in \mathbb{Z}[x]$ of degree $n > 0$, squarefree factorization computes squarefree and pairwise coprime polynomials $f_1, \ldots, f_n \in \mathbb{Z}[x]$ such that

$$f = f_1 f_2^2 \cdots f_n^n \ ,$$

i.e., the roots of f_i are precisely the roots of f of multiplicity i. This is a subtask in algorithms for rational function integration.

Difference Algebra and Symbolic Summation. Difference algebra is the discrete analog of differential algebra, where the *difference operator* Δ plays the role of the differential operator D. The difference operator is defined by $(\Delta f)(x) = f(x+1) - f(x)$ for a unary function f. If E denotes the *shift operator* satisfying $(Ef)(x) = f(x+1)$ and I is the identity operator, then $\Delta = E - I$. A *linear difference equation* of order n has the form

$$b_n \Delta^n y + \cdots + b_1 \Delta y + b_0 = g \ ,$$

where the *coefficients* $b_n, \ldots, b_0$, the *perturbation function* g, and the unknown function y are unary functions in the independent variable x. Such difference equations occur naturally as discretizations of differential equations. Using the relation $\Delta = E - I$, each such difference equation can be equivalently written as a *recurrence equation*

$$a_n E^n y + \cdots + a_1 E y + a_0 = g \ ,$$

or, even shorter,

$$Ly = g \, ,$$

where L is the *linear difference operator* or *linear recurrence operator*

$$L = b_n \Delta^n + \cdots + b_1 \Delta + b_0 = a_n E^n + \cdots + a_1 E + a_0 \, .$$

The discrete analog of the notion of differential field is the *difference field*. This is a field F together with an automorphism E. A *difference extension field* of F is an algebraic extension field K with an automorphism extending E, which is usually also denoted by E. An element $f \in K$ is *hypergeometric over* F if $Ef = a \cdot f$ for some $a \in F$, and it is *hypergeometric* if $a \in \mathbb{Q}(x)$. For example, if E is the shift operator, then $f = \Gamma(x)$ and $f = 2^x$ are both hypergeometric, since $(E\Gamma)(x) = \Gamma(x + 1) = x \cdot \Gamma(x)$ and $(E2^x) = 2^{x+1} = 2 \cdot 2^x$. On the other hand, $f = 2^{x^2}$ is not hypergeometric, since the ratio $(Ef)/f = 2^{2x+1}$ is not a rational function. Trivially, every element of F is hypergeometric over F. The class of hypergeometric terms includes products of rational functions, factorials, binomial coefficients, and exponentials.

The discrete analog of symbolic integration is *symbolic summation*, where an element g in a difference field F is given and we look for an *antidifference*, or *indefinite sum*, f in F or some difference extension field of F. This is a solution for y of the special difference equation

$$\Delta y = (E - I)y = g \, .$$

The name "indefinite sum" comes from the following elementary fact. When E is the shift operator and f satisfies this difference equation, then

$$\sum_{0 \le k < n} g(k) = f(n) - f(0)$$

for all $n \in \mathbb{N}$, so f provides a closed form for the sum on the left hand side. This is the discrete analog of the fundamental theorem of calculus, which says that

$$\int_a^b g(x)dx = f(b) - f(a)$$

holds for all $a, b \in \mathbb{R}$ if f is an antiderivative of g.

In this work, we consider algorithms for *hypergeometric summation*, more precisely, for finding hypergeometric antidifferences of hypergeometric elements over $\mathbb{Q}(x)$, where E is the shift operator on $\mathbb{Q}(x)$. As in the differential case, this also involves the computation of a polynomial solution of a linear first order difference equation with polynomial coefficients, also known as *key equation*: given $a, b, c \in \mathbb{Z}[x]$, find a polynomial $y \in \mathbb{Q}[x]$ satisfying

$$(aE + b)y = c \, .$$

We also discuss the *greatest factorial factorization*, introduced by Paule (1995), which is the discrete analog of squarefree factorization. The goal is to write a polynomial $f \in \mathbb{Z}[x]$ of degree $n > 1$ as a product

$$f = f_1 f_2^{\underline{2}} \cdots f_n^{\underline{n}}$$

of *falling factorial powers* of polynomials $f_1, \ldots, f_n \in \mathbb{Z}[x]$ in a unique way. Here, the ith falling factorial power is defined as $g^{\underline{i}} = g \cdot (E^{-1} g) \cdots (E^{1-i} g)$ for $g \in \mathbb{Z}[x]$. To achieve uniqueness, additional conditions analogous to $f_1, \ldots, f_n$ being squarefree and pairwise coprime in the case of squarefree factorization are necessary; see Sect. 5.2.

Modular Algorithms. Exact computation with symbolic objects of arbitrary precision, such as integers, rational numbers, or multivariate polynomials, often faces the severe problem of *intermediate expression swell*: the coefficients of intermediate results of such a computation, and often also of the final result, tend to be very large and often involve fractions even if the input does not contain fractions (see, e.g., Chap. 6 in von zur Gathen & Gerhard 1999). Most algorithms normalize their intermediate results by rewriting the coefficients as fractions of two coprime polynomials or integers whenever possible. Experience shows that these normalizations make arithmetic with fractions computationally costly.

There is an important paradigm for overcoming both the problem of intermediate expression swell and arithmetic with fractions: *homomorphic imaging* or *modular computation* (see, e.g., Lipson 1981). The idea is to transform the original task, such as the computation of a greatest common divisor of two polynomials, the solution of a system of linear equations, or the factorization of a polynomial into a product of irreducible polynomials, into one or several tasks over coefficient domains where the two problems above do not exist, namely finite fields. Elements of a finite field can be represented as univariate polynomials of bounded degree with integer coefficients of bounded size, and the representation of the result of an arithmetic operation of two elements can be computed very efficiently. In this respect, symbolic computations with coefficients from a finite field are comparable to numerical computations with floating point numbers of a fixed precision. The transformation works by substituting values for indeterminates and reducing integral coefficients modulo prime numbers.

There are two main schemes of modular computation. The first one uses several moduli, i.e., evaluation points or prime numbers, independently performs the computation for each modulus, and reconstructs the result via the *Chinese Remainder Algorithm* or *interpolation*. Following Chap. 6 in von zur Gathen & Gerhard (1999), we call this the *small primes modular computation* scheme. The second scheme uses a single modulus, performs the computation for that modulus, and then lifts the result modulo powers of the modulus by techniques known as *Newton iteration* and *Hensel lifting*. We call this the *prime power modular computation* scheme. There are also mixed forms of both schemes.

Both modular computation schemes have the advantage that fractions occur only in the very last step of the computation, namely the reconstruction of the final result from its modular image(s), and that the size of the intermediate results never exceeds the size of the final result. Often modular algorithms are asymptotically faster than the corresponding non-modular variants. In the small primes scheme, the intuitive reason is that it is cheaper to solve many problems with "small" coefficients instead of one problem with "big" coefficients.

Modular techniques are successfully employed in many areas of symbolic computation and are implemented in many computer algebra systems. The most important applications have already been mentioned: polynomial gcd computation (Brown 1971), linear system solving (McClellan 1973; Moenck & Carter 1979; Dixon 1982), and polynomial factorization (Zassenhaus 1969; Loos 1983). However, up to now the modular paradigm appears not very often in algorithms for symbolic solutions of differential and difference equations, symbolic integration, and symbolic summation in the literature; some examples are Horowitz (1971) and Li & Nemes (1997). The goal of this work is to bring the two worlds closer together.

We only discuss modular algorithms for the basic case of integral coefficients and give asymptotic O-estimates for their running times in the worst case. It is usually not difficult to adapt our algorithms to problems where the coefficients are polynomials with coefficients in a finite field or an algebraic number field. However, with the exception of the case of univariate polynomials over a finite field, the analysis of such an adaption is usually more complicated. For reasons of practical efficiency, we choose our moduli as primes that fit into one machine word of our target computer whenever possible.

The algorithms that we discuss in this book work on polynomials. We mostly give two kinds of cost analysis: a high-level running time estimate in terms of arithmetic operations in the coefficient ring, and a refined estimate in terms of word operations in case the coefficients are integers. In principle, however, most of our modular algorithms can be easily adapted to other coefficient rings as well, in particular, to rings of (multivariate) polynomials. In the important case where the coefficients are univariate polynomials over an infinite field, we can choose linear polynomials as our prime moduli, and the cost analyses become much simpler than in the case of integer coefficients, due to the absence of carries.

At first thought, solving differential and difference equations modulo prime numbers may seem strange, at least in the differential case, since the concepts of derivative and integral are defined in terms of limits, which are analytical objects that do not make sense in a finite field. However, as stated above, the notion of derivative can be defined in a purely algebraic way, without any limit. Nevertheless, algorithms in differential algebra and difference algebra are often restricted to coefficient rings of characteristic zero. This is due to the fact that unexpected things may happen in positive characteristic: for example, the nonconstant polynomial x^p has derivative zero modulo a prime number p, and similarly the difference of the pth falling factorial power, namely $\Delta(x^{\underline{p}})$, vanishes modulo p. However, these phenomenons can only happen when the degree of the polynomials involved exceeds the characteristic. In our modular algorithms, we choose the primes so as to be larger than the degrees of all input polynomials, and then the algorithms for characteristic zero usually work literally modulo all prime moduli.

Often the dominant cost of a modular algorithm is the reconstruction of the final result from its modular images by Chinese remaindering and rational number reconstruction. We usually prove a worst case upper bound on the size of the coefficients of the output, and our algorithm uses as many single precision primes as to cover this guaranteed upper bound. Often our cost estimate with fast arithmetic agrees – up to logarithmic factors – with the bound on the output size. Then we say that our algorithm is *asymptotically optimal* up to logarithmic factors.

Moreover, most of our algorithms are *output sensitive*, in the following statical sense. If it is possible to improve the order of magnitude of the upper bound on the output size, then a better cost estimate, which is asymptotically optimal with respect to the improved bound, follows immediately.

Sometimes it is also possible to make our algorithms output sensitive in a dynamical sense, by choosing the primes in an adaptive fashion, checking correctness of the final result, and adding more primes in case of failure, say by doubling their number. This guarantees that for each individual input, never than twice as many primes as needed are used, at the expense of logarithmically many additional correctness checks.

2.1 Outline

Chap. 3 collects some technical results for later use and may be skipped at first reading.

In Chap. 4, we discuss and analyze several algorithms for polynomial basis conversion. The main applications are the conversion between the *monomial basis* $1, x, x^2, x^3, \ldots$, the *shifted monomial basis* $1, x - b, (x - b)^2, (x - b)^3, \ldots$, for some constant b, and the *falling factorial basis* $1, x, x^{\underline{2}}, x^{\underline{3}}, \ldots$, where $x^{\underline{i}} = x(x - 1)(x - 2) \cdots (x - i + 1)$ denotes the ith falling factorial, for all $i \in \mathbb{N}$. These conversions will be put to use in Chap. 9. We also present and analyze new modular variants of these methods.

In Chap. 5, we discuss and analyze new modular algorithms for squarefree factorization and its discrete analog, the greatest factorial factorization. We also discuss a new asymptotically fast algorithm for greatest factorial factorization, which is an adaption of Yun's (1976) squarefree factorization algorithm.

The main results in Chap. 6 through 10 are new modular algorithms, including a cost analysis in terms of word operations, for symbolic integration of rational functions in $\mathbb{Q}(x)$, hypergeometric summation over $\mathbb{Q}(x)$, and hyperexponential integration over $\mathbb{Q}(x)$. These algorithms and their analysis form the core of this book. To our knowledge, this is the first time that a complete cost estimate for any of these problems is given.

In Chap. 6, we discuss and analyze two new modular variants of Hermite's (1872) integration algorithm for rational functions in $\mathbb{Q}(x)$.

An important subtask in the algorithms for rational and hyperexponential integration and for hypergeometric summation, whose modular variants we discuss, is to compute the squarefree decomposition or all integral roots of a certain resultant. This problem is addressed in Chap. 7. Moreover, we give and analyze a new modular variant of the method by Lazard & Rioboo (1990) and Trager (unpublished; according to Bronstein (1997), Trager implemented the algorithm in SCRATCHPAD, but did not publish it) for the integration of rational functions with only simple poles. Together with the results from Chap. 6, we obtain a complete cost estimate for modular integration of rational functions in $\mathbb{Q}(x)$.

Gosper's (1978) algorithm for hypergeometric summation and its continuous analog, Almkvist & Zeilberger's (1990) algorithm for hyperexponential integration, each comprise essentially two steps, namely the computation of the denominator and of the numerator of a rational function that satisfies a certain linear first order difference or differential equation with rational coefficients, respectively. In Chap. 8, we employ the results from the previous chapter to obtain new modular algorithms for computing the denominator.

Chap. 9 tackles the second step of Gosper's algorithm and of Almkvist & Zeilberger's algorithm, namely the computation of the numerator. This amounts to computing a polynomial solution $y \in F[x]$ of a linear first order difference equation $(aE + b)y = c$ or differential equation $(aD + b)y = c$ with given polynomial coefficients $a, b, c \in F[x]$, where F is an arbitrary field. We discuss and analyze six algorithms for the latter problem, among them a new asymptotically fast variant of Newton's method of indeterminate coefficients. Moreover, we present and analyze new modular variants of these algorithms.

Finally, Chap. 10 collects the results of the preceding two chapters and gives a complete cost analysis of our modular variants of Gosper's algorithm for hypergeometric summation over $\mathbb{Q}(x)$ and Almkvist & Zeilberger's algorithm for hyperexponential integration over $\mathbb{Q}(x)$.

Fig. 2.1 illustrates the dependencies between the various algorithms.

2.2 Statement of Main Results

Chap. 4 discusses algorithms for polynomial basis conversion. The conversion between the monomial basis and the shifted monomial basis is known as *Taylor shift*, since it amounts to computing the coefficients of the Taylor expansion of a given polynomial f around the point b, which in turn is equivalent to computing the coefficients of the "shifted" polynomial $f(x + b)$ with respect to the monomial basis. In Sect. 4.1, which follows closely von zur Gathen & Gerhard (1997), we discuss six known algorithms for the Taylor shift (due to Horner 1819; Shaw & Traub 1974; Paterson & Stockmeyer 1973; von zur Gathen 1990; and Aho, Steiglitz & Ullman 1975) and analyze them in terms of arithmetic operations in the coefficient ring, and also in terms of word operations if the coefficients are integers. It turns out that the asymptotically fastest method, due to Aho, Steiglitz & Ullman (1975), takes $O(\mathsf{M}(n))$ arithmetic operations for polynomials of degree n

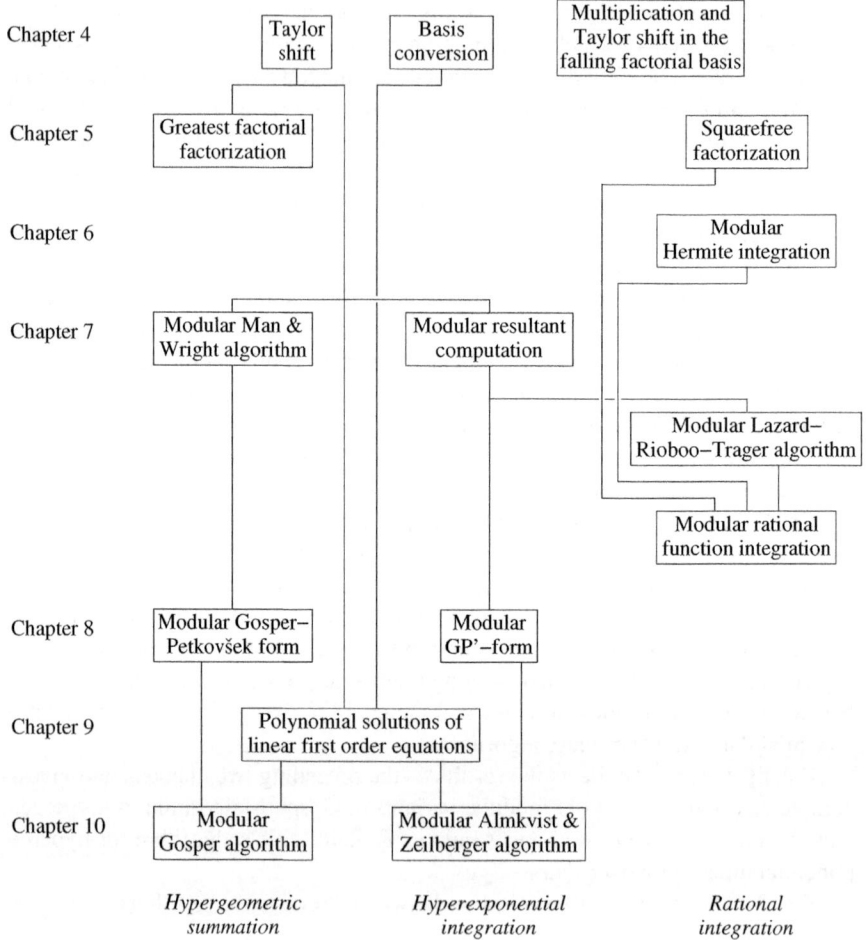

Fig. 2.1. Algorithm dependency graph

(Theorem 4.5). (Here and in what follows, M denotes a multiplication time for integers and polynomials, such that two polynomials of degrees at most n or two integers of length at most n can be multiplied with $O(M(n))$ arithmetic operations or word operations, respectively. Classical arithmetic has $M(n) = n^2$, and the asymptotically fastest of the currently known algorithms, by Schönhage & Strassen (1971), yields $M(n) = n \log n \log\log n$. See also Sect. 8.3 in von zur Gathen & Gerhard 1999.) When the coefficients and b are integers bounded by 2^λ in absolute value, then we obtain a cost estimate of $O(M(n^2(\lambda + \log n)))$ word operations for Aho, Steiglitz & Ullman's algorithm with fast arithmetic (Theorem 4.5). Moreover, we present and analyze new small primes modular algorithms for the Taylor shift (Algorithm 4.7). The modular variant of Aho, Steiglitz & Ullman's algorithm takes $O(n M(n\lambda) \log(n\lambda))$ word operations with fast arithmetic (Theorem 4.8). Both the non-modular algorithm of Aho, Steiglitz & Ullman and our new modular variant are – up to logarithmic factors – asymptotically optimal, since the output size for the Taylor shift is $\Theta(n^2\lambda)$ machine words. We have implemented various Taylor shift algorithms and give tables of running time experiments.

In Sect. 4.2, we discuss and analyze two variants of Horner's rule for converting polynomials from the monomial basis to a common generalization of the shifted monomial basis and the falling factorial basis, and also vice versa. The asymptotically fast variant seems to be new, but is closely related to the methods of Borodin & Moenck (1974) (see also Strassen 1973, and Strassen 1974 and §4.5 in Borodin & Munro (1975) for a survey) for evaluation of a polynomial at many points and for interpolation. With fast arithmetic, the cost is $O(\mathsf{M}(n) \log n)$ arithmetic operations for polynomials of degree n (Theorems 4.15 and 4.16), so the estimate for the general case is slower by a factor of $\log n$ than the estimate for the special case of a Taylor shift. We also show that the conversion between the monomial basis and the falling factorial basis takes $O(\mathsf{M}(n^2 \log n + n\lambda) \log n)$ word operations if the integer coefficients are bounded by 2^λ in absolute value (Corollary 4.17). Moreover, we present and analyze new small primes modular algorithms for this basis conversion (Algorithms 4.18 and 4.19), taking $O(n\,\mathsf{M}(n \log n + \lambda) \log(n + \lambda) + \lambda\,\mathsf{M}(n) \log n)$ word operations with fast arithmetic (Corollary 4.22). Both the non-modular algorithms and their modular variants with fast arithmetic are – up to logarithmic factors – asymptotically optimal for those inputs where the upper bound $O(n^2 \log n + n\lambda)$ on the output size is reached.

Sect. 4.3 discusses asymptotically fast algorithms for multiplication and for Taylor shift of polynomials, when both the input and the output are represented with respect to the falling factorial basis. These algorithms are not needed later, but may be of independent interest. In principle, both problems can be solved by converting the input polynomial(s) into the monomial basis, applying the corresponding algorithm for this representation, and converting the result again into the falling factorial basis. However, the cost for this approach for polynomials of degree at most n is dominated by the $O(\mathsf{M}(n) \log n)$ arithmetic operations for the two basis conversions, while our algorithms take only $O(\mathsf{M}(n))$ operations (Theorems 4.27 and 4.28). The multiplication algorithm seems to be new, and the algorithm for the Taylor shift is an adaption of Aho, Steiglitz & Ullman's algorithm for the monomial basis. The material of Sect. 4.2 and 4.3 first appeared in Gerhard (2000).

In Chap. 5, we discuss and analyze new modular algorithms for squarefree factorization and greatest factorial factorization. The small primes modular squarefree factorization algorithm 5.6 in Sect. 5.1, from Gerhard (2001), is a modular variant of Yun's (1976) method, which takes $O(\mathsf{M}(n) \log n)$ arithmetic operations for a polynomial of degree n and is the asymptotically fastest of the currently known squarefree factorization algorithms. We show that our modular variant uses

$$O(n\,\mathsf{M}(n + \lambda) \log(n + \lambda) + \lambda\,\mathsf{M}(n) \log n)$$

word operations with fast arithmetic if the polynomial has coefficients bounded by 2^λ in absolute value (Theorem 5.10). We also analyze a prime power modular algorithm due to Yun (1976) (Algorithm 5.12), and show that it takes

$$O((\mathsf{M}(n)\log n + n\log \lambda)\mathsf{M}(n + \lambda))$$

word operations with fast arithmetic (Theorem 5.14). In the diagonal case where $n \approx \lambda$, both algorithms are – up to logarithmic factors – asymptotically optimal, since then the input size is $\Theta(n^2)$ machine words.

We discuss and analyze a new algorithm for the greatest factorial factorization in Sect. 5.2 (Algorithm 5.20), which is an adaption of Yun's method, and show that it also takes $O(\mathsf{M}(n)\log n)$ arithmetic operations with fast arithmetic (Theorem 5.21). This is faster by a factor of n than the algorithm given by Paule (1995). Moreover, we present and analyze a small primes modular variant of this algorithm (Algorithm 5.24) and show that it takes $O(n\,\mathsf{M}(n + \lambda)\log(n + \lambda) + \lambda\,\mathsf{M}(n)\log n)$ word operations with fast arithmetic (Theorem 5.28). This is the same estimate as for our small primes modular algorithm for the squarefree factorization. Again, this algorithm is – up to logarithmic factors – asymptotically optimal in the diagonal case where $n \approx \lambda$ and the input size is $\Theta(n^2)$ machine words. The presentation in this section follows Gerhard (2000).

In Chap. 6, we discuss and analyze two new modular variants of Hermite's (1872) integration algorithm for rational functions in $\mathbb{Q}(x)$ (Algorithms 6.4 and 6.9). Given two nonzero coprime polynomials $f, g \in \mathbb{Z}[x]$ with $n = \deg g \geq 1$, plus a squarefree decomposition $g = g_1 g_2^2 \cdots g_n^n$, with all $g_i \in \mathbb{Z}[x]$ nonzero, squarefree, and pairwise coprime, this algorithm computes polynomials h, c_{ij}, and a_i in $\mathbb{Q}[x]$ for $1 \leq j < i \leq n$ such that

$$\frac{f}{g} = h' + \sum_{1 \leq j < i \leq n}\left(\frac{c_{ij}}{g_i^j}\right)' + \sum_{1 \leq i \leq n}\frac{a_i}{g_i}.$$

Thus it reduces the problem of integrating an arbitrary rational function to the integration of rational functions with squarefree denominator. If $n \leq m$, $\deg f < m$, and the coefficients of f and g are absolutely less than 2^λ, then our modular algorithms take $O(m^3(n^2 + \log^2 m + \lambda^2))$ word operations with classical arithmetic. The cost with fast arithmetic is

$$O(m\,\mathsf{M}(m(n + \log m + \lambda))\log(m(n + \lambda)))\ \text{or}\ O^\sim(m^2(n + \lambda))$$

for the small primes modular variant and $O^\sim(m^2(n^2 + \lambda^2))$ for the prime power modular variant, where the $O^\sim$ notation suppresses logarithmic factors (Corollary 6.7 and Theorem 6.10). The small primes modular algorithm is from Gerhard (2001). Horowitz (1971) has also given a small primes modular algorithm, based on linear algebra. Our estimate for classical arithmetic is better by about two orders of magnitude than the estimate of Horowitz. Our estimate for the small primes modular algorithm with fast arithmetic is – up to logarithmic factors – asymptotically optimal for those inputs where the upper bound $O(m^2(n + \log m + \lambda))$ on the output size is reached. The prime power modular variant with fast arithmetic is slower by about one order of magnitude. We also report on an implementation of both modular algorithms and give some running time experiments in Sect. 6.3.

Chap. 7 contains algorithms for computing integral roots of certain resultants. In Sect. 7.1 and 7.2, we discuss two quite different modular approaches to compute all integral roots of the resultant $r = \text{res}_x(f(x), g(x+y)) \in \mathbb{Z}[y]$ for given nonconstant polynomials $f, g \in \mathbb{Z}[x]$. This is major subtask for hypergeometric summation.

The idea of the first approach, which seems to be new, is as follows. The roots of r are precisely the differences of the roots of f and the roots of g, and they are considerably smaller than the coefficients of r. In order to compute all integral roots of r, it would therefore be too costly to compute its coefficients, and we compute them only modulo a prime power that is just slightly larger than the roots. This leads to a new probabilistic algorithm of Monte Carlo type (Algorithm 7.12), i.e., it may return a wrong result, with small probability. We show that for polynomials of degree at most n with integer coefficients absolutely bounded by 2^λ, the cost is $O(n^4(\lambda^2 + \log n))$ word operations with classical arithmetic and

$$O((n^2 \, \mathsf{M}(n) + \mathsf{M}(n^2) \log n)(\log n)\mathsf{M}(\lambda) \log \lambda) \text{ or } O^\sim(n^3\lambda)$$

with fast arithmetic (Theorem 7.18).

The second approach is a modular variant of an algorithm by Man & Wright (1994), which does not compute the resultant r at all, but instead computes the irreducible factorizations of f and g. Our modular variant only computes the irreducible factorizations modulo a sufficiently large prime power (Algorithm 7.20). We show that it takes $O(n^3\lambda + n^2\lambda^2)$ word operations with classical arithmetic and $O^\sim(n^2\lambda)$ with fast arithmetic (Theorem 7.21). These estimates are faster than the corresponding estimates for the first approach by at least one order of magnitude. This method also appears in Gerhard, Giesbrecht, Storjohann & Zima (2003).

In Sect. 7.3, we analyze a small primes modular algorithm for computing all integral roots of the resultant $r = \text{res}_x(g, f - yg') \in \mathbb{Z}[y]$ for given nonzero polynomials $f, g \in \mathbb{Z}[x]$ (Algorithm 7.25). This is a subtask for hyperexponential integration. In this case, the roots of r have about the same length as its coefficients in the worst case, and we therefore compute the coefficients of r exactly by a modular approach. We show that for polynomials of degree at most n with integer coefficients absolutely bounded by 2^λ, the cost is $O(n^4(\lambda^2 + \log^2 n))$ word operations with classical arithmetic and $O^\sim(n^3\lambda)$ with fast arithmetic (Theorem 7.26).

Sect. 7.4 contains a new small primes modular variant of the algorithm of Lazard, Rioboo and Trager for integrating a rational function $f/g \in \mathbb{Q}(x)$ with only simple poles, where $f, g \in \mathbb{Z}[x]$ are nonzero coprime and g is squarefree (Algorithm 7.29). If the degrees of f and g are at most n and their coefficients are bounded by 2^λ in absolute value, then our algorithm takes $O(n^4(\lambda^2 + \log n))$ word operations with classical arithmetic and

$$O(n^2 \, \mathsf{M}(n(\lambda + \log n)) \log(n\lambda)) \text{ or } O^\sim(n^3\lambda)$$

with fast arithmetic (Theorem 7.31). The output size is $O(n^3(\lambda + \log n))$ machine words in the worst case, and hence our modular algorithm with fast arithmetic is – up to logarithmic factors – asymptotically optimal for those inputs where this upper bound is reached. Together with the results from Chap. 6, we obtain a small primes modular algorithm for integrating a rational function in $\mathbb{Q}(x)$ (Theorem 7.32):

Theorem. *Let* $f, g \in \mathbb{Z}[x]$ *be nonzero polynomials with* $\deg f, \deg g \leq n$ *and* $\|f\|_\infty, \|g\|_\infty < 2^\lambda$. *We can compute a symbolic integral of* f/g *using*

$$O(n^8 + n^6 \lambda^2)$$

word operations with classical arithmetic and

$$O(n^2 \, \mathsf{M}(n^3 + n^2 \lambda) \log(n\lambda)) \text{ or } O^\sim(n^5 + n^4 \lambda)$$

with fast arithmetic.

A cost estimate of a non-modular variant of this algorithm in terms of arithmetic operations also appears in the 2003 edition of von zur Gathen & Gerhard (1999), Theorem 22.11.

In Chap. 8, we discuss modular variants of the first step of the algorithms of Gosper (1978) for hypergeometric summation and of Almkvist & Zeilberger (1990) for hyperexponential integration. In each case, the first step can be rephrased as the computation of a certain "normal form" of a rational function in $\mathbb{Q}(x)$. In the difference case, this is the *Gosper-Petkovšek form*: Gosper (1978) and Petkovšek (1992) showed that for a field F of characteristic zero and nonzero polynomials $f, g \in F[x]$, there exist unique nonzero polynomials $a, b, c \in F[x]$ such that b, c are monic and

$$\frac{f}{g} = \frac{a}{b} \cdot \frac{Ec}{c}, \quad \gcd(a, E^i b) = 1 \text{ for } i \in \mathbb{N}, \quad \gcd(a, c) = \gcd(b, Ec) = 1 .$$

We discuss and analyze two new small primes modular variants of Gosper's and Petkovšek's algorithm (Algorithm 8.2) for computing this normal form in the case $F = \mathbb{Q}$. Our algorithms employ the methods from Sect. 7.2 to compute all integral roots of the resultant $\mathrm{res}_x(f(x), g(x + y))$. It is well-known that the degree of the polynomial c in general is exponential in the size of the coefficients of f and g. For example, if $f = x$ and $g = x - e$ for some positive integer e, then the Gosper-Petkovšek form is $(a, b, c) = (1, 1, (x - 1)^e)$, and $e = \deg c$ is exponential in the word size $\Theta(\log e)$ of the coefficients of f and g. Thus there is no polynomial time algorithm for computing this normal form. However, if we do not explicitly need the coefficients of c in the usual dense representation with respect to the monomial basis, but are satisfied with a product representation of the form

$$c = h_1^{z_1} \cdots h_t^{z_1} ,$$

where $h_1, \ldots, h_1 \in \mathbb{Q}(x)$ are monic and $z_1, \ldots, z_t$ are nonnegative integers, then it is possible to compute a, b, and c in polynomial time. If $\deg f, \deg g \leq n$ and the coefficients of f and g are absolutely bounded by 2^λ, then both of our two modular algorithms take $O(n^4 + n^2 \lambda^2)$ word operations with classical arithmetic, and the faster variant uses

$$O(n^2 \, \mathsf{M}(n + \lambda) \log(n + \lambda) + n\lambda \, \mathsf{M}(n) \log n + \mathsf{M}(n^2) \log n)$$

or $O^{\sim}(n^2(n+\lambda))$ word operations with fast arithmetic (Theorem 8.18). The additional cost for explicitly computing the coefficients of c is polynomial in n, λ, and the *dispersion*

$$e = \mathrm{dis}(f, g) = \max\{i \in \mathbb{N}: i = 0 \text{ or } \mathrm{res}(f, E^i g) = 0\}$$

of f and g, which was introduced by Abramov (1971). If nonzero, the dispersion is the maximal positive integer distance between a root of f and a root of g. Using a standard small primes modular approach, we show that the cost is $O(e^3(n^3 + n\lambda)^2)$ word operations with classical arithmetic and

$$O((en\,\mathsf{M}(e(n+\lambda)) + e\lambda\,\mathsf{M}(en))\log(e(n+\lambda)))$$

or $O^{\sim}(e^2 n(n+\lambda))$ with fast arithmetic. The latter estimate is – up to logarithmic factors – asymptotically optimal for those inputs where the upper bounds $O(e^2 n(n+\lambda))$ on the size of c are achieved (Theorem 8.18). As a corollary, we obtain an algorithm for computing rational solutions of homogeneous linear first order difference equations with polynomial coefficients within the same time bound (Corollary 8.19).

In Sect. 8.1, we introduce the continuous analog of the Gosper-Petkovšek form, which we call the *GP'-form*. Given nonzero polynomials $f, g \in F[x]$, where F is a field of characteristic zero, we show that there exist unique polynomials $a, b, c \in F[x]$ such that b, c are nonzero monic and

$$\frac{f}{g} = \frac{a}{b} + \frac{c'}{c}, \quad \gcd(b, a - ib') = 1 \text{ for } i \in \mathbb{N}, \quad \gcd(b, c) = 1$$

(Lemma 8.23). Moreover, we discuss two new modular algorithms for computing this normal form when $F = \mathbb{Q}$. As in the difference case, the degree of c is exponential in the size of the coefficients of f and g in general. For example, if $f = e$ and $g = x$ for a positive integer e, the GP'-form is $(a, b, c) = (0, 1, x^e)$. However, we can compute a, b and a product representation of c of the form

$$c = h_1^{z_1} \cdots h_t^{z_t}$$

in polynomial time. If $\deg f, \deg g \le n$ and the coefficients of f and g are less than 2^λ in absolute value, then both our modular algorithms take $O(n^4(\lambda^2 + \log^2 n))$ word operations with classical arithmetic and $O^{\sim}(n^3 \lambda)$ with fast arithmetic (Theorem 8.42). The additional cost for explicitly computing the coefficients of c is polynomial in n, λ, and the continuous analog of the dispersion

$$e = \varepsilon(f, g) = \max\{i \in \mathbb{N}: i = 0 \text{ or } \mathrm{res}(g, f - ig') = 0\}\,.$$

If nonzero, $\varepsilon(f, g)$ is the maximal positive integer residue of the rational function f/g at a simple pole. Using a small primes modular approach, we obtain the same cost estimates for computing the coefficients of c and the same bounds on the output size as in the difference case, and the estimate for fast arithmetic is – up to logarithmic factors – asymptotically optimal for those inputs where the upper bounds on the output size are achieved (Theorem 8.42). As in the difference case, we also obtain an algorithm computing rational solutions of homogeneous linear first order differential equations with polynomial coefficients within the same time bound (Corollary 8.43).

Chap. 9 tackles the second step of Gosper's algorithm and of Almkvist & Zeilberger's algorithm. It amounts to computing a polynomial solution $y \in F[x]$ of a linear first order difference equation $(aE + b)y = c$ or differential equation $(aD + b)y = c$ with given polynomial coefficients $a, b, c \in F[x]$, where F is an arbitrary field. We discuss and analyze six algorithms for the latter problem: Newton's well-known method of undetermined coefficients (Sect. 9.1) together with a new asymptotically fast divide-and-conquer variant (Sect. 9.5 and 9.6), taken from von zur Gathen & Gerhard (1997), and the algorithms of Brent & Kung (1978) (Sect. 9.2); Rothstein (1976) (Sect. 9.3); Abramov, Bronstein & Petkovšek (1995) (Sect. 9.4); and Barkatou (1999) (Sect. 9.7). It turns out that the cost for all algorithms in terms of arithmetic operations is essentially the same, namely quadratic in the sum of the input size and the output size when using classical arithmetic, and softly linear with fast arithmetic. More precisely, if $\deg a, \deg b, \deg c \leq n$ and $\deg y = d$, then the cost for the algorithms employing classical arithmetic is $O(n^2 + d^2)$ arithmetic operations, and the asymptotically fast algorithms take $O(\mathsf{M}(n+d) \log(n+d))$. There are, however, minor differences, shown in Table 9.2.

In Sect. 9.8, we discuss new small primes modular algorithms in the case $F = \mathbb{Q}$. Under the assumptions that $d \in O(n)$ and the coefficients of a, b, c are bounded by 2^λ in absolute value, the cost estimates in the differential case are $O(n^3(\lambda^2 + \log^2 n))$ word operations with classical arithmetic and

$$O(n \, \mathsf{M}(n(\lambda + \log n)) \log(n\lambda)) \text{ or } O^\sim(n^2\lambda)$$

with fast arithmetic (Corollaries 9.59 and 9.69). The cost estimates in the difference case are $O(n^3(\lambda^2 + n^2))$ word operations with classical arithmetic and $O(n \, \mathsf{M}(n(\lambda+n)) \log(n\lambda))$ or $O^\sim(n^2(\lambda+n))$ with fast arithmetic (Corollaries 9.63 and 9.66).

We can associate to a linear difference operator $L = aE + b$ a "number" $\delta_L \in \mathbb{Q} \cup \{\infty\}$, depending only on the coefficients of a and b, such that either $\deg y = \deg c - \max\{1 + \deg a, \deg(a+b)\}$ or $\deg y = \delta_L$ (Gosper 1978). (The element δ_L is the unique root of the indicial equation of L at infinity.) This δ_L plays a similar role as the dispersion in the preceding chapter: there are examples where the degree of the polynomial y is equal to δ_L and is exponential in the size of the coefficients of a and b; some are given in Sect. 10.1. One can define a similar δ_L for a linear differential operator $L = aD + b$, and then analogous statements hold.

Chap. 10 collects the results of the preceding two chapters and gives a complete cost analysis of our modular variants of Gosper's algorithm (Algorithm 10.1) for hypergeometric summation over $\mathbb{Q}(x)$ and Almkvist & Zeilberger's algorithm (Algorithm 10.4) for hyperexponential integration over $\mathbb{Q}(x)$. Gosper (1978) showed that a hypergeometric element u in an extension field of $\mathbb{Q}(x)$ has a hypergeometric antidifference if and only if the linear first order difference equation

$$\frac{f}{g} \cdot E\sigma - \sigma = 1 \,,$$

where $f, g \in \mathbb{Z}[x]$ are nonzero such that $(Eu)/u = f/g$, has a rational solution $\sigma \in \mathbb{Q}(x)$, and then σu is an antidifference of u. The algorithms of Chap. 8 and 9 essentially compute the denominator and the numerator of σ, respectively. The following is proven as Theorem 10.2:

Theorem. *Let* $L = fE - g = f\Delta + f - g$, *and assume that* f, g *have degree at most* n *and coefficients absolutely bounded by* 2^λ. *If* $e = \mathrm{dis}(f, g)$ *is the dispersion of* f *and* g *and* $\delta = \max(\{0, \delta_L\} \cap \mathbb{N})$, *then* $e < 2^{\lambda+2}$, $\delta < 2^{\lambda+1}$, *and Algorithm 10.1 takes*

$$O(e^5 n^5 + e^3 n^3 \lambda^2 + \delta^5 + \delta^3 \lambda^2) \text{ and } O^\sim(e^3 n^3 + e^2 n^2 \lambda + \delta^3 + \delta^2 \lambda)$$

word operations with classical and fast arithmetic, respectively.

Similarly, a hyperexponential element u in an extension field of $\mathbb{Q}(x)$ has a hyperexponential antiderivative if and only if the linear first order differential equation

$$D\sigma + \frac{f}{g}\sigma = 1$$

has a rational solution $\sigma \in \mathbb{Q}(x)$, where $f, g \in \mathbb{Z}[x]$ are nonzero such that $(Du)/u = f/g$, and then σu is an antiderivative of u. The following is proven as Theorem 10.5:

Theorem. *Let* $L = gD + f$, *and assume that* f, g *have degree at most* n *and coefficients absolutely bounded by* 2^λ. *If* $e = \varepsilon(f, g)$ *and* $\delta = \max(\{0, \delta_L\} \cap \mathbb{N})$, *then* $e \leq (n+1)^n 2^{2n\lambda}$, $\delta \leq 2^\lambda$, *and Algorithm 10.4 takes*

$$O((e^3 n^5 + \delta^3 n^2)(\lambda^2 + \log^2 n)) \text{ and } O^\sim(e^2 n^3 \lambda + \delta^2 n \lambda)$$

word operations with classical and fast arithmetic, respectively.

In Sect. 10.1, we give some examples where a hypergeometric antidifference or a hyperexponential antiderivative, respectively, of a non-rational element u exists and the degree of the numerator or the denominator of σ are exponential in the input size. Some of these examples appear in Gerhard (1998). We also exhibit a subclass of the hyperexponential elements for which the bounds δ and e are polynomial in the input size.

2.3 References and Related Works

Standard references on ordinary differential equations are, e.g., Ince (1926) and Kamke (1977). *Differential Galois theory* classifies linear differential operators in terms of the algebraic group of differential automorphisms, called the *differential*

Galois group, of the corresponding solution spaces, and also provides algorithms for computing solutions from some partial knowledge about this group. Classical texts on differential Galois theory and differential algebra are Ritt (1950); Kaplansky (1957); and Kolchin (1973); see also van der Put & Singer (2003).

There are many algorithms for computing symbolic solutions of higher order linear differential equations, e.g., by Abramov (1989a, 1989b); Abramov & Kvansenko (1991); Singer (1991); Bronstein (1992); Petkovšek & Salvy (1993); Abramov, Bronstein & Petkovšek (1995); Pflügel (1997); Bronstein & Fredet (1999); or Fakler (1999). The special case of the first order equation, also known as *Risch differential equation* since it plays a prominent role in Risch's algorithm, is discussed, e.g., by Rothstein (1976); Kaltofen (1984); Davenport (1986); and Bronstein (1990, 1991).

Classical works on rational function integration are due to Johann Bernoulli (1703); Ostrogradsky (1845); and Hermite (1872). The latter two algorithms write a rational function as the sum of a rational function with only simple poles plus the derivative of a rational function. Horowitz (1969, 1971); Mack (1975); and Yun (1977a) stated and analyzed modern variants of these algorithms. Rothstein (1976, 1977); Trager (1976); and Lazard & Rioboo (1990) and Trager (unpublished) gave algorithms for the remaining task of integrating a rational function with only simple poles.

Rational and hyperexponential integration are special cases of Risch's (1969, 1970) famous algorithm for symbolic integration of elementary functions. Variants of his algorithm are implemented in nearly any general purpose computer algebra system. See Bronstein (1997) for a comprehensive treatment and references.

Already Gauß (1863), article 368, contains an algorithm for computing the squarefree part of a polynomial. The first "modern" works on squarefree factorization are – among others – Tobey (1967); Horowitz (1969, 1971); Musser (1971); and Yun (1976, 1977a, 1977b). The latter papers contain the fastest currently known algorithm when counting arithmetic operations in the coefficient field. Most of these algorithms only work in characteristic zero; see Gianni & Trager (1996) for a discussion of the case of positive characteristic. Bernardin (1999) discusses a variant of Yun's algorithm for multivariate polynomials. Diaz-Toca & Gonzales-Vega (2001) give an algorithm for parametric squarefree factorization.

Classical references on difference algebra are, e.g., Boole (1860); Jordan (1939); or Cohn (1965). *Difference Galois theory* studies the algebraic group of difference automorphisms of the solution space of such a linear difference operator; see van der Put & Singer (1997) for an overview.

The first algorithms for rational summation are due to Abramov (1971, 1975) and Moenck (1977), and Gosper (1978) first solved the hypergeometric summation problem. Karr (1981, 1985) presented an analog of Risch's integration algorithm for summation in $\Sigma\Pi$-fields, which correspond to the Liouvillian fields in the case of integration; see Schneider (2001). More recent works on rational and hypergeometric summation and extensions of Gosper's algorithm are due to Lisoněk, Paule & Strehl (1993); Man (1993); Petkovšek (1994); Abramov (1995b);

Koepf (1995); Pirastu & Strehl (1995); Paule (1995); Pirastu (1996); Abramov & van Hoeij (1999); and Bauer & Petkovšek (1999). Paule & Strehl (1995) give an overview. Algorithms for solving higher order linear difference equations are given, e.g., by Abramov (1989a, 1989b, 1995a); Petkovšek (1992); van Hoeij (1998, 1999); Hendriks & Singer (1999); and Bronstein (2000).

The problem of *definite summation* is, given a bivariate function g such that $g(n, \cdot)$ is summable for each $n \in \mathbb{N}$, to compute a "closed form" f such that

$$\sum_{k \in \mathbb{Z}} g(n, k) = f(n) \text{ for } n \in \mathbb{N} .$$

If g is hypergeometric with respect to both arguments, such that both

$$\frac{g(n+1, k)}{g(n, k)} \text{ and } \frac{g(n, k+1)}{g(n, k)}$$

are rational functions of n and k, then in many cases the algorithm of Zeilberger (1990a, 1990b, 1991), employing a variant of Gosper's (1978) algorithm for indefinite hypergeometric summation as a subroutine, computes a linear difference operator L with polynomial coefficients that annihilates f. Then any algorithm for solving linear difference equations with polynomial coefficients can be used to find a closed form for f. For example, the algorithms of Petkovšek (1992) or van Hoeij (1999) decide whether there is a hypergeometric element f such that $Lf = 0$. The related method of Wilf & Zeilberger (1990) is able to produce routinely short proofs of all kinds of combinatorial identities, among them such famous ones as Dixon's theorem (Ekhad 1990) and the Rogers-Ramanujan identities (Ekhad & Tre 1990; Paule 1994). Recent work related to Zeilberger's method includes Abramov (2002); Abramov & Le (2002); Abramov & Petkovšek (2002b); Le (2002); Le (2003b); and Le (2003a).

There is also a continuous variant of Zeilberger's algorithm, due to Almkvist & Zeilberger (1990). Given a bivariate function g that is hyperexponential with respect to both arguments, such that both ratios

$$\frac{(\partial g / \partial x)(x, y)}{g(x, y)} \text{ and } \frac{(\partial g / \partial y)(x, y)}{g(x, y)}$$

are rational functions of x and y, this algorithm employs a variant of the continuous analog of Gosper's algorithm for hyperexponential integration to find a linear differential operator with polynomial coefficients annihilating the univariate function f defined by

$$f(x) = \int_{y \in \mathbb{R}} g(x, y) .$$

Then any algorithm for solving linear differential equations with polynomial coefficients can be used to find a closed form for f. Generalizations of these algorithms are discussed in Wilf & Zeilberger (1992); Chyzak (1998a, 1998b, 2000); and Chyzak & Salvy (1998). Graham, Knuth & Patashnik (1994) and Petkovšek,

Wilf & Zeilberger (1996) give nice introductions to Gosper's and Zeilberger's algorithm, and Koepf (1998) discusses also their continuous variants.

Besides the shift operator E with $(Ef)(x) = f(x + 1)$, another interesting automorphism in difference algebra is known as the *q-shift operator* Q. For a fixed element q and a unary function f, it is defined by $(Qf)(x) = f(qx)$. See Koepf (1998) for an overview and references on q-analogs of various symbolic summation algorithms.

There are many analogies between differential algebra and difference algebra, and also between the corresponding algorithms. There are several attempts at providing a unified framework. *Ore rings*, invented by Ore (1932a, 1932b, 1933), cover the similarities between the algebraic properties of linear differential operators and linear difference operators (see also Bronstein & Petkovšek 1994). *Pseudo-linear algebra* (Jacobson 1937; Bronstein & Petkovšek 1996) focuses on the solution space of such operators. *Umbral calculus* (Rota 1975; Roman & Rota 1978; Roman 1984) studies the connection between linear operators, comprising the differential operator and the difference operator as special cases, and sequences of polynomials that are uniquely characterized by certain equations involving the linear operator and the elements of the sequence.

Differential equations in positive characteristic occur also in a different area of computer algebra: Niederreiter (1993a, 1993b, 1994a, 1994b); Niederreiter & Göttfert (1993, 1995); Göttfert (1994); and Gao (2001) employ differential equations to factor polynomials over finite fields.

Throughout this book, we often refer to von zur Gathen & Gerhard (1999), where appropriate pointers to the original literature are provided.

2.4 Open Problems

We conclude this introduction with some problems that remain unsolved in this book.

- Give a cost analysis for a modular variant of a complete algorithm for rational summation, i.e., including the computation of the transcendental part of the anti-difference of a rational function.
- Give a cost analysis for a modular variant of Zeilberger's algorithm for definite hypergeometric summation and its continuous analog for definite hyperexponential integration, due to Almkvist & Zeilberger.
- The methods presented in this book essentially provide rational function solutions of linear first order differential or difference equations with polynomial coefficients. Generalize the modular algorithms in this book to equations of higher order.
- A major open problem in the area of symbolic integration and summation is to provide a complete cost analysis of Risch's and Karr's algorithms. In light of the fact that a general version of the symbolic integration problem is known to be unsolvable (see, e.g., Richardson 1968), it is not even clear that these algorithms

are primitive recursive, i.e., can be written in a programming language using just for loops but not general while loops.

- An open question related to the problems discussed in this book is whether the resultant of two bivariate polynomials over and abstract field can be computed in pseudo-linear time, when counting only arithmetic operations in the field. If the degrees of the input polynomials in the two variables are at most n and d, respectively, then the degree of the resultant is at most nd, so that is conceivable that there is an $O^\sim(nd)$ algorithm for computing the resultant. However, the best known algorithms so far take time $O^\sim(n^2d)$; see Theorem 7.7.

3. Technical Prerequisites

This chapter summarizes some technical ingredients for later use and may be skipped at first reading. More background on these can also be found in von zur Gathen & Gerhard (1999).

We say that a nonzero polynomial in $\mathbb{Z}[x]$ is *normalized* if it is primitive, so that its coefficients have no nontrivial common divisor in $\mathbb{Z}$, and its leading coefficient is positive. Then each nonzero polynomial $f \in \mathbb{Z}[x]$ can be uniquely written as $f = cg$ with a nonzero constant $c \in \mathbb{Z}$ and a normalized polynomial $g \in \mathbb{Z}[x]$. We call $c = \mathrm{lu}(f) = \pm\mathrm{cont}(f)$, which is a unit in $\mathbb{Q}$, the *leading unit* and $g = \mathrm{normal}(f) = \pm\mathrm{pp}(f)$ the *normal form* of f. Then the following properties hold for all nonzero polynomials $f, g \in \mathbb{Z}[x]$ and all nonzero constants $c \in \mathbb{Z}$.

- normal and lu are both multiplicative: $\mathrm{normal}(fg) = \mathrm{normal}(f)\mathrm{normal}(g)$ and $\mathrm{lu}(fg) = \mathrm{lu}(f)\mathrm{lu}(g)$.
- $\mathrm{normal}(c) = 1$ and $\mathrm{lu}(c) = c$, and hence $\mathrm{normal}(cf) = \mathrm{normal}(f)$ and $\mathrm{lu}(cf) = c\,\mathrm{lu}(f)$.
- f is normalized if and only if $f = \mathrm{normal}(f)$, or equivalently, $\mathrm{lu}(f) = 1$.
- There exist nonzero constants $a, b \in \mathbb{Q}$ such that $af = bg$ if and only if $\mathrm{normal}(f) = \mathrm{normal}(g)$.

We use the following norms for a vector $f = (f_i)_{0 \le i < n} \in \mathbb{C}^n$ or a polynomial $f = \sum_{0 \le i < n} f_i x^i \in \mathbb{C}[x]$:

- the *max-norm* $\|f\|_\infty = \max\limits_{0 \le i < n} |f_i|$,

- the *one-norm* $\|f\|_1 = \sum\limits_{0 \le i < n} |f_i|$, and

- the *two-norm* or *Euclidean norm* $\|f\|_2 = \left(\sum\limits_{0 \le i < n} |f_i|^2 \right)^{1/2}$.

Each of the three norms satisfies

$$\|f + g\| \le \|f\| + \|g\|, \quad \|cf\| \le |c| \cdot \|f\|$$

for all vectors (polynomials) $f, g \in \mathbb{C}^n$ and all scalars $c \in \mathbb{C}$. The three norms are related by the well-known inequalities

$$\|f\|_\infty \le \|f\|_2 \le \|f\|_1 \le n\|f\|_\infty, \quad \|f\|_2 \le n^{1/2}\|f\|_\infty .$$

J. Gerhard: Modular Algorithms, LNCS 3218, pp. 27–40, 2004.
© Springer-Verlag Berlin Heidelberg 2004

The one-norm has the distinguished property of being sub-multiplicative:

$$\|fg\|_1 \leq \|f\|_1 \cdot \|g\|_1$$

for all vectors (polynomials) $f, g \in \mathbb{C}^n$.

3.1 Subresultants and the Euclidean Algorithm

Lemma 3.1 (Determinant bounds). *(i) Let $A = (a_{ij})_{1 \leq i,j \leq n} \in \mathbb{C}^{n \times n}$ and $a^1, \ldots, a^n \in \mathbb{C}^n$ be the columns of A. Then $|\det A| \leq \|a^1\|_2 \cdots \|a^n\|_2 \leq n^{n/2} \|A\|_\infty$.*

(ii) Let R be a ring, $A \in R[y_1, y_2, \ldots]^{n \times n}$, and $a^1, \ldots, a^n \in \mathbb{C}^n$ be the columns of A. Moreover, for $1 \leq i \leq n$ let $\deg A$ and $\deg a^i$ denote the maximal total degree of a coefficient of A or a^i, respectively, and define $\deg_{y_j} A$ and $\deg_{y_j} a^i$ similarly for $j \geq 1$. Then $\deg(\det A) \leq \deg a^1 + \cdots + \deg a^n \leq n \deg A$, and similarly for $\deg_{y_j}(\det A)$.

(iii) With $R = \mathbb{C}$ and A as in (ii), we have $\|\det A\|_\infty \leq \|\det A\|_1 \leq n! B^n$, where $B \in \mathbb{R}$ is the maximal one-norm of an entry of A.

Proof. (i) is Hadamard's well-known inequality (see, e.g., Theorem 16.6 in von zur Gathen & Gerhard 1999) and (ii) is obvious from the definition of the determinant as a sum of $n!$ terms:

$$\det A = \sum_{\pi \in S_n} \mathrm{sign}(\pi) \cdot a_{1\pi_1} \cdots a_{n\pi_n} \, ,$$

where S_n denotes the symmetric group of all permutations of $\{1, \ldots, n\}$ and $\mathrm{sign}(\pi) = \pm 1$ is the sign of the permutation π. Finally, (iii) follows from

$$\|\det A\|_1 \leq \sum_{\pi \in S_n} \|a_{1\pi_1}\|_1 \cdots \|a_{n\pi_n}\|_1 \leq n! B^n \, . \quad \square$$

Of course, all these bounds are valid for rows instead of columns as well.

Let R be a ring, $f = \sum_{0 \leq i \leq n} f_i x^i$ and $g = \sum_{0 \leq i \leq m} g_i x^i$ in $R[x]$ two nonzero polynomials of degree n and m, respectively, and $0 \leq d \leq \min\{n, m\}$. The dth *subresultant* $\sigma_d(f, g)$ is the determinant of the submatrix of the Sylvester matrix depicted in Figure 3.1 (coefficients with negative indices are considered to be zero). (See Sect. 6.10 in von zur Gathen & Gerhard 1999.) In particular, for $d = 0$ this matrix is the Sylvester matrix of f and g and $\sigma_0(f, g) = \mathrm{res}(f, g)$ is their *resultant*. If $n = m = d$, then the matrix above is the empty 0×0 matrix, and $\sigma_d(f, g) = 1$. For convenience, we define $\sigma_d(f, g) = 0$ for $\min\{n, m\} < d < \max\{n, m\}$, and $\sigma_d(f, 0) = 0$ and $\sigma_d(0, g) = 0$ if $d < n$ and $d < m$, respectively.

Corollary 3.2 (Subresultant bounds). *(i) Let $f, g \in \mathbb{Z}[x]$ be nonzero polynomials with degree at most n and max-norm at most A. Moreover, let $0 \leq d \leq \min\{\deg f, \deg g\}$ and $\sigma \in \mathbb{Z}$ be the dth subresultant of f, g. Then $|\sigma| \leq (\|f\|_2 \|g\|_2)^{n-d} \leq ((n+1)A^2)^{n-d}$.*

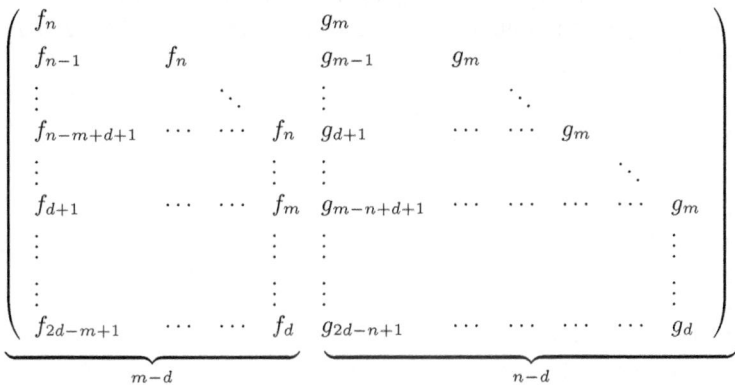

Fig. 3.1. The dth submatrix of the Sylvester matrix

(ii) *Let R be a ring, $f, g \in R[y_1, y_2, \ldots][x]$ with $0 \leq d \leq \min\{\deg_x g, \deg_x f\}$ and $\max\{\deg_x f, \deg_x g\} \leq n$, and let $\sigma \in R[y_1, y_2, \ldots]$ be the dth subsultant of f and g with respect to x. If $\deg_y$ denotes the total degree with respect to all variables y_j, then $\deg_y \sigma \leq (n - d)(\deg_y f + \deg_y g)$, and similarly for $\deg_{y_j} \sigma$.*

(iii) *With $R = \mathbb{Z}$ and f, g as in (ii), we have $\|\sigma\|_\infty \leq \|\sigma\|_1 \leq (2n - 2d)! \, B^{2n-2d}$, where $B \in \mathbb{N}$ is the maximal one-norm of a coefficient in $\mathbb{Z}[y_1, y_2, \ldots]$ of f or g.*

The following well-known fact is due to Mignotte (1989), Théorème IV.4.4.

Fact 3.3 (Mignotte's bound). *Let $f, f_1, \ldots, f_t \in \mathbb{Z}[x]$ be nonconstant such that $(f_1 \cdots f_t) \mid f$ and $\deg(f_1 \cdots f_t) = n$. Then*

$$\|f_1\|_1 \cdots \|f_t\|_1 \leq 2^n \|f\|_2 .$$

The famous Euclidean Algorithm computes the greatest common divisor of two univariate polynomials with coefficients in a field (see, e.g., von zur Gathen & Gerhard 1999, Chap. 4). In Chap. 7, we use the following variant, which works over an arbitrary commutative ring. In contrast to the usual notation, the indices of the intermediate results are the degrees of the corresponding remainder.

We use the convention that the zero polynomial has degree $-\infty$ and leading coefficient 1. If R is a ring and $f, g \in R[x]$ are nonzero polynomials such that the leading coefficient $\mathrm{lc}(f)$ is invertible, then there exist unique polynomials $q, r \in R[x]$ with $f = qg + r$ and $\deg r < \deg g$, and we denote them by $q = f$ quo g and $r = f$ rem g.

Algorithm 3.4 (Monic Extended Euclidean Algorithm). ▬▬▬▬▬
Input: Polynomials $f, g \in R[x]$, where R is a commutative ring, such that $f \neq 0$, $\deg f, \deg g \leq n$, and $\mathrm{lc}(f), \mathrm{lc}(g) \in R^\times$.

Output: The Euclidean length ℓ of f, g, their degree sequence $\deg f \geq d_1 > d_2 > \cdots > d_\ell \geq 0$, and the values $\gamma_{d_i} \in R$ and $r_{d_i}, s_{d_i}, t_{d_i} \in R[x]$ for $2 \leq i \leq \ell$ as computed below.

1. $d_0 \longleftarrow \infty$, $\qquad d_1 \longleftarrow \deg g$
   $\rho_{d_0} \longleftarrow \mathrm{lc}(f)$, $\qquad \rho_{d_1} \longleftarrow \mathrm{lc}(g)$
   $r_{d_0} \longleftarrow \rho_{d_0}^{-1} f$, $\qquad r_{d_1} \longleftarrow \rho_{d_1}^{-1} g$
   $s_{d_0} \longleftarrow \rho_{d_0}^{-1}$, $\qquad s_{d_1} \longleftarrow 0$
   $t_{d_0} \longleftarrow 0$, $\qquad t_{d_1} \longleftarrow \rho_{d_1}^{-1}$
   $\beta_{d_1} \longleftarrow \rho_{d_0}\rho_{d_1}$, $\qquad \gamma_{d_1} \longleftarrow \rho_{d_1}^{\deg f - \deg g}$
   $i \longleftarrow 0$

2. **while** $r_{d_{i+1}} \neq 0$ **do**
   $\qquad i \longleftarrow i + 1$
   $\qquad d_{i+1} \longleftarrow \deg(r_{d_{i-1}} \text{ rem } r_{d_i})$
   $\qquad a_{d_{i+1}} \longleftarrow r_{d_{i-1}} \text{ rem } r_{d_i}, \qquad \rho_{d_{i+1}} \longleftarrow \mathrm{lc}(a_{d_{i+1}})$
   $\qquad$ **if** $\rho_{d_{i+1}} \notin R^\times$ **then goto 3**
   $\qquad q_{d_i} \longleftarrow r_{d_{i-1}} \text{ quo } r_{d_i}$
   $\qquad r_{d_{i+1}} \longleftarrow \rho_{d_{i+1}}^{-1} a_{d_{i+1}}$
   $\qquad s_{d_{i+1}} \longleftarrow \rho_{d_{i+1}}^{-1}(s_{d_{i-1}} - q_{d_i}s_{d_i})$
   $\qquad t_{d_{i+1}} \longleftarrow \rho_{d_{i+1}}^{-1}(t_{d_{i-1}} - q_{d_i}t_{d_i})$
   $\qquad \beta_{d_{i+1}} \longleftarrow \beta_{d_i}\rho_{d_{i+1}}$
   $\qquad \gamma_{d_{i+1}} \longleftarrow (-1)^{(d_i - d_{i+1})(\deg f - d_{i+1} + i + 1)}\beta_{d_{i+1}}^{d_i - d_{i+1}} \cdot \gamma_{d_i}$

3. $\ell \longleftarrow i$

In step 2, we let $d_{i+1} = -\infty$ and $\rho_{d_{i+1}} = 1$ if $a_{d_{i+1}} = 0$. If R is a field, then the above algorithm always terminates with $r_{d_\ell} \mid r_{d_{\ell-1}}$, and r_{d_ℓ} is the monic gcd of f and g in $R[x]$. Otherwise, it may happen that the algorithm terminates prematurely when the "if" condition in step 2 is true. If this is not the case, then we say that the monic EEA terminates *regularly*.

Example 3.5. Let $R = \mathbb{Z}/8\mathbb{Z}$, $f = x^3 + 2x + 1$, and $g = x^2$. Then $r_\infty = f, r_2 = g$, $a_1 = r_\infty \text{ rem } r_2 = 2x + 1$, and $\rho_1 = \mathrm{lc}(a_1) = 2$. The latter is not invertible in R, and the monic EEA does not terminate regularly.

Remark 3.6. *One of our main applications of Algorithm 3.4 in Chap. 7 is when $R = \mathbb{Z}/p^k\mathbb{Z}$ for a prime $p \in \mathbb{N}$ and some $k \geq 2$. Suppose that the condition in step 2 is true, so that $\mathrm{lc}(a_{d_{i+1}})$ is divisible by p, and assume additionally that $a_{d_{i+1}}$ is not divisible by p. Using Hensel lifting, one can compute a unique unit $u \in R[x]$ with $\deg u \leq d_{i+1}$ such that $a_{d_{i+1}}/u$ is monic of degree $d_{i+1} - \deg u$ (see Corollary 3.17 in von zur Gathen & Hartlieb 1998). One might then proceed with $a_{d_{i+1}}/u$ instead of $a_{d_{i+1}}$. For example, we might choose $u = a_1$ in Example 3.5. However, we are not aware of such workaround if $a_{d_{i+1}}$ is divisible by p.*

Proposition 3.7 (Invariants of the monic EEA). *Let f, g be as in Algorithm 3.4 and σ_d be the dth subresultant of f, g, for $0 \leq d \leq \min\{n, m\}$. The following hold for $0 \leq i \leq \ell$.*

(i) $d_i < d_{i-1}$ if $i \geq 1$,

(ii) $s_{d_i} f + t_{d_i} g = r_{d_i}$,

(iii) $s_{d_i} t_{d_{i-1}} - s_{d_{i-1}} t_{d_i} = (-1)^i (\rho_{d_0} \cdots \rho_{d_i})^{-1}$ if $i \geq 1$,

(iv) $\deg r_{d_i} = d_i$ if $i \geq 1$, $\deg s_{d_i} = \deg g - d_{i-1}$ if $i \geq 2$, and $\deg t_{d_i} = \deg f - d_{i-1}$ if $i \geq 2$,

(v) $\beta_{d_i} = (\rho_{d_0} \cdots \rho_{d_i})$, and $\gamma_{d_i} = \sigma_{d_i}$ if $i \geq 2$.

With the convention that $d_{\ell+1} = -\infty$, $r_{d_{\ell+1}} = 0$, and $\rho_{d_{\ell+1}} = 1$, claims (i) through (iv) are also valid for $i = \ell + 1$ in case the monic EEA terminates regularly.

Proof. All proofs can be found in von zur Gathen & Gerhard (1999) (where they are stated only for the case when R is a UFD or even a field): (i) through (iii) are from Lemma 3.8, (iv) is Lemma 3.10, and (v) follows from Theorem 11.13 (ii); the numbering corresponds to the 1999 edition. □

Definition 3.8. *Let R be a ring and $s, t \in R$. We say that s and t are* coprime *if there exist $a, b \in R$ such that $sa + tb = 1$.*

The following lemma shows the relation between subresultants and the Euclidean Algorithm.

Lemma 3.9. *Let R be a ring and $f, g \in R[x]$ nonzero polynomials of degrees n and m. Moreover, let $0 \leq d \leq \min\{n, m\}$ and $\sigma \in R$ be the dth subresultant of f and g.*

(i) *If R is a field, then d occurs in the degree sequence of the monic Euclidean Algorithm of f, g in $R[x]$ if and only if $\sigma \neq 0$.*

(ii) *Suppose that $d < \max\{m, n\}$. If σ is a unit, then there are unique polynomials $s, t \in R[x]$ such that $\deg s < m - d$, $\deg t < n - d$, and $sf + tg$ is a monic polynomial of degree d.*

(iii) *There exist polynomials $s, t \in R[x]$, not both zero, such that $\deg s < m - d$, $\deg t < n - d$, and $\deg(sf + tg) < d$, if and only if σ is a zero divisor in R.*

Proof. Let $S \in R^{(n+m-2d) \times (n+m-2d)}$ be the square submatrix of the Sylvester matrix of f and g whose determinant is σ (Figure 3.1).

(i) is Corollary 6.49 (i) in von zur Gathen & Gerhard (1999).

(ii) For polynomials s, t with the given degree constraints, $sf + tg$ is a monic polynomial of degree d if and only if the coefficient vector v of $sx^{n-d} + t$ satisfies $Sv = (0, \ldots, 0, 1)^T$. Since $\sigma = \det S$ is a unit, there is a unique such vector v, and hence there are unique such polynomials s, t.

(iii) Let s, t be as assumed, $v \in R^{n+m-2d}$ the coefficient vector of $sx^{n-d} + t$, and $T \in R^{(n+m-2d) \times (n+m-2d)}$ the adjoint of S. Then $\sigma v = T S v = 0$ and σ is a zero divisor. Conversely, if σ is a zero divisor, then McCoy's theorem (see, e.g., Theorem 3.5.1 in Balcerzyk & Józefiak 1989 or Satz 49.2 in Scheja & Storch 1980) implies that there exists a nonzero vector $v \in R^{n+m-2d}$ such that $Sv = 0$, and if we let $s, t \in R[x]$ with $\deg s < m - d$ and $\deg t < n - d$ be such that v is the coefficient vector of $sx^{n-d} + t$, then $\deg(sf + tg) < d$. □

Corollary 3.10. *Let R be a UFD with field of fractions K and $f, g \in R[x]$ nonzero polynomials.*

(i) $\gcd(f, g)$ is nonconstant if and only if $\mathrm{res}(f, g) = 0$ in R.
(ii) If K is a perfect field, then g is squarefree (in $K[x]$) $\iff \gcd(g, g')$ is constant $\iff \mathrm{res}(g, g') \neq 0$.

The next lemma investigates the relation between the Extended Euclidean Algorithm for two polynomials on the one hand and for a homomorphic image of them on the other hand.

Lemma 3.11. *Let R be an integral domain with field of fractions K, $f, g \in R[x]$ nonzero with $\deg g \leq \deg f = n$, and $I \subseteq R$ an ideal coprime to $\mathrm{lc}(f)$. Moreover, let a bar denote reduction modulo I and let $\sigma_d \in R$ be the dth subresultant of f and g, for $0 \leq d \leq \deg g$.*

(i) Let $0 \leq d \leq \deg g$, $d < \deg f$, and $d \geq d_\ell$ if the monic Euclidean Algorithm for $\bar{f}, \bar{g}$ in $\overline{R}[x]$ does not terminate regularly, where ℓ is its length. Then d occurs in the degree sequence of $\bar{f}, \bar{g}$ if and only if σ_d is a unit modulo I. In that case, the denominators of the polynomials $r_d^*, s_d^*, t_d^* \in K[x]$ in the monic EEA of f, g in $K[x]$ divide σ_d, so that they are invertible modulo I, and their images modulo I are equal to the polynomials r_d, s_d, t_d in the monic EEA of $\bar{f}, \bar{g}$ in $\overline{R}[x]$.
(ii) If the monic Euclidean Algorithm for $\bar{f}, \bar{g}$ in $\overline{R}[x]$ does not terminate regularly, then there is some $d \in \{0, \ldots, d_\ell - 1\}$ such that σ_d is nonzero and $\overline{\sigma_d}$ is not a unit.

Proof. (i) Let $m = \deg \bar{g}$, and suppose that $d = d_i$ for some $i \in \{1, \ldots, \ell\}$. Then $d \geq 0$ implies that $\bar{g} \neq 0$. We have $\deg s_d = m - d_{i-1} < m - d$, $\deg t_d = n - d_{i-1} < n - d$, the polynomial $r_d = s_d f + t_d g$ is monic of degree d, and s_d, t_d are coprime, by Proposition 3.7. Since $\overline{\mathrm{lc}(f)}$ is invertible, the dth subresultant of $\bar{f}, \bar{g}$ is $\overline{\sigma_d}$ divided by a power of $\overline{\mathrm{lc}(f)}$, and Proposition 3.7 (v) implies that $\overline{\sigma_d}$ is a unit in $\overline{R}$. The denominators of the polynomials r_d^*, s_d^*, t_d^* in the monic EEA of f, g in $K[x]$, which exist by Lemma 3.9 (i), are divisors of σ_d, by Corollary 6.49 (ii) in von zur Gathen & Gerhard (1999) and Cramer's rule. Moreover, we have $\deg \overline{t_d^*} < n - d$, and $\overline{s_d^*} \bar{f} + \overline{t_d^*} \bar{g} = \overline{r_d^*}$ is monic of degree d. Thus $\deg \overline{s_d^*} < m - d$, and Lemma 3.9 (ii) yields $\overline{r_d^*} = r_d$, $\overline{s_d^*} = s_d$, and $\overline{t_d^*} = t_d$.
Conversely, assume that $d_{i-1} > d > d_i$ for some $i \in \{1, \ldots, \ell + 1\}$, with the convention that $d_{\ell+1} = -\infty$ and $r_{d_{\ell+1}} = 0$ if the monic Euclidean Algorithm terminates regularly. If $i \geq 2$, then $\deg s_{d_i} = m - d_{i-1} < m - d$, $\deg t_{d_i} = n - d_{i-1} < n - d$, and $\deg(s_{d_i} \bar{f} + t_{d_i} \bar{g}) = \deg r_{d_i} = d_i < d$. By Proposition 3.7 (iv), t_{d_i} is nonzero. As before, the dth subresultant of $\bar{f}, \bar{g}$ is $\overline{\sigma_d}$ divided by a power of $\overline{\mathrm{lc}(f)}$, and Lemma 3.9 (iii) implies that $\overline{\sigma_d}$ is a zero divisor in $\overline{R}$, so that σ_d is not invertible modulo I.
If $i = 1$, then

$$\deg f = \deg \bar{f} \geq \deg g \geq d > \deg \bar{g} = d_1 .$$

If $d < \deg g$, then all g-columns in the matrix from Figure 3.1, whose determinant is σ_d, vanish modulo I on and above the diagonal, and hence $\overline{\sigma_d} = 0$. If $d = \deg g$, then $d < \deg f = \deg \overline{f}$ and $\sigma_d = \mathrm{lc}(g)^{\deg f - \deg g}$. Since $d > \deg \overline{g}$, we have $\overline{\mathrm{lc}(g)} = 0$. Thus $\overline{\sigma_d} = 1$ if $\deg f = d = \deg g$ and $\overline{\sigma_d} = 0$ if $\deg f > d = \deg g$, and the claims follow in all cases.

(ii) Since $\sigma_{d_{\ell-1}}$ and σ_{d_ℓ} are nonzero in R, by (i), the degrees $d_{\ell-1}$ and d_ℓ occur (not necessarily consecutively) in the degree sequence of the monic EEA of f, g in $K[x]$, by Lemma 3.9 (i). So let $r^*_{d_{\ell-1}}, r^*_{d_\ell} \in K[x]$ be the corresponding remainders of degrees $d_{\ell-1}$ and d_ℓ, respectively. Division with remainder yields $q, r \in K[x]$ with $\deg r < d_\ell$ and $r^*_{d_{\ell-1}} = q r^*_{d_\ell} + r$. Now $r^*_{d_{\ell-1}}$ and $r^*_{d_\ell}$ are monic and defined modulo I and their images modulo I are equal to $r_{d_{\ell-1}}$ and r_{d_ℓ}, respectively, again by (i), so that we can take this equation modulo I, and the uniqueness of division with remainder implies that $\overline{r} = a_{d_{\ell+1}} \neq 0$. Hence $r^*_{d_\ell} \nmid r^*_{d_{\ell-1}}$, and Proposition 3.7 (ii) implies that $r^*_{d_\ell}$ is not the gcd of f and g. Thus there exists a nonzero remainder of degree $d < d_\ell$ in the monic EEA of f and g in $K[x]$, and Lemma 3.9 (i) yields $\sigma_d \neq 0$.

Since the "if" condition in step 2 is true, $\rho_{d_{\ell+1}}$ is not a unit in $\overline{R}$. Let $J \subseteq \overline{R}$ be a maximal ideal containing $\rho_{d_{\ell+1}}$. Then $\overline{\mathrm{lc}(f)}$ is a unit modulo J and d does not occur in the degree sequence of the monic Euclidean Algorithm over the field $\overline{R}/J$. Now Lemma 3.9 (i) with R replaced by $\overline{R}/J$ implies that $\overline{\sigma_d} \in J \neq \overline{R}$, and hence it is not a unit. $\square$

Corollary 3.12. *Let R be a UFD, $f, g \in R[x]$ nonzero, $d = \deg \gcd(f, g)$, and $I \subseteq R$ a maximal ideal not containing $\mathrm{lc}(f)$. Moreover, let $\sigma_d \in R$ be the dth subresultant of f and g. Then $\deg \gcd(f \bmod I, g \bmod I) \geq d$, with equality if and only if $\sigma_d \notin I$.*

Proof. Let $h = \gcd(f, g)$ and denote reduction modulo I by a bar. Then $\overline{h} \mid \overline{f}$ and $\overline{h} \mid \overline{g}$, and hence $\overline{h} \mid \gcd(\overline{f}, \overline{g})$. Now $\mathrm{lc}(h) \mid \mathrm{lc}(f)$, which implies that $\overline{\mathrm{lc}(h)} \neq 0$ and $\deg \overline{h} = \deg h$, and the first claim follows. Lemma 3.11 (i) yields the subsequent claims. $\square$

3.2 The Cost of Arithmetic

In the remainder of this chapter, we collect some cost estimates for basic arithmetic tasks. We denote by ω the word size of our processor, i.e., the number of bits that a machine word occupies; common values are 32 or 64. The *word length* of a nonzero integer $a \in \mathbb{Z}$ is the number of machine words that it occupies in memory, namely $\lfloor (\log_2 |a|)/\omega \rfloor + 1$. A nonnegative integer a is called *single precision* if it fits into one machine word, so that $0 \leq a < 2^\omega$. In our cost estimates, we count *word operations*, which the reader may imagine as processor instructions, operating on single precision data. A *single precision prime* is a prime $p \in \mathbb{N}$ fitting precisely into one machine word, so that $2^{\omega-1} < p < 2^\omega$. Thus an arithmetic operation in the field $\mathbb{F}_p$ takes $O(1)$ word operations.

We let M be a multiplication time for integers and polynomials, so that integers of word length at most n and polynomials of degree at most n can be multiplied using $O(M(n))$ word operations or coefficient additions and multiplications, respectively. Using classical arithmetic corresponds to $M(n) = n^2$, Karatsuba's (1962) algorithm has $M(n) = n^{1.59}$, and the asymptotically fastest currently known algorithm of Schönhage & Strassen (1971), which is based on the Fast Fourier Transform, leads to $M(n) = n \log n \log\log n$. We assume that M is *sub-additive*, so that $M(m) + M(n) \leq M(m + n)$, and that $m\, M(n) \leq M(mn) \leq m^2 M(n)$, for all $m, n \in \mathbb{N}$.

We will freely use the following facts, whose proofs or references can be found in Aho, Hopcroft & Ullman (1974) or von zur Gathen & Gerhard (1999).

Fact 3.13 (Basic arithmetic). *For integers of word length at most n or polynomials of degree at most n, we have the following cost estimates in word operations or arithmetic operations on coefficients, respectively.*

(i) *Division with remainder:* $O(M(n))$,
(ii) *Extended Euclidean Algorithm:* $O(M(n) \log n)$,
(iii) *Reduction modulo n single precision integers or simultaneous evaluation at n points:* $O(M(n) \log n)$,
(iv) *Chinese Remainder Algorithm for n pairwise coprime single precision integers or interpolation at n points:* $O(M(n) \log n)$.

In fact, when $M(n) \in \Omega(n^{1+\varepsilon})$ for some positive ε, as for the classical and Karatsuba's (1962) algorithm, then the cost for (ii), (iii), and (iv) is $O(M(n))$ as well. In particular, the cost for all four tasks is $O(n^2)$ when using classical arithmetic.

We note that the cost estimates above are also valid for Algorithm 3.4. In particular, there is an asymptotically fast variant of this algorithm that computes an arbitrary row of the output with running time $O(M(n) \log n)$ (see Reischert 1997).

Remark 3.14 (Kronecker substitution). *Let $f, g \in \mathbb{Z}[x]$ be of degree at most n and max-norm less than 2^λ. Then we have $\|fg\|_\infty < (n + 1)2^{2\lambda}$. Let $t = 2\lambda + \lceil \log_2(n+1) \rceil$. Using the Kronecker substitution $x = 2^t$, we can read off the coefficients of the product polynomial fg from the integer $f(2^t)g(2^t)$. Thus we can compute fg at a cost of $O(M(n(\lambda + \log n)))$ word operations with fast arithmetic.*

Kronecker (1882) introduced in §4 a closely related concept to reduce the factorization of multivariate polynomials to factorization of univariate polynomials, by substituting suitable powers of the first variable for all other variables.

Lemma 3.15. *Let R be a ring, $f_1, \ldots, f_r \in R[x]$ nonconstant polynomials, $f = f_1 \cdots f_r$, and $n = \deg f$.*

(i) *Given $f_1, \ldots, f_r$, we can compute f at a cost of $O(n^2)$ additions and multiplications in R with classical arithmetic and $O(M(n) \log n)$ with fast arithmetic.*
(ii) *If $R = \mathbb{Z}$ and $\prod_{1 \leq i \leq r} \|f_i\|_1 < 2^\lambda$, then the cost is $O(n^2\lambda^2)$ word operations with classical arithmetic and $O(M(n\lambda) \log r)$ or $O^\sim(n\lambda)$ with fast arithmetic.*

(iii) By using a modular algorithm, the cost drops to $O(n^2\lambda + n\lambda^2)$ word operations with classical arithmetic and $O(\lambda\,\mathsf{M}(n)\log r + n\,\mathsf{M}(\lambda)\log\lambda)$ with fast arithmetic, if we assume that there are sufficiently many single precision primes and ignore the cost for prime finding.

Proof. Let $n_i = \deg f_i$ for $1 \le i \le r$. With classical arithmetic, we successively compute $g_i = f_1 \cdots f_i$, for $2 \le i \le r$. This takes $O(n_i n)$ additions and multiplications in R to compute g_i from g_{i-1}, in total $O(n^2)$. If $R = \mathbb{Z}$, then

$$\|g_i\|_\infty \le \|g_i\|_1 \le \prod_{1 \le j \le i} \|f_i\|_1 < 2^\lambda \,,$$

so that the coefficients of all intermediate results have word size at most λ, and hence one multiplication in $\mathbb{Z}$ takes $O(\lambda^2)$ word operations.

With fast arithmetic, we think of a binary tree of depth $\lceil \log_2 r \rceil$ with $f_1, \ldots, f_r$ at the leaves, f at the root, and such that each inner vertex is the product of its two children. The arithmetic cost estimate is due to Strassen (1973) and Borodin & Moenck (1974) in the case when all f_i are linear; see Lemma 10.4 in von zur Gathen & Gerhard (1999) for the general case. Now let $g \in \mathbb{Z}[x]$ be an inner vertex and $m = \deg g$. By a similar argument as above, we have $\|g\|_\infty < 2^\lambda$. By using the Kronecker substitution $x = 2^\lambda$, we can reduce the computation of g to the multiplication of two integers of word size $O(m\lambda)$, taking $O(\mathsf{M}(m\lambda))$ word operations. The sum of the degrees at one fixed level of the tree is at most n, so that the overall cost for that level is $O(\mathsf{M}(n\lambda))$, by the sub-additivity of M, and the claim follows since there are $O(\log r)$ levels. This concludes the proof of (i) and (ii).

The modular algorithm reduces all f_i modulo $O(\lambda)$ single precision primes, taking $O(n\lambda^2)$ word operations with classical arithmetic and $O(n\,\mathsf{M}(\lambda)\log\lambda)$ with fast arithmetic. Then it computes the modular products for each prime independently, at a cost of λn^2 word operations with classical arithmetic and $O(\lambda\,\mathsf{M}(n)\log r)$ with fast arithmetic, by (i). Finally, the coefficients of f are reconstructed by Chinese remaindering from their $O(\lambda)$ modular images. This takes $O(\lambda^2)$ word operations per coefficient with classical arithmetic, in total $O(n\lambda^2)$, and $O(n\,\mathsf{M}(\lambda)\log\lambda)$ with fast arithmetic. Now (iii) follows by adding up costs. $\square$

In fact, by balancing the binary tree with respect to the degree, the factor $\log r$ in the above lemma can be replaced by the entropy

$$H(\deg f_1, \ldots, \deg f_r) = \sum_{1 \le i \le r} -\frac{\deg r_i}{n} \log_2 \frac{\deg r_i}{n} \tag{3.1}$$

(Theorem 2.2 in Strassen 1983).

The following fact can be found in von zur Gathen & Gerhard (1999), Theorem 5.31 and Exercise 10.18.

Fact 3.16 (Partial fraction decomposition). *Let F be a field, $f, g \in F[x]$ nonzero and coprime with $\deg f < \deg g = n$, and $g = g_1^{e_1} \cdots g_t^{e_t}$, with nonconstant and* *pairwise coprime polynomials $g_1, \ldots, g_t \in F[x]$ and positive integers $e_1, \ldots, e_t$. Then we can compute the partial fraction decomposition*

$$\frac{f}{g} = \sum_{1 \leq i \leq t} \sum_{1 \leq j \leq e_i} \frac{\gamma_{ij}}{g_i^j},$$

i.e., the unique polynomials $\gamma_{ij} \in F[x]$ of degree less than $\deg g_i$ for all i, j, using $O(n^2)$ arithmetic operations in F with classical arithmetic and $O(\mathsf{M}(n) \log n)$ with fast arithmetic.

Lemma 3.17. *Let R be a ring and $n \in \mathbb{N}$.*

(i) *Let $g_1, \ldots, g_r, a_1, \ldots, a_r \in R[x]$, $g = g_1 \cdots g_r$, and assume that $\deg g = n \geq r$ and $\deg a_i < \deg g_i$ for $1 \leq i \leq r$. Then we can compute*

$$f = \sum_{1 \leq i \leq r} a_i \frac{g}{g_i}$$

using $O(\mathsf{M}(n) \log r)$ arithmetic operations in R. If $R = \mathbb{Z}$, $\|a_i\|_1 \leq A$ for all i, and $\|h\|_1 \leq B$ for all divisors h of g, then $\|f\|_1 \leq rAB$, and the computation of f takes $O(\mathsf{M}(n \log(nAB)) \log r)$ word operations.

(ii) *Let $h, c_0, \ldots, c_{r-1} \in R[x]$ and assume that $r \deg h = n$ and $\deg c_i < \deg h$ for all i. Then we can compute*

$$p = \sum_{0 \leq i < r} c_i h^i$$

using $O(\mathsf{M}(n) \log r)$ arithmetic operations in R. If $R = \mathbb{Z}$, $\|c_i\|_1 \leq A$ and $\|h^i\|_1 \leq B$ for all i, then $\|p\|_1 \leq rAB$, and the computation of p takes $O(\mathsf{M}(n \log(nAB)) \log r)$ word operations.

Proof. (i) For the arithmetic cost estimate see, e.g., Theorem 10.21 in von zur Gathen & Gerhard (1999). If $R = \mathbb{Z}$, then

$$\|f\|_1 \leq \sum_{1 \leq i \leq r} \|a_i\|_1 \cdot \left\| \frac{g}{g_i} \right\|_1 \leq rAB.$$

The intermediate results in the divide-and-conquer algorithm described in von zur Gathen & Gerhard (1999) are either divisors of g or of a similar form as f, so that rAB is an upper bound on the one-norm of any intermediate result. The cost estimate now follows from Remark 3.14.

(ii) The arithmetic cost estimate follows, e.g., from Exercise 9.20 in von zur Gathen & Gerhard (1999). The estimates for $R = \mathbb{Z}$ follow by similar arguments as in the proof of (i). $\square$

For a proof of the following fact and references, see von zur Gathen & Gerhard (1999), Sect. and Notes 5.10, and Wang & Pan (2003).

Fact 3.18 (Rational number reconstruction). *Given n pairwise coprime single precision primes $p_1, \ldots, p_s$, an integer c with $2|c| < p_1 \cdots p_s$, and a bound $C < p_1 \cdots p_s$, we can decide whether there exist coprime integers a, b such that $|a| < C$, $0 < b < p_1 \cdots p_s / 2C$, none of the primes divides b, and $ab^{-1} \equiv c \bmod p_i$ for $1 \leq i \leq s$, and if so, compute these unique a, b. This amounts to essentially one application of the Chinese Remainder Algorithm and one application of the Extended Euclidean Algorithm, so that it takes $O(s^2)$ word operations with classical arithmetic, and $O(\mathsf{M}(s) \log s)$ with fast arithmetic.*

Remark 3.19. *We often want to reconstruct a vector of rational coefficients that is the unique solution of a square system of linear equations with integer coefficients. In this case Cramer's (1750) rule implies that the determinant of the coefficient matrix is a common denominator of all coefficients of the solution vector. Heuristically, one may then expect that the denominator b of the first reconstructed coefficient is already very close to a common denominator. If we multiply any other coefficient by b, then this is likely to be a rational number with a "small" denominator, for which rational reconstruction via the Extended Euclidean Algorithm works faster than for a rational number with "large" denominator. This heuristic is sometimes useful to speed up the whole rational reconstruction process.*

Fact 3.20 (Hensel lifting). *Let $f \in \mathbb{Z}[x]$ of degree $n \geq 1$ be primitive and squarefree, $k, r \in \mathbb{N}_{\geq 1}$, $p \in \mathbb{N}$ a prime not dividing $\mathrm{lc}(f)$ such that $f \bmod p$ is squarefree, and*

$$f \equiv \mathrm{lc}(f) g_1 \cdots g_r \bmod p$$

a factorization of f modulo p into nonconstant monic polynomials $g_1, \ldots, g_r \in \mathbb{Z}[x]$ that are pairwise coprime modulo p. Then there exist nonconstant monic polynomials $f_1, \ldots, f_r \in \mathbb{Z}[x]$ with

$$f \equiv \mathrm{lc}(f) f_1 \cdots f_r \bmod p^k, \quad f_i \equiv g_i \bmod p \text{ for } 1 \leq i \leq r .$$

These polynomials are unique modulo p^k. If p is single precision and $\|f\|_\infty < p^k$, then they can be computed using $O(n^2 k^2)$ word operations with classical arithmetic and $O(\mathsf{M}(n) \log r \cdot (\mathsf{M}(k) + \log n))$ with fast arithmetic.

Proof. The uniqueness follows from Theorem 15.14 in von zur Gathen & Gerhard (1999), and Theorem 15.18 in the 1999 edition gives a cost estimate of $O(\mathsf{M}(n) \mathsf{M}(k) \log r)$ word operations. This does not include the cost for setting up a factor tree modulo p, which amounts to $O(\mathsf{M}(n)(\log n) \log r)$ and $O(n^2 \log r)$ word operations with fast and classical arithmetic, respectively, and the estimate for fast arithmetic follows. As in Lemma 3.15, a finer analysis shows that the factor $\log r$ can be replaced by the entropy (3.1). For classical arithmetic, where $\mathsf{M}(n) = n^2$, the above estimate leads to a bound of $O(n^2 k^2 \log r)$ word operations. We now show how to shave off the factor $\log r$.

We consider one lifting step from a factor tree modulo $p^{2^{j-1}}$ to a factor tree modulo p^{2^j}, for some $j \geq 0$. For each vertex $v \in \mathbb{Z}[x]$ in the factor tree, let $p_v = (\deg v)/n$. We assume that the factor tree is balanced with respect to the degree, so that $i < 1 - \log_2 p_v$ holds for all leaves at level i (see Exercise 10.5 in von zur Gathen & Gerhard 1999). Topological induction along the factor tree shows that this inequality is valid for the inner vertices as well. With classical arithmetic, the cost at an inner vertex v is at most $c(\deg v)^2$ arithmetical operations modulo p^{2^j}, for some positive real constant c. Summing over all inner vertices, we find that the cost for lifting the complete tree is at most

$$
\begin{aligned}
\sum_v c(\deg v)^2 &= cn^2 \sum_v p_v^2 = cn^2 \sum_{i \geq 0} \sum_{v \text{ at level } i} p_v^2 < cn^2 \sum_{i \geq 0} \sum_{v \text{ at level } i} 2^{2-2i} \\
&\leq 4cn^2 \sum_{i \geq 0} 2^i \cdot 2^{-2i} = 4cn^2 \sum_{i \geq 0} 2^{-i} = 8cn^2
\end{aligned}
$$

arithmetic operations modulo p^{2^j} or $O(n^2 2^{2j})$ word operations. Summing over all $j \leq \lceil \log_2 k \rceil$ gives a total cost estimate of $O(n^2 k^2)$. By a similar argument, the cost for setting up the factor tree modulo p is $O(n^2)$ word operations, and the estimate for classical arithmetic follows. $\square$

See von zur Gathen (1984) and Bernardin (1999) for other Hensel lifting algorithms and Notes 15.4 in von zur Gathen & Gerhard (1999) for historical notes and more references.

Fact 3.21 (Factorization over finite fields). *Let $p \in \mathbb{N}$ be a prime and $f \in \mathbb{F}_p[x]$ of degree $n \geq 1$.*

(i) *We can compute all roots of f in $\mathbb{F}_p$ with an expected number of $O(n^2 \log n \cdot \log p)$ arithmetic operations in $\mathbb{F}_p$ with classical arithmetic and $O(\mathsf{M}(n) \cdot \log(pn) \log n)$ or $O^\sim(n \log p)$ with fast arithmetic.*

(ii) *We can compute the irreducible factorization of f in $\mathbb{F}_p[x]$ with an expected number of $O(n^3 + n^2 \log p)$ arithmetic operations in $\mathbb{F}_p$ with classical arithmetic and $O(\mathsf{M}(n^2) \log n + \mathsf{M}(n) \log n \cdot \log p)$ or $O^\sim(n^2 + n \log p)$ with fast arithmetic.*

See Corollaries 14.16 and 14.30 and Theorem 14.32 in von zur Gathen & Gerhard (1999) for proofs and the notes to Chap. 14 for references.

The following lemma says that there are sufficiently many single precision primes for all practical purposes. A similar result is in Theorem 1.8 of Giesbrecht (1993).

Lemma 3.22. *For an integer $\omega \geq 7$, the number π_ω of primes between $2^{\omega-1}$ and 2^ω is at least*

$$
\pi_\omega > \frac{2^\omega \log_2 e}{2\omega} \left(1 - \frac{1 + \log_2 e}{\omega - 1} \right) .
$$

For example, we have $\pi_{32} > 8.9 \cdot 10^8$ and $\pi_{64} > 1.99 \cdot 10^{18}$.

Proof. By the prime number theorem, the number $\pi(x)$ of primes less than $x \in \mathbb{R}_{>0}$ is about $x / \ln x$, and more precisely

$$\frac{x}{\ln x}\left(1 + \frac{1}{2\ln x}\right) < \pi(x) < \frac{x}{\ln x}\left(1 + \frac{3}{2\ln x}\right) \text{ if } x \geq 59$$

(Rosser & Schoenfeld 1962). Using this for $x = 2^\omega$ and $x = 2^{\omega-1}$, we obtain, after some calculation, the claim about $\pi_\omega = \pi(2^\omega) - \pi(2^{\omega-1})$. □

Remark 3.23. • *We stress that the cost estimates for all our modular algorithms are not strictly asymptotic estimates, since they require a certain number of single precision primes and there are only finitely many of them for a fixed ω. It is possible to derive proper asymptotic estimates by allowing for primes of arbitrary precision and using the prime number theorem to bound their word length. The corresponding estimates are usually slower by a logarithmic factor, and we do not state them in what follows.*

• *There are several possible ways how to find random single precision primes for our modular algorithms. In our analyses, we do not discuss this and also ignore the cost. Suppose that we need s distinct random single precision primes. In principle, we can find them by repeatedly choosing random integers between $2^{\omega-1}$ and 2^ω, subjecting them to a probabilistic primality test (like the ones of Solovay & Strassen 1977 or Miller 1976 and Rabin 1976, 1980), and stop as soon as sufficiently many "probable" primes are found. Since we are looking for primes of a fixed precision, the cost for one primality test is $O(1)$ word operations, and the expected number of trials until a prime is found is also $O(1)$, provided that there are sufficiently many of them, say at least $2s$ (one has to be a bit careful about getting s distinct primes here). Thus the cost for randomly and independently choosing s single precision primes is $O(s)$ word operations. The prime number theorem (see Lemma 3.22) guarantees that there are sufficiently many single precision primes for all practical purposes.*

A second possibility is to use the well-known sieve of Eratosthenes to compute all primes below 2^ω, at a cost of $O(2^\omega \cdot \omega \log \omega)$ word operations (see, e.g., von zur Gathen & Gerhard (1999), Theorem 18.10). When ω is fixed, then this is $O(1)$, at least in theory, but in practice this is probably too expensive for $\omega \geq 32$, in particular too memory-consuming.

In practice, however, it is useful to maintain a list of precomputed single precision primes and then to choose randomly from this list if there are sufficiently many of them or simply to take the first s primes from the list otherwise. In the last case, we no longer have a guaranteed bound on the error probability of our algorithms; in principle, it may happen that all primes from our list are "unlucky".

Nevertheless, using such a precomputed list is very attractive, for two reasons. Firstly, the prime list can be reused for many different kinds of modular algorithms, such as gcd computations and linear algebra. Secondly, we may choose all primes p to be Fourier primes, so that $p - 1$ is divisible by a "large" power of two, say $2^{\omega/2}$. This allows for very efficient implementations of FFT-based polynomial arithmetic, as has been shown by Shoup (1995) with his software package NTL, at the expense of reducing the number of suitable primes.

The final lemma in this chapter is about probability and will be put to use in Chap. 5.

Lemma 3.24. *Consider the following random experiment. An urn contains b black and w white balls, and we draw $k \le w + b$ balls without replacement.*

 (i) If $w \ge b$, then the probability that at most $k/2$ balls are white is at most $1/2$.
(ii) If $w \ge 4b$ and $k \ge 8$, then the probability that at most $k/2$ balls are white is at most $1/4$.

Proof. We only prove (ii); see, e.g., Exercise 6.31 in von zur Gathen & Gerhard (1999) for a proof of (i). Let X denote the random variable counting the number of white balls after k trials. Then X has a hypergeometric distribution

$$\operatorname{prob}(X = i) = \frac{\binom{w}{i}\binom{b}{k-i}}{\binom{w+b}{k}}$$

for $0 \le i \le k$, with mean

$$\mu = \mathcal{E}X = \sum_{0 \le i \le k} k \cdot \operatorname{prob}(X = k) = \frac{kw}{w+b} \ge \frac{4}{5}k .$$

We want to prove that $\operatorname{prob}(X \le k/2) \le 1/4$. For $0 < t \le \mu/k$, the following tail inequality holds:

$$\operatorname{prob}(X - \mu \le -tk) \le e^{-2kt^2} \tag{3.2}$$

(see, e.g., Chvátal 1979). We let $c = \sqrt{\ln 4}$ and $t = \sqrt{c/2k}$. Then

$$\frac{k}{2} - \mu \le -\frac{3}{10}k \le -\frac{3\sqrt{2}}{5}\sqrt{k} < -\sqrt{\frac{c}{2}}\sqrt{k} = -tk ,$$

since $k \ge 8$, and (3.2) implies that

$$\begin{aligned}
\operatorname{prob}\left(X \le \frac{k}{2}\right) &= \operatorname{prob}\left(X - \mu \le \frac{k}{2} - \mu\right) \le \operatorname{prob}(X - \mu \le -tk) \le e^{-2kt^2} \\
&= e^{-\ln 4} = \frac{1}{4} . \ \square
\end{aligned}$$

4. Change of Basis

In this chapter, we discuss conversion algorithms for univariate polynomials in $R[x]$, where R is a commutative ring with 1. These algorithms convert between a basis of the form

$$\mathcal{M}_{\mathbf{b}} = \Big((x - b_0)(x - b_1) \cdots (x - b_{i-1}) \Big)_{i \in \mathbb{N}} ,$$

where $\mathbf{b} = (b_0, b_1, \ldots) \in R^{\mathbb{N}}$ is an infinite sequence of arbitrary constants from R, and the usual *monomial basis* $\mathcal{M} = \mathcal{M}_{\mathbf{0}} = (x^i)_{i \in \mathbb{N}}$. This comprises important special cases such as the *shifted monomial basis* $((x - b)^i)_{i \in \mathbb{N}}$, for some $b \in R$, the *falling factorial basis* $\mathcal{F} = (x^{\underline{i}})_{i \in \mathbb{N}}$, where $x^{\underline{i}} = x(x - 1) \cdots (x - i + 1)$ is the *ith falling factorial*, and the *rising factorial basis* $(x^{\overline{i}})_{i \in \mathbb{N}}$, where $x^{\overline{i}} = x(x + 1) \cdots (x + i - 1)$ is the *ith rising factorial*. We give cost estimates for our algorithms in terms of arithmetic operations in R and, in the case $R = \mathbb{Z}$, also in word operations. For the shifted monomial basis, the conversion is simply a *Taylor shift*, which we discuss in the following section.

Umbral calculus (Rota 1975; Roman & Rota 1978; Roman 1984) covers the similarities between various polynomial bases, comprising all of the special cases mentioned above, and relates monomial bases and linear operators on polynomials that commute with the differential operator. For example, the relations $Dx^n = nx^{n-1}$ and $\Delta(x^{\underline{n}}) = nx^{\underline{n-1}}$ for $n \in \mathbb{N}$ say that the monomial basis $\mathcal{M}$ and the falling factorial basis $\mathcal{F}$ are the *associated sequences* of the differential operator D and the difference operator Δ, respectively. One of the nice results of umbral calculus is the following. Given a sequence $(p_n)_{n \in \mathbb{N}}$ of monic polynomials with $\deg p_n = n$ for all n and $p_n(0) = 0$ if $n > 0$, the linear operator T defined by $Tp_n = np_{n-1}$ for $n \in \mathbb{N}$, such that $(p_n)_{n \in \mathbb{N}}$ is the associated sequence of T, commutes with the differential operator D if and only if the sequence satisfies a binomial theorem of the form

$$p_n(x + y) = \sum_{0 \le i \le n} \binom{n}{i} p_i(x) p_{n-i}(y) \text{ for } n \in \mathbb{N} .$$

(See Chap. 2 in Roman 1984.) We use this binomial theorem for the falling factorial basis $\mathcal{F}$ in Sect. 4.3 below.

J. Gerhard: Modular Algorithms, LNCS 3218, pp. 41–60, 2004.
© Springer-Verlag Berlin Heidelberg 2004

4.1 Computing Taylor Shifts

In this section, we discuss algorithms for converting between $\mathcal{M}$ and $\mathcal{M}_{(b,b,\ldots)}$ for some fixed $b \in R$. Given the coefficients $f_0, \ldots, f_n \in R$ of the polynomial $f = \sum_{0 \le i \le n} f_i x^i \in R[x]$, we want to find $g_0, \ldots, g_n \in R$ such that $f(x) = \sum_{0 \le i \le n} g_i \cdot (x - b)^i$, or vice versa. If we let $g = \sum_{0 \le i \le n} g_i x^i$, then $f(x) = g(x - b)$, or equivalently,

$$g(x) = \sum_{0 \le k \le n} g_k x^k = f(x + b) = \sum_{0 \le i \le n} f_i \cdot (x + b)^i , \qquad (4.1)$$

so that the conversions in both directions can be reduced to computing *Taylor shifts* by b and $-b$, respectively. This is a basic operation in many computer algebra systems (e.g., `translate` in MAPLE).

Problem 4.1 (Taylor shift). *Let R be a ring. Given $b \in R$ and the coefficients of a nonzero polynomial $f \in R[x]$ of degree n, compute the coefficients of $f(x + b)$ with respect to the monomial basis.*

Writing out (4.1) explicitly, for $0 \le k \le n$ we have

$$g_k = \sum_{k \le i \le n} \binom{i}{k} f_i b^{i-k} . \qquad (4.2)$$

An important special case is $b = \pm 1$. The following lemma says how the coefficient size of a polynomial increases at most by a Taylor shift in the case $R = \mathbb{Z}$.

Lemma 4.2. *Let $f \in \mathbb{Z}[x]$ be nonzero of degree $n \in \mathbb{N}$ and $b \in \mathbb{Z}$. Then*

$$\|f(x + b)\|_\infty \le \|f(x + b)\|_1 \le (|b| + 1)^n \|f\|_1 \le (n + 1)(|b| + 1)^n \|f\|_\infty .$$

For $b = \pm 1$, the following sharper bound is valid:

$$\|f(x \pm 1)\|_\infty \le \|f(x \pm 1)\|_1 \le 2^{n+1} \|f\|_\infty .$$

Proof. Let $f = \sum_{0 \le i \le n} f_i x^i$. Then

$$\|f(x + b)\|_1 = \left\| \sum_{0 \le i \le n} f_i \cdot (x + b)^i \right\|_1 \le \sum_{0 \le i \le n} |f_i|(1 + |b|)^i \le (|b| + 1)^n \|f\|_1 .$$

Moreover, we have

$$\|f(x + b)\|_1 \le \|f\|_\infty \sum_{0 \le i \le n} (1 + |b|)^i = \|f\|_\infty \frac{(1 + |b|)^{n+1} - 1}{|b|} ,$$

and the claim for $|b| = 1$ follows. $\square$

For all $b, B, n \in \mathbb{N}_{>0}$, the polynomial $f = B \sum_{0 \leq i \leq n} x^i$ achieves the first bound within a factor of at most $n + 1$:

$$\|f(x + b)\|_1 = B \sum_{0 \leq i \leq n} (b + 1)^i \geq \|f\|_\infty (b + 1)^n .$$

We now discuss several computational methods for computing Taylor shifts and analyze their costs. The presentation follows closely von zur Gathen & Gerhard (1997). We start with algorithms employing classical arithmetic.

A. *Horner's (1819 method)*: We compute

$$g(x) = f_0 + (x + b) \left(f_1 + \cdots + (x + b) \left(f_{n-1} + (x + b) f_n \right) \cdots \right)$$

in n steps

$$g^{(n)} = f_n, \quad g^{(i)} = (x + b) \cdot g^{(i+1)} + f_i \text{ for } i = n - 1, \ldots, 0 ,$$

and obtain $g = g^{(0)}$.

B. *Shaw & Traub's (1974) method* (if b is not a zero divisor): Compute $f^*(x) = f(bx)$, $g^*(x) = f^*(x + 1)$, using A, and then $g(x) = g^*(x/b)$. (See also de Jong & van Leeuwen 1975; Schönhage 1998; and §4.6.4 in Knuth 1998.)

C. *Multiplication-free method* (if $R = \mathbb{Z}$): Suppose that $b > 0$. We denote by E^b the shift operator acting as $(E^b f)(x) = f(x + b)$ on polynomials. We write $b = \sum_{0 \leq j < d} b_j 2^j$ in binary, with $b_j \in \{0, 1\}$ for all j. Then $E^b = E^{b_{d-1} 2^{d-1}} \circ \cdots \circ E^{b_1 \cdot 2} \circ E^{b_0}$. For $j = 0, 1, \ldots, d - 1$, successively apply method A with $b_j 2^j$. The case $b < 0$ is handled by noting that $E^b = M E^{-b} M$, where M denotes the operator $(M f)(x) = f(-x)$.

Method C seems to be new. We note that both methods B and C boil down to Horner's method if $b = \pm 1$.

Theorem 4.3. *Let R be a ring, $f \in R[x]$ of degree $n \geq 1$, and $b \in R$ nonzero. Then method A solves the Taylor shift problem 4.1 with $O(n^2)$ additions and multiplications in R, method B takes $O(n^2)$ additions plus $O(n)$ multiplications and divisions in R, and method C uses $O(n^2 d)$ additions in $R = \mathbb{Z}$.*

More precisely, if $R = \mathbb{Z}$, $\|f\|_\infty < 2^\lambda$, and $|b| < 2^d$, then the cost in word operations to compute $(E^b f)(x) = f(x + b) \in \mathbb{Z}[x]$ for the three methods above is

A: $O(n^2 d(nd + \lambda))$ *with classical and* $O(n^2 \mathsf{M}(nd + \lambda))$ *with fast integer arithmetic,*

B: $O(n^2 d(nd + \lambda))$ *with classical and* $O(n^2(nd + \lambda) + n \mathsf{M}(nd + \lambda))$ *with fast integer arithmetic,*

C: $O(n^2 d(nd + \lambda))$.

Proof. A. In step $n - i$, we have at most $i - 1$ additions and i multiplications by b, in total $O(n^2)$ additions and multiplications in R each. In the integer case, the word size of the integers involved is $O(nd + \lambda)$, by Lemma 4.2. Thus the cost for one addition is $O(nd + \lambda)$, and the cost for one multiplication by b is $O(d(nd + \lambda))$ with classical multiplication and $O(\mathsf{M}(nd + \lambda))$ with fast multiplication. Thus we get a total cost of $O(n^2 d(nd + \lambda))$ and $O(n^2\,\mathsf{M}(nd + \lambda))$ word operations, respectively.

B. We have $n - 1$ multiplications, of size $O(d) \times O(nd)$ in the integer case, for the computation of $b^2, \ldots, b^n$, plus n multiplications, of size $O(\lambda) \times O(nd)$ in the integer case, for the computation of $f_1 b, \ldots, f_n b^n$ which yields f^*. In using method A to compute $f^*(x + 1)$, no multiplications are required, and the cost is $O(n^2)$ additions in R. In the case $R = \mathbb{Z}$, Lemma 4.2 implies that all additions involve integers of size $O(nd + \lambda)$, and hence the cost is $O(n^2(nd + \lambda))$ word operations. Finally, we have n exact divisions by $b, b^2, \ldots, b^n$ in R. In the integer case, the dividends are of size $O(nd + \lambda)$ and the divisors of size $O(nd)$. The cost for such a division is $O(nd(nd + \lambda))$ word operations with classical integer arithmetic and $O(\mathsf{M}(nd + \lambda))$ with fast arithmetic. We obtain a total of $O(n^2 d(nd+\lambda))$ word operations with classical integer arithmetic and $O(n^2(nd + \lambda) + n\,\mathsf{M}(nd + \lambda))$ with fast arithmetic.

C. When $\pm b$ is a power of two, a multiplication by $\pm b$ is – up to sign – just a shift in the binary representation, yielding a total cost of $O(n^2)$ additions or $O(n^2(nd + \lambda))$ word operations for A. Applying this special case at most d times gives a total running time of $O(n^2 d)$ additions or $O(n^2 d(nd + \lambda))$ word operations. $\square$

Corollary 4.4. *With the assumptions of Theorem 4.3 in the case $R = \mathbb{Z}$, the cost for the three algorithms in word operations is*

(i) $O(n^2(n + \lambda))$ *if* $b = \pm 1$,
(ii) $O(n^3 \lambda^2)$ *with classical integer arithmetic if* $d \in O(\lambda)$,
(iii) $O^\sim(n^3 \lambda)$ *for A and* $O(n^3 \lambda)$ *for B, respectively, with fast integer arithmetic if* $d \in O(\lambda)$.

Table 4.1. Running times in CPU seconds with method A for $b = 1$, degree $n - 1$, "small" coefficients between $-n$ and n, and "large" coefficients between -2^n and 2^n

n	small	large
128	0.001	0.002
256	0.005	0.010
512	0.030	0.068
1024	0.190	0.608
2048	2.447	8.068
4096	22.126	65.758
8192	176.840	576.539

Tables 4.1, 4.2, and 4.3 show the performances of methods A, B, and C in our experiments. Running times are given in average CPU seconds for 10 pseudorandomly chosen inputs on a Linux PC with an 800 MHz Pentium III CPU. Our software

Table 4.2. Running times in CPU seconds with methods A,B,C for degree $n-1$ and "small" coefficients and values of b between $-n$ and n

n	A	B	C
128	0.009	0.005	0.013
256	0.067	0.044	0.102
512	0.541	0.494	0.802
1024	5.897	7.053	9.786
2048	48.386	57.504	98.722
4096	499.149	601.716	1078.450

Table 4.3. Running times in CPU seconds with methods A,B,C for degree $n-1$ and "large" coefficients and values of b between -2^n and 2^n

n	A	B	C
128	0.302	0.298	1.653
256	7.840	8.931	65.840
512	279.988	259.658	2232.080

is written in C++. For arithmetic in $\mathbb{Z}$, we have used Victor Shoup's highly optimized C++ library NTL version 5.0c for integer and polynomial arithmetic, parts of which are described in Shoup (1995). It uses Karatsuba's (Karatsuba & Ofman 1962) method for multiplying large integers.

Next we discuss three methods employing fast polynomial arithmetic.

D. *Paterson & Stockmeyer's (1973) method*: We assume that $(n+1) = m^2$ is a square (padding f with leading zeroes if necessary), and write $f = \sum_{0 \leq i < m} f^{(i)} x^{mi}$, with polynomials $f^{(i)} \in R[x]$ of degree less than m for $0 \leq i < m$.
   1. Compute $(x+b)^i$ for $1 \leq i \leq m$.
   2. For $0 \leq i < m$, compute $f^{(i)}(x+b)$ as a linear combination of $1, (x+b), (x+b)^2, \ldots, (x+b)^{m-1}$.
   3. Compute
$$g(x) = \sum_{0 \leq i < m} f^{(i)}(x+b) \cdot (x+b)^{mi}$$
   in a Horner-like fashion.
E. *Divide & conquer method* (von zur Gathen 1990; see also Bini & Pan 1994): We assume that $n + 1 = 2^m$ is a power of two. In a precomputation stage, we compute $(x+b)^{2^i}$ for $0 \leq i < m$. In the main stage, we write $f = f^{(0)} + x^{(n+1)/2} f^{(1)}$, with polynomials $f^{(0)}, f^{(1)} \in R[x]$ of degree less than $(n+1)/2$. Then
$$g(x) = f^{(0)}(x+b) + (x+b)^{(n+1)/2} f^{(1)}(x+b) \, ,$$
where we compute $f^{(0)}(x+b)$ and $f^{(1)}(x+b)$ recursively.
F. *Convolution method* (Aho, Steiglitz & Ullman 1975; see also Schönhage, Grotefeld & Vetter 1994, §9.3): This only works if $n!$ is not a zero divisor in R. After multiplying both sides of (4.2) by $n! \, k!$, we obtain

$$n!k!\,g_k = \sum_{k \le i \le n} (i!\,f_i) \cdot \frac{n!\,b^{i-k}}{(i-k)!}$$

in R. If we let $u = \sum_{0 \le i \le n} i!\,f_i x^{n-i}$ and $v = n! \sum_{0 \le j \le n} b^j x^j / j!$ in $R[x]$, then $n!k!\,g_k$ is the coefficient of x^{n-k} in the product polynomial uv.

Theorem 4.5. *Let R be a ring, $f \in R[x]$ of degree $n \ge 1$, and $b \in R$ nonzero. Then method D solves the Taylor shift problem 4.1 with $O(n^{1/2}\mathsf{M}(n))$ additions and multiplications in R, method E takes $O(\mathsf{M}(n) \log n)$ additions and multiplications in R, and method F uses $O(\mathsf{M}(n))$ additions and multiplications plus $O(n)$ divisions in R.*

More precisely, if $R = \mathbb{Z}$, $\|f\|_\infty < 2^\lambda$, and $|b| < 2^d$, then the cost in word operations to compute $(E^b f)(x) = f(x+b) \in \mathbb{Z}[x]$ for the three methods above is

D: $O(n\,\mathsf{M}(n^{3/2}d + n^{1/2}\lambda))$ *or* $O^\sim(n^{2.5}\lambda)$,
E: $O(\mathsf{M}(n^2 d + n\lambda) \log n)$ *or* $O^\sim(n^2\lambda)$,
F: $O(\mathsf{M}(n^2 \log n + n^2 d + n\lambda))$ *or* $O^\sim(n^2\lambda)$,

where the $O^\sim$-estimates are valid if $d \in O^\sim(\lambda)$.

Proof. We first note that by using Kronecker substitution (Remark 3.14), we can multiply two polynomials of degree at most n and with coefficients of word size at most k using $O(\mathsf{M}(n(k + \log n)))$ word operations.

D. In step 1, we have $O(m^2)$ multiplications and additions, of size $O(md)$ in the integer case, by Lemma 4.2, or $O(n\,\mathsf{M}(n^{1/2}d))$ word operations. The computation of each $f^{(i)}(x + b)$ for $0 \le i < m$ in step 2 uses $O(m^2)$ multiplications and additions, of size $O(md + \lambda)$ in the case $R = \mathbb{Z}$, and the total cost of step 2 is $O(n^{3/2})$ additions and multiplications in R or $O(n^{3/2}\,\mathsf{M}(n^{1/2}d + \lambda))$ word operations. Finally, we have at most m polynomial multiplications and additions of degree $m \times O(n)$, with coefficients of size $O(nd + \lambda)$ in the integer case. By dividing the larger polynomial into blocks of size m, each such multiplication can be performed with $O(m\,\mathsf{M}(m))$ arithmetic operations, or $O(m\,\mathsf{M}(m(nd + \lambda)))$ word operations in the integer case. Thus the overall cost for step 3 is $O(n\,\mathsf{M}(n^{1/2}))$ additions and multiplications in R or $O(n\,\mathsf{M}(n^{3/2}d + n^{1/2}\lambda))$ word operations if $R = \mathbb{Z}$. This dominates the cost of the other two steps.

E. The cost of the precomputation stage is at most

$$\sum_{1 \le i < m} \mathsf{M}(2^{i-1}) \le \mathsf{M}(n)$$

additions and multiplications in R. If $R = \mathbb{Z}$, then the size of the coefficients of the polynomials in the precomputation stage is $O(nd)$, by Lemma 4.2, and the cost is

$$\sum_{1 \le i < m} O(\mathsf{M}(2^{i-1}nd)) \subseteq O(\mathsf{M}(n^2 d))$$

word operations. Let $T(n + 1)$ denote the cost of the main stage in ring operations for polynomials of degree less than $n + 1$. Then

$$T(1) = 0 \text{ and } T(n + 1) \in 2T((n + 1)/2) + O(\mathsf{M}(n)) \,,$$

and unraveling the recursion yields $T(n+1) \in O(\mathsf{M}(n) \log n)$, dominating the cost for the precomputation stage. Similarly, if we let $T(n + 1)$ denote the cost of the main stage in word operations for polynomials of degree less than $n + 1$ and with coefficients of word size at most k, with k independent of n, we find

$$T(1) = 0 \text{ and } T(n + 1) \in 2T((n + 1)/2) + O(\mathsf{M}(n(k + \log n))) \,,$$

and unraveling the recursion yields $T(n+1) \in O(\mathsf{M}(n(k+\log n)) \log n)$. The result follows from $k \in O(nd + \lambda)$.

F. Computing uv takes $O(\mathsf{M}(n))$ additions and multiplications in R. The cost for calculating $i!$ and b^i for $2 \leq i \leq n$, for computing u, v from f and these data, and for determining the g_k's from uv amounts to $O(n)$ multiplications and divisions in R.

In the integer case, the size of the coefficients of u and v is $O(n \log n + \lambda)$ and $O(n(\log n + d))$, respectively. Hence the word size of the coefficients of uv is $O(n(\log n + d) + \lambda)$, and computing uv takes $O(\mathsf{M}(n^2 \log n + n^2 d + n\lambda))$ word operations. Using $O(n \, \mathsf{M}(n(\log n + d) + \lambda))$ word operations, the coefficients of u and v can be computed, and the same number suffices to recover the g_k from the coefficients of uv. Thus the total cost is $O(\mathsf{M}(n^2 \log n + n^2 d + n\lambda))$ word operations. $\square$

We note that the input size is $\Theta(n\lambda + d)$, and by Lemma 4.2 and the discussion following it, the size of the output $f(x + b)$ is $\Theta(n(nd + \lambda))$ words, or $\Theta(n^2\lambda)$ if $d \in \Theta(\lambda)$. Thus Algorithms E and F are – up to logarithmic factors – asymptotically optimal. For $b = \pm 1$, the output size is $\Theta(n(n + \lambda))$.

If we want to compute integral shifts of the same polynomial for several $b_1, \ldots, b_k \in \mathbb{Z}$ of absolute value $2^{\Theta(d)}$, then the output size is $\Theta(kn(nd + \lambda))$, and hence the simple idea of applying method E or F k times independently is – up to logarithmic factors – asymptotically optimal.

Corollary 4.6. *Let $f \in \mathbb{Z}[x]$ be of degree $n \geq 1$ with $\|f\|_\infty < 2^\lambda$. Then the cost in word operations for computing $(Ef)(x) = f(x + 1)$ or $(E^{-1}f)(x) = f(x - 1)$ using the above algorithms is*

D: $O(n^{1/2}\mathsf{M}(n^2 + n\lambda))$ or $O^\sim(n^{1.5}(n + \lambda))$,
E: $O(\mathsf{M}(n^2 + n\lambda) \log n)$ or $O^\sim(n^2 + n\lambda)$,
F: $O(\mathsf{M}(n^2 \log n + n\lambda))$ or $O^\sim(n^2 + n\lambda)$.

Tables 4.4 and 4.5 give running times of methods D, E, and F in our experiments in average CPU seconds for 10 pseudorandomly chosen inputs on a Linux PC with an 800 MHz Pentium III CPU. Integer and polynomial arithmetic is again taken from NTL, which implements FFT-multiplication modulo Fermat numbers $2^{2^k} + 1$,

Table 4.4. Running times in CPU seconds with methods D,E,F for degree $n - 1$ and "small" coefficients and values of b between $-n$ and n

n	D	E	F
128	0.010	0.010	0.044
256	0.072	0.072	0.290
512	0.602	0.432	2.007
1024	6.364	2.989	13.958
2048	57.744	16.892	98.807
4096	722.757	125.716	787.817

Table 4.5. Running times in CPU seconds with methods D,E,F for degree $n - 1$ and "large" coefficients and values of b between -2^n and 2^n

n	D	E	F
128	0.700	0.489	0.524
256	14.894	9.566	13.262
512	420.562	166.138	234.087

as used by Schönhage & Strassen (1971), for polynomials with large coefficients, and Karatsuba's (Karatsuba & Ofman 1962) algorithm as well as a modular Chinese remaindering approach, as described by Pollard (1971), for polynomials with moderately sized coefficients.

Fig. 4.6 compares the timings from Tables 4.2 and 4.4. The conclusion is that in our computing environment method B is the best choice for small problems, and method E for large ones.

Our final algorithm that we discuss is a new modular method for computing Taylor shifts in $\mathbb{Z}[x]$. We recall that ω is the word size of our processor.

Algorithm 4.7 (Small primes modular Taylor shift).
Input: A polynomial $f \in \mathbb{Z}[x]$ of degree $n \geq 1$ and max-norm $\|f\|_\infty < 2^\lambda$, and $b \in \mathbb{Z}$ with $0 < |b| < 2^d$.
Output: The coefficients of $f(x + b) \in \mathbb{Z}[x]$.

1. $r \longleftarrow \lceil \log_2((n+1)2^{nd+\lambda+1})/(\omega - 1) \rceil$
   choose odd single precision primes $p_1 < \cdots < p_r$
2. **for** $1 \leq j \leq r$ compute $g_j \in \mathbb{Z}[x]$ of max-norm less than $p_j/2$ such that $g_j \equiv f(x + b) \bmod p_j$
3. use the Chinese Remainder Algorithm to compute $g \in \mathbb{Z}[x]$ of max-norm less than $(\prod_{1 \leq j \leq r} p_j)/2$ such that $g \equiv g_j \bmod p_j$ for $1 \leq j \leq r$
4. **return** g

Theorem 4.8. *Algorithm 4.7 solves the Taylor shift problem 4.1 correctly as specified. Steps 2 and 3 take $O(n^3 d^2 + n^2 d\lambda + n\lambda^2)$ word operations with classical arithmetic and $O(\lambda \mathsf{M}(n) + n \mathsf{M}(nd + \lambda) \log(nd + \lambda))$ or $O^\sim(n^2 d + n\lambda)$ with fast arithmetic.*

Proof. Let $m = \prod_{1 \leq j \leq r} p_j > 2^{(\omega-1)r} \geq (n+1)2^{nd+\lambda+1}$. Then $f(x + b) \equiv g \bmod m$, both sides of the congruence have max-norms less than $m/2$, by Lemma 4.2, and hence they are equal. In step 2, we first reduce f and b modulo

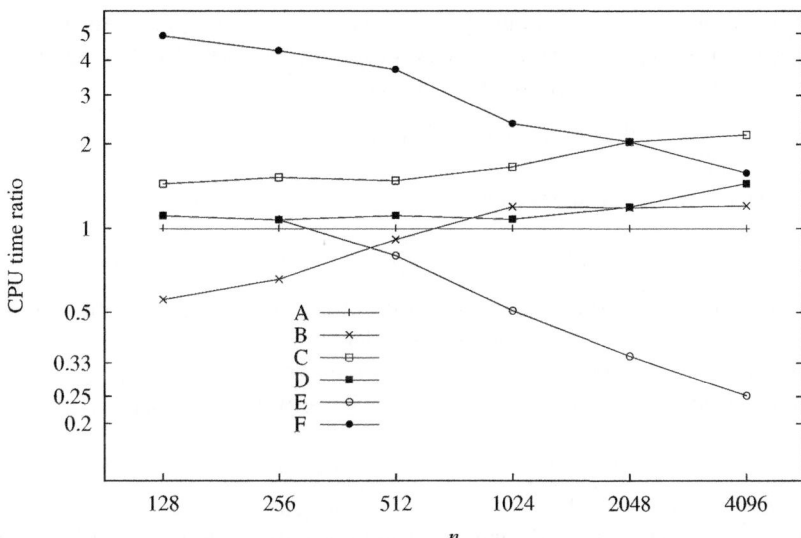

Fig. 4.6. Comparison of the running times of methods A through F for degree $n - 1$ and "small" coefficients and values of b between $-n$ and n. The vertical axis corresponds to the ratio of the computing time by the computing time of method A.

all primes, taking $O(r(n\lambda + d))$ word operations with classical arithmetic and $O(n\,\mathsf{M}(r)\log r)$ with fast arithmetic. Then we perform the Taylor shift modulo each prime, taking $O(rn^2)$ word operations with classical arithmetic and $O(r\,\mathsf{M}(n))$ with fast arithmetic, by Theorems 4.3 and 4.5. Finally, the cost for the Chinese remaindering in step 3 is $O(r^2)$ per coefficient with classical arithmetic and $O(\mathsf{M}(r)\log r)$ with fast arithmetic, in total $O(nr^2)$ and $O(n\,\mathsf{M}(r)\log r)$, respectively. The estimates now follow from $r \in O(nd + \lambda)$. $\square$

Corollary 4.9. *Let $f \in \mathbb{Z}[x]$ of degree $n \geq 1$ and $b \in \mathbb{Z} \setminus \{0\}$ with $\|f\|_\infty, |b| < 2^\lambda$. If we ignore the cost for prime finding, then the cost in word operations for computing $(E^b f)(x) = f(x + b)$ using Algorithm 4.7 is*

(i) *$O(n^3\lambda^2)$ with classical and $O(n\,\mathsf{M}(n\lambda)\log(n\lambda))$ or $O^\sim(n^2\lambda)$ with fast arithmetic,*

(ii) *$O(n^3 + n\lambda^2)$ with classical arithmetic and $O(\lambda\,\mathsf{M}(n) + n\,\mathsf{M}(n+\lambda)\log(n+\lambda))$ or $O^\sim(n^2 + n\lambda)$ with fast arithmetic if $b = \pm 1$.*

Theorem 4.8 and Corollary 4.9 indicate that the modular algorithm 4.7 with classical arithmetic is slower than methods A, B, and C.

4.2 Conversion to Falling Factorials

In this section, we address algorithms for converting between the usual monomial basis $\mathcal{M}$ and the falling factorial basis $\mathcal{F} = (x^{\underline{i}})_{i\in\mathbb{N}}$. Since the algorithms are the same in the general case of arbitrary interpolation points, we discuss them in the more general setting and give corollaries for the special case of falling factorials. Other special cases are the conversion between $\mathcal{M}$ and

- the rising factorial basis $(x^{\overline{i}})_{i \in \mathbb{N}}$, where $x^{\overline{i}} = x(x+1) \cdots (x+i-1) = (-1)^i (-x)^{\underline{i}}$, or
- the generalized factorial basis $(x(x-h)(x-2h) \cdots (x-(i-1)h))_{i \in \mathbb{N}}$, where $h \in R$ is arbitrary, or
- the shifted falling factorial basis $((x-b)^{\underline{i}})_{i \in \mathbb{N}}$, where $b \in R$ is arbitrary,

or more generally, the shifted generalized factorial basis. The presentation in this and the following section follows closely Gerhard (2000).

Given the coefficients $f_0, \ldots, f_n \in R$ of a polynomial $f = \sum_{0 \leq i \leq n} f_i x^i \in R[x]$, and arbitrary elements $b_0, \ldots, b_{n-1} \in R$, we want to find $g_0, \ldots, g_n \in R$ such that

$$f = \sum_{0 \leq i \leq n} g_i \cdot (x - b_0) \cdots (x - b_{i-1}) , \qquad (4.3)$$

or vice versa. If all b_i's are distinct, then the right hand side in (4.3) is the representation of f that we obtain from *Newton interpolation* at the sample points $b_0, \ldots, b_{n-1}$, and the g_i's are the divided differences with respect to these points. If all b_i's are equal, then this is exactly the problem of Taylor shift discussed in the previous section.

Problem 4.10 (Change of basis: forward conversion). *Let R be a ring. Given b_0, $\ldots, b_{n-1} \in R$ and the coefficients of a nonzero polynomial $f \in R[x]$, compute $g_0, \ldots, g_n \in R$ satisfying (4.3).*

Problem 4.11 (Change of basis: backward conversion). *Let R be a ring. Given $b_0, \ldots, b_{n-1} \in R$ and $g_0, \ldots, g_n \in R$, compute the coefficients of the polynomial (4.3) with respect to the monomial basis.*

The main idea of the following algorithms is to employ the following variant of *Horner's rule* (method A in Sect. 4.1):

$$f = g_0 + (x - b_0) \left(g_1 + \cdots + (x - b_{n-2}) \Big(g_{n-1} + (x - b_{n-1}) g_n \Big) \cdots \right) . \quad (4.4)$$

To analyze the bit cost of the algorithms, we need some upper bounds. If h is a polynomial and $i \in \mathbb{Z}$, then we denote the coefficient of x^i in h by $[x^i]h$, with the convention that it be zero if $i < 0$ or $i > \deg h$.

Lemma 4.12. *Let $R = \mathbb{Z}$, $n \in \mathbb{N}_{\geq 1}$, and $b_0, \ldots, b_{n-1} \in \mathbb{Z}$ of absolute value at most b. Moreover, let $u_n = (x - b_0) \cdots (x - b_{n-1})$, $u_n^* = \mathrm{rev}(u_n) = (1 - b_0 x) \cdots (1 - b_{n-1} x)$, and $v_n \in \mathbb{Z}[x]$ of degree at most n such that $u_n^* v_n \equiv 1 \bmod x^{n+1}$.*

(i) $\|u_n^*\|_\infty = \|u_n\|_\infty \leq \|u_n\|_1 \leq (b+1)^n$ and $\|v_n\|_\infty \leq (4b)^n$.
(ii) If $f_0, \ldots, f_n, g_0, \ldots, g_n \in \mathbb{Z}$ are such that (4.3) holds, $A_\infty = \max\{|f_i|: 0 \leq i \leq n\}$, $B_\infty = \max\{|g_i| : 0 \leq i \leq n\}$, and $B_1 = \sum_{0 \leq i \leq n} |g_i|$, then $A_\infty \leq (b+1)^n B_1 \leq (n+1)(b+1)^n B_\infty$ and $B_\infty \leq (n+1)(4b)^n A_\infty$.

Proof. The claim about $\|u_n\|_1$ follows since the one-norm is sub-multiplicative. We have $v_n \equiv \prod_{0 \le i < n} h_i \mod x^{n+1}$, where $h_i = 1/(1 - b_i x) = \sum_{j \ge 0} b_i^j x^j \in R[[x]]$ for $0 \le i < n$. Let $v_n = \sum_{0 \le k \le n} w_k x^k$, with all $w_k \in \mathbb{Z}$. Then

$$|w_k| = \left| \sum_{j_0 + \cdots + j_{n-1} = k} b_0^{j_0} \cdots b_{n-1}^{j_{n-1}} \right| \le \binom{n + k - 1}{n - 1} b^k \le \binom{2n - 1}{n - 1} b^n \le (4b)^n$$

for $0 \le k \le n$. This concludes the proof of (i).

For (ii), we obtain from (4.3) and (i) that

$$A_\infty = \|f\|_\infty \le \sum_{0 \le i \le n} |g_i| \, \|u_i\|_\infty \le (b+1)^n B_1 \le (n+1)(b+1)^n B_\infty \, .$$

Let $f^* = \mathrm{rev}(f)$ and $0 \le j < n$. To prove the second inequality, we claim that $g_j = [x^{n-j}](f^* v_{j+1})$. If $n \ge i > j$, then

$$u_i^* v_{j+1} \equiv \frac{(1 - b_0 x) \cdots (1 - b_{i-1} x)}{(1 - b_0 x) \cdots (1 - b_j x)} = (1 - b_{j+1} x) \cdots (1 - b_{i-1} x) \mod x^{n+1} \, ,$$

and hence $[x^{i-j}](u_i^* v_{j+1}) = 0$. If $i = j$, then $[x^{i-j}](u_i^* v_{j+1}) = u_j^*(0) v_{j+1}(0) = 1$, and if $i < j$, then $[x^{i-j}](u_i^* v_{j+1}) = 0$, by definition. Thus (4.3) yields

$$[x^{n-j}](f^* v_{j+1}) = [x^{n-j}] \sum_{0 \le i \le n} g_i u_i^* x^{n-i} v_{j+1} = \sum_{0 \le i \le n} g_i [x^{n-j}](u_i^* v_{j+1} x^{n-i})$$

$$= \sum_{0 \le i \le n} g_i [x^{i-j}](u_i^* v_{j+1}) = g_j \, ,$$

and the claim is proved. Finally,

$$|g_j| \le \|f^* v_{j+1}\|_\infty \le (n+1)\|f\|_\infty \, \|v_{j+1}\|_\infty \le (n+1)(4b)^n A_\infty \, ,$$

by (i). The second inequality in (ii) now follows since $|g_n| = |f_n| \le A_\infty$. $\square$

Theorem 4.13. *Let R be a ring, $f \in R[x]$ of degree $n \ge 1$, and $\mathbf{b} = (b_0, b_1, \ldots)$ in $R^\mathbb{N}$. Using classical arithmetic, we can solve both conversion problems 4.10 and 4.11 with $O(n^2)$ additions and multiplications in R.*

If $R = \mathbb{Z}$, $|b_i| < 2^d$ for all $i \le n$, and $\|f\|_\infty < 2^\lambda$, then the cost for solving the forward conversion problem is $O(n^2 d(nd + \lambda))$ word operations. The same estimate is valid for the backward conversion problem if $|g_i| < 2^\lambda$ for all i.

Proof. For computing the usual coefficients from $g_0, \ldots, g_n$, we employ the Horner-like scheme (4.4). In n steps, we compute

$$f^{(n)} = g_n, \quad f^{(i)} = (x - b_i) f^{(i+1)} + g_i \text{ for } i = n - 1, \ldots, 0 \, , \qquad (4.5)$$

and obtain $f = f^{(0)}$. Then the invariant

$$f = f^{(i)} \cdot (x - b_0) \cdots (x - b_{i-1}) + \sum_{0 \leq j < i} g_j \cdot (x - b_0) \cdots (x - b_{j-1}) \qquad (4.6)$$

holds for $0 \leq i \leq n$. The cost for step $n - i$ is i multiplications and $i - 1$ additions in R, and summing up yields a total cost of $O(n^2)$ ring operations.

For the other conversion, we reverse the scheme (4.5), by letting $f^{(0)} = f$, obtaining $f^{(i+1)}$ and g_i as quotient and remainder in the division of $f^{(i)}$ by $x + b_{i+1}$, and finally $g_n = f^{(n)}$. This takes exactly the same number of arithmetic operations as the Horner-like scheme (each addition in the former corresponds to a subtraction in the latter).

If $R = \mathbb{Z}$, then we have to estimate the size of the coefficients of the $f^{(i)}$. By (4.6), we have

$$f^{(i)} = \sum_{i \leq j \leq n} g_j \cdot (x - b_i) \cdots (x - b_{j-1}) \qquad (4.7)$$

for all i. If we start with $g_0, \ldots, g_n$ of absolute value less than 2^λ, then Lemma 4.12 (ii) shows that $\|f^{(i)}\|_\infty < (n + 1)2^{nd+\lambda}$, and hence the word length of the coefficients of all $f^{(i)}$ is $O(nd + \lambda)$. Thus in step $n - i$, we have i integer multiplications of size $O(d) \times O(nd+\lambda)$ and $i-1$ additions of integers of length $O(nd+\lambda)$, for $0 \leq i \leq n$. This yields a total cost of $O(n^2d(nd + \lambda))$ word operations.

For the reverse transformation, when the usual coefficients of f are given and of absolute value less than 2^λ, then Lemma 4.12 (ii) implies that $|g_i| < (n + 1)2^{n(d+2)+\lambda}$ for $0 \leq i \leq n$. Now (4.7) together with another application of the same lemma shows that $\|f^{(i)}\|_\infty \leq (n+1)^2 2^{2n(d+1)+\lambda}$, and hence the coefficients of all $f^{(i)}$ are of word length $O(nd+\lambda)$. This yields an overall cost estimate of $O(n^2d(nd + \lambda))$ word operations. $\square$

For the falling (or rising) factorial coefficients, we have $b_i = \pm i$ for all i, so that may take $d = \lceil \log_2(n - 1) \rceil$.

Corollary 4.14. *Let $f \in R[x]$ of degree $n \geq 1$. Using classical arithmetic, we can compute the falling (or rising) factorial coefficients $g_0, \ldots, g_n$ of f from its usual coefficients, and also vice versa, with $O(n^2)$ additions and multiplications in R. If $R = \mathbb{Z}$ and $\|f\|_\infty < 2^\lambda$, then the cost for computing the falling (or rising) factorial coefficients from the usual coefficients is $O(n^2 \log n \cdot (n \log n + \lambda))$ word operations. The same estimate is valid for computing the usual coefficients from the falling (or rising) factorial coefficients if $|g_i| < 2^\lambda$ for all i.*

The following two theorems analyze new conversion algorithms applying the divide-and-conquer technique from the fast evaluation and interpolation algorithms by Borodin & Moenck (1974) (see also Strassen 1973; and Strassen 1974 and §4.5 in Borodin & Munro (1975) for a survey) to the Horner-like scheme (4.4). They are analogous to method E in Sect. 4.1.

Theorem 4.15. *Let R be a ring, $f \in R[x]$ of degree less than $n \geq 2$, and $\mathbf{b} = (b_0, b_1, \ldots)$ in $R^\mathbb{N}$. Using fast arithmetic, we can solve the forward conversion problem 4.10 – with n replaced by $n + 1$ – using $O(\mathsf{M}(n) \log n)$ additions and multiplications in R.*

If $R = \mathbb{Z}$, $\|f\|_\infty < 2^\lambda$, and $|b_i| < 2^d$ for $i < n - 1$, then the cost in word operations is $O(\mathsf{M}(n^2 d + n\lambda)\log n)$ or $O^\sim(n^2 d + n\lambda)$, where the $O^\sim$ notation suppresses logarithmic factors.

Proof. We assume that $n = 2^k$ is a power of two. In a precomputation stage, we first compute recursively the subproducts

$$m_{ij} = (x - b_{j2^i}) \cdots (x - b_{(j+1)2^i - 1}) \tag{4.8}$$

of $(x - b_0) \cdots (x - b_{n-1})$, for $0 \le i < k$ and $0 \le j < 2^{k-i}$. This takes $O(\mathsf{M}(n)\log n)$ ring operations, by (the proof of) Lemma 3.15. In the main stage, we divide f with remainder to obtain $f^{(0)}, f^{(1)} \in R[x]$ of degrees less than $n/2$ such that $f = f^{(0)} + m_{k-1,0} \cdot f^{(1)}$, and proceed recursively to compute the coefficients with respect to $\mathcal{M}_{\mathbf{b}}$ of

$$\begin{aligned}
f^{(0)} &= \sum_{0 \le i < n/2} g_i \cdot (x - b_0) \cdots (x - b_{i-1}) , \\
f^{(1)} &= \sum_{0 \le i < n/2} g_{n/2+i} \cdot (x + b_{n/2}) \cdots (x + b_{n/2+i-1}) .
\end{aligned} \tag{4.9}$$

Then

$$\begin{aligned}
f &= f^{(0)} + (x - b_0) \cdots (x - b_{n/2-1}) \cdot f^{(1)} \\
&= \sum_{0 \le i < n/2} g_i \cdot (x - b_0) \cdots (x - b_{i-1}) \\
&\quad + \sum_{0 \le i < n/2} g_{n+i} \cdot (x - b_0) \cdots (x - b_{n/2+i-1}) \\
&= \sum_{0 \le i < n} g_i \cdot (x - b_0) \cdots (x - b_{i-1}) .
\end{aligned}$$

The cost for the division with remainder is $O(\mathsf{M}(n))$ ring operations. If $T(n)$ denotes the cost of the main stage, then we have $T(1) = 0$ and $T(n) \in 2T(n/2) + O(\mathsf{M}(n))$, and unraveling the recursion yields $T(n) \in O(\mathsf{M}(n)\log n)$. This dominates the cost for the precomputation stage.

If $R = \mathbb{Z}$, then $\|m_{ij}\|_1 \le 2^{nd}$, by Lemma 4.12 (i), for all i, j. Thus we can compute all m_{ij} with $O(\mathsf{M}(n^2 d)\log n)$ word operations, by Lemma 3.15. Lemma 4.12 (ii) shows that $|g_i| < n2^{n(d+2)+\lambda}$ for all i. As in the proof of Theorem 4.13, another application of the same lemma yields $\|f^{(0)}\|_\infty, \|f^{(1)}\|_\infty < n^2 2^{2n(d+1)+\lambda}$, and hence the coefficients of $f^{(0)}$ and $f^{(1)}$ are of word size $O(nd + \lambda)$. We perform the division with remainder by first computing $\mathrm{rev}(m_{k-1,0})^{-1}$ modulo $x^{n/2}$ using Newton iteration, then multiplying the result by $\mathrm{rev}(f)$ modulo $x^{n/2}$ to get $\mathrm{rev}(f^{(1)})$, and finally obtain $f^{(0)} = f - m_{k-1,0}f^{(1)}$, as described, for example, in Sect. 9.1 of von zur Gathen & Gerhard (1999). Lemma 4.12 (i) implies that the coefficients of $\mathrm{rev}(m_{k-1,0})^{-1} \bmod x^{n/2}$ have word length $O(nd)$, and hence the coefficients of all intermediate results in the computation have word length $O(nd + \lambda)$. Thus the cost for the division with remainder is $O(\mathsf{M}(n^2 d + n\lambda))$ word operations. If $T(n)$ denotes the cost in word operations for the main stage, then $T(1) = 0$ and $T(n) \in 2T(n/2) + O(\mathsf{M}(n^2 d + n\lambda))$, and we obtain $T(n) \in O(\mathsf{M}(n^2 d + n\lambda) \log n)$. $\square$

Theorem 4.16. *Let R be a ring, $n \geq 2$, and $b_0, \ldots, b_{n-1}, g_0, \ldots, g_{n-1} \in R$. We can solve the backward conversion problem 4.11 – with n replaced by $n + 1$ – with $O(\mathsf{M}(n) \log n)$ additions and multiplications in R, using fast arithmetic.*

If $R = \mathbb{Z}$, $|g_i| < 2^{\lambda}$, and $|b_i| < 2^d$ for all $i < n - 1$, then the cost in word operations is $O(\mathsf{M}(n^2 d + n\lambda) \log n)$ or $O^{\sim}(n^2 d + n\lambda)$.

Proof. The proof parallels the proof of Theorem 4.15. Again, we assume that $n = 2^k$ is a power of two, and precompute the coefficients of the subproducts m_{ij} for $0 \leq i < k$ and $0 \leq j < 2^{k-i}$, as defined in (4.8), using $O(\mathsf{M}(n) \log n)$ ring operations. In the main stage, we recursively compute the coefficients of $f^{(0)}, f^{(1)}$ as in (4.9), and then obtain $f = f^{(0)} + m_{k-1,0} \cdot f^{(1)}$. Multiplying $m_{k-1,0}$ by $f^{(1)}$ and adding $f^{(0)}$ takes $O(\mathsf{M}(n))$ additions and multiplications in R. If $T(n)$ denotes the cost of the main stage, then we have $T(1) = 0$ and $T(n) \in 2T(n/2) + O(\mathsf{M}(n))$, and unraveling the recursion yields $T(n) \in O(\mathsf{M}(n) \log n)$. This dominates the cost for the precomputation stage.

If $R = \mathbb{Z}$, then $\|m_{ij}\|_1 \leq 2^{nd}$ for all i, j, and computing all m_{ij} takes $O(\mathsf{M}(n^2 d) \log n)$ word operations, as in the proof of Theorem 4.15. Now $\|f^{(0)}\|_{\infty}$ and $\|f^{(1)}\|_{\infty}$ are less than $n2^{nd+\lambda}$, by Lemma 4.12 (ii), and their word size is $O(nd + \lambda)$. Thus multiplying $m_{k-1,0}$ by $f^{(1)}$ and adding $f^{(0)}$ uses $O(\mathsf{M}(n^2 d + n\lambda))$ word operations. If now $T(n)$ denotes the cost of the main stage in word operations, then $T(1) = 0$ and $T(n) \in 2T(n/2) + O(\mathsf{M}(n^2 d + n\lambda))$, and, as usual, we find $T(n) \in O(\mathsf{M}(n^2 d + n\lambda) \log n)$. $\square$

The output size for both conversion problems is $O(n^2 d + n\lambda)$ if the input size is $\Theta(n\lambda)$, by Lemma 4.12 (ii), and hence the estimates of Theorems 4.15 and 4.16 are – up to logarithmic factors – optimal for those inputs where the output size is close to the upper bound.

Corollary 4.17. *Let $f \in R[x]$ of degree less than $n \geq 2$. Using fast arithmetic, we can compute the falling (or rising) factorial coefficients of f from the usual ones using $O(\mathsf{M}(n) \log n)$ ring operations. If $R = \mathbb{Z}$ and $\|f\|_{\infty} < 2^{\lambda}$, then the cost is $O(\mathsf{M}(n^2 \log n + n\lambda) \log n)$ or $O^{\sim}(n^2 + n\lambda)$ word operations. The same estimates are valid for computing the usual coefficients from the falling (or rising) factorial coefficients if the latter are absolutely less than 2^{λ}.*

The arithmetic cost estimate of $O(\mathsf{M}(n) \log n)$ from Corollary 4.17 is the same as for arbitrary interpolation points. Although the interpolation points $0, 1, 2, \ldots$ are very special in the case of the falling factorials, it seems that there is neither a faster algorithm known for this special case, nor a nontrivial lower bound. We mention some related interesting results. In the proofs of Theorems 4.15 and 4.16, we have seen that the coefficients of the polynomial $x^{\underline{n}}$, i.e., the Stirling numbers of the first kind, can be computed with $O(\mathsf{M}(n)\log n)$ or $O^{\sim}(n)$ coefficient additions and multiplications. If we do not need the coefficients of this polynomial explicitly, but only want an algorithm for evaluating it at some point $x = u \in R$, the this can be done with only $O^{\sim}(n^{1/2})$ additions and multiplications in R, by Strassen (1976). Shub & Smale (1995) obtain the following surprising result. Let $S \subseteq \mathbb{Z}[x]^{\mathbb{N}}$ be the set of all sequences $(f_n)_{n \in \mathbb{N}}$ such that $f_n \neq 0$ and $x^{\underline{n}} \mid f_n$ for all $n \in \mathbb{N}$. If for all sequences $(f_n)_{n \in \mathbb{N}} \in S$ there is a lower bound of $\Omega(n^{\varepsilon})$, with $\varepsilon \in \mathbb{R}_{>0}$, for evaluating f_n at a point, then $\mathcal{P}_{\mathbb{C}} \neq \mathcal{NP}_{\mathbb{C}}$. These complexity classes are defined similarly to the well-known classes $\mathcal{P}$ and $\mathcal{NP}$, but they allow for algorithms to work exactly with arbitrary complex numbers and perform arithmetic operations on them with unit cost (Blum, Shub & Smale 1989; Blum, Cucker, Shub & Smale 1998). In fact, there is a relation between the two types of complexity classes: Cucker, Karpinski, Koiran, Lickteig & Werther (1995) prove that $\mathcal{P}_{\mathbb{C}} = \mathcal{NP}_{\mathbb{C}}$ implies $\mathcal{NP} \subseteq \mathcal{BPP}$. Heintz & Morgenstern (1993) show in their Theorem 18 that evaluating the polynomial $f_n = \prod_{0 \leq i < n}(x - i^{1/2}) \in \mathbb{R}[x]$ at a point requires at least $\Omega(\sqrt{n}/\log n)$ nonscalar multiplications. Now $x^{\underline{n}} = (-1)^n \cdot f_n(x^{1/2}) \cdot f_n(-x^{1/2})$, but this does of course not yield a lower bound for evaluating $x^{\underline{n}}$.

We now present new modular algorithms for converting between $\mathcal{M}$ and $\mathcal{M}_{\mathrm{b}}$. We recall that ω is the word size of our hypothetical processor.

Algorithm 4.18 (Modular conversion from $\mathcal{M}$ to $\mathcal{M}_{\mathrm{b}}$).
Input: A polynomial $f \in \mathbb{Z}[x]$ of degree $n \geq 1$ and max-norm less than 2^{λ}, and $b_0, \ldots, b_{n-1} \in \mathbb{Z}$ of absolute value less than 2^d.
Output: Integers $g_0, \ldots, g_n$ such that $f(x) = \sum_{0 \leq i \leq n} g_i \cdot (x + b_0) \cdots (x + b_{i-1})$.

1. $C \longleftarrow (n+1)2^{n(d+2)+\lambda}$, $r \longleftarrow \lceil (\log_2 2C)/(\omega - 1) \rceil$
   choose odd single precision primes $p_1 < \cdots < p_r$
2. **for** $1 \leq j \leq r$ compute $g_{0j}, \ldots, g_{nj} \in \mathbb{Z}$ of absolute value less than $p_j/2$ such that $f(x) \equiv \sum_{0 \leq i \leq n} g_{ij} \cdot (x - b_0) \cdots (x - b_{i-1}) \bmod p_j$
3. **for** $0 \leq i \leq n$ use the Chinese Remainder Algorithm to compute $g_i \in \mathbb{Z}$ of absolute value less than $(\prod_{1 \leq j \leq r} p_j)/2$ such that $g_i \equiv g_{ij} \bmod p_j$ for $1 \leq j \leq r$
4. **return** $g_0, \ldots, g_n$

Algorithm 4.19 (Modular conversion from $\mathcal{M}_{\mathrm{b}}$ to $\mathcal{M}$).
Input: Integers $g_0, \ldots, g_n$ of absolute value less than 2^{λ}, for some $n \geq 1$, and $b_0, \ldots, b_{n-1} \in \mathbb{Z}$ of absolute value less than 2^d.
Output: The coefficients of the polynomial $\sum_{0 \leq i \leq n} g_i \cdot (x - b_0) \cdots (x - b_{i-1})$.

1. $C \longleftarrow (n+1)2^{nd+\lambda}$, $r \longleftarrow \lceil (\log_2 2C)/(\omega - 1) \rceil$
   choose odd single precision primes $p_1 < \cdots < p_r$
2. **for** $1 \leq j \leq r$ compute $f_j \in \mathbb{Z}[x]$ of max-norm less than $p_j/2$ such that $f_j \equiv \sum_{0 \leq i \leq n} g_i \cdot (x - b_0) \cdots (x - b_{i-1}) \bmod p_j$
3. use the Chinese Remainder Algorithm to compute $f \in \mathbb{Z}[x]$ of max-norm less than $(\prod_{1 \leq j \leq r} p_j)/2$ such that $f \equiv f_j \bmod p_j$ for $1 \leq j \leq r$
4. **return** f

Theorem 4.20. *Algorithms 4.18 and 4.19 solve the forward conversion problem 4.10 and the backward conversion problem 4.11, respectively, correctly as specified. Steps 2 and 3 take $O(n^3 d^2 + n^2 d\lambda + n\lambda^2)$ word operations with classical arithmetic and $O(\lambda\, \mathsf{M}(n) \log n + n\, \mathsf{M}(nd + \lambda) \log(nd + \lambda))$ or $O^{\sim}(n^2 d + n\lambda)$ with fast arithmetic.*

Proof. Let $m = \prod_{1 \le i \le r} p_i > 2^{(\omega-1)r} \ge 2C$. In Algorithm 4.19, we have $f(x) \equiv \sum_{0 \le i \le n} g_i \cdot (x - b_0) \cdots (x - b_{i-1}) \bmod m$, both sides of the congruence have max-norms less than $m/2$, by Lemma 4.12 (ii), and hence they are equal. The proof that Algorithm 4.18 works correctly is completely analogous.

In step 2 of both algorithms, we first reduce the g_i or the coefficients of f, respectively, and the b_i modulo all primes, taking $O(nr(\lambda + d))$ word operations with classical arithmetic and $O(n\, \mathsf{M}(r) \log r)$ with fast arithmetic. Then we perform the corresponding conversion modulo each prime, taking $O(rn^2)$ word operations with classical arithmetic and $O(r\, \mathsf{M}(n) \log n)$ with fast arithmetic, by Theorems 4.14, 4.15, and 4.16. Finally, the cost for the Chinese remaindering in step 3 is $O(r^2)$ per coefficient with classical arithmetic and $O(\mathsf{M}(r) \log r)$ with fast arithmetic, in total $O(nr^2)$ and $O(n\, \mathsf{M}(r) \log r)$, respectively. The estimates now follow from $r \in O(nd + \lambda)$. $\square$

We note that for particular bases $\mathcal{M}_b$, the bound C in Algorithms 4.18 and 4.19, arising from the estimate in Lemma 4.12, may be much too large. For example, for the falling factorials, we may take $d \approx \log_2 n$, and then the bound C is about $(n+1)n^n 2^{2n+\lambda}$ for Algorithm 4.18 and about $(n+1)n^n 2^{\lambda}$ for Algorithm 4.19. In fact, the following lemma shows that $C = (n+1)!\, 2^{\lambda}$ is sufficient in both cases.

Lemma 4.21. *Let $R = \mathbb{Z}$ and $n \ge 1$. If $f_0, \dots, f_n, g_0, \dots, g_n \in \mathbb{Z}$ are such that*

$$\sum_{0 \le i \le n} f_i x^i = \sum_{0 \le i \le n} g_i x^i, \tag{4.10}$$

$A = \max\{|f_i| : 0 \le i \le n\}$, *and* $B = \max\{|g_i| : 0 \le i \le n\}$, *then* $A \le (n+1)!\, B$ *and* $B \le (n+1)!\, A$.

Proof. For $0 \le j \le i$, we denote by $\left[\begin{smallmatrix} i \\ j \end{smallmatrix}\right]$ and $\left\{\begin{smallmatrix} i \\ j \end{smallmatrix}\right\}$ the Stirling numbers of the first and second kind, counting the number of permutations on i elements having exactly j cycles and the number of partitions of a set of i elements having exactly j blocks, respectively. We use the following obvious inequalities:

$$\left\{ \begin{matrix} i \\ j \end{matrix} \right\} \le \left[\begin{matrix} i \\ j \end{matrix} \right] \le \sum_{0 \le k \le i} \left[\begin{matrix} i \\ k \end{matrix} \right] = i! \, .$$

It is well-known that

$$x^{\underline{i}} = \sum_{0 \le j \le i} (-1)^{i-j} \left[\begin{matrix} i \\ j \end{matrix} \right] x^j, \quad x^i = \sum_{0 \le j \le i} \left\{ \begin{matrix} i \\ j \end{matrix} \right\} x^{\underline{j}}$$

for all $i \in \mathbb{N}$. Plugging the two equalities into (4.10), we find

$$|f_j| = \left| \sum_{j \leq i \leq n} (-1)^{i-j} \begin{bmatrix} i \\ j \end{bmatrix} g_i \right| \leq \sum_{j \leq i \leq n} \begin{bmatrix} i \\ j \end{bmatrix} |g_i| \leq (n-j+1)n!B ,$$

$$|g_j| = \left| \sum_{j \leq i \leq n} \begin{Bmatrix} i \\ j \end{Bmatrix} f_i \right| \leq \sum_{j \leq i \leq n} \begin{Bmatrix} i \\ j \end{Bmatrix} |f_i| \leq (n-j+1)n!A$$

for $0 \leq j \leq n$, and the claim follows. $\square$

Corollary 4.22. *Let $f \in \mathbb{Z}[x]$ of degree $n \geq 1$. If we ignore the cost for prime finding, then we have a modular algorithm for computing the falling (or rising) factorial coefficients of f from the usual coefficients. If $\|f\|_\infty < 2^\lambda$, then it takes $O(n^3 \log^2 n + n^2 \log n \cdot \lambda + n\lambda^2)$ word operations with classical arithmetic and $O(\lambda \mathsf{M}(n) \log n + n \mathsf{M}(n \log n + \lambda) \log(n+\lambda))$ or $O^\sim(n^2 + n\lambda)$ with fast arithmetic. The same estimate is valid for computing the usual coefficients from the falling (or rising) factorial coefficients if the latter are absolutely less than 2^λ.*

Table 4.7 summarizes the cost estimates for the basis conversion algorithms that we have presented so far. The costs are to be read as O-estimates for polynomials of degree n with λ bit coefficients, and d is the bit size of the b_i. For the modular algorithms, we have neglected the cost for prime finding. We note that the estimates for the modular algorithms are not asymptotic estimates; see Remark 3.23. The last two rows are only valid for the Taylor shift, where all b_i's are equal.

Table 4.7. Cost estimates for polynomial basis conversion

Algorithm	Cost	$n = d = \lambda$
classical (Theorem 4.13)	$n^3 d^2 + n^2 d\lambda$	n^5
classical modular (Theorem 4.20)	$n^3 d^2 + n^2 d\lambda + n\lambda^2$	n^5
fast (Theorems 4.15, 4.16)	$\mathsf{M}(n^2 d + n\lambda) \log n$	$\mathsf{M}(n^3) \log n$
fast modular (Theorem 4.20)	$\lambda \mathsf{M}(n) \log n +$ $n \mathsf{M}(nd + \lambda) \log(nd + \lambda)$	$n \mathsf{M}(n^2) \log n$
fast Taylor shift (Theorem 4.5)	$\mathsf{M}(n^2 \log n + n^2 d + n\lambda)$	$\mathsf{M}(n^3)$
fast modular Taylor shift (Theorem 4.8)	$\lambda \mathsf{M}(n) +$ $n \mathsf{M}(nd + \lambda) \log(nd + \lambda)$	$n \mathsf{M}(n^2) \log n$

4.3 Fast Multiplication in the Falling Factorial Basis

In this section, we give some new fast algorithms for polynomials in the *falling factorial basis*

$$\mathcal{F} = (x^{\underline{i}})_{i \in \mathbb{N}} = \mathcal{M}_{(0,1,2,3,\ldots)} ,$$

namely for multiplication and Taylor shift. The algorithms are not used later, but may be of independent interest. In principle, both problems can be solved by first converting the input polynomial(s) to the monomial basis $\mathcal{M}$, applying the corresponding fast algorithm for the monomial basis, and finally converting back to the falling factorial basis. However, the following algorithms are more direct and faster by a logarithmic factor. We note that the algorithms can be easily generalized to the basis $\mathcal{M}_{\mathbf{b}}$, where $\mathbf{b} = (b_0, b_0 + h, b_0 + 2h, \ldots)$ is an arbitrary arithmetic progression, so in particular to the rising factorial basis $\mathcal{M}_{(0,-1,-2,-3,\ldots)}$. Our algorithms assume that the characteristic of the coefficient ring R is zero or "big enough". We give cost estimates in arithmetic operations.

We start with the fast multiplication algorithm. The idea is to use the well-known evaluation-interpolation scheme: evaluate the input polynomials at a suitable set of points, multiply the results pointwise, and interpolate the product polynomial. The most profitable use of this scheme is in FFT multiplication algorithms, where the points are the powers of a primitive root of unity (see, e.g., Chap. 8 in von zur Gathen & Gerhard 1999). In our case of the falling factorial basis representation, it is not surprising that the integers $0, 1, 2, 3, \ldots$ (in the more general case: the ring elements $b_0, b_0 + h, b_0 + 2h, \ldots$) are a suitable set of evaluation points.

Algorithm 4.23 (Evaluation in the falling factorial basis).
Input: $f_0, \ldots, f_n \in R$, where $n \geq 1$ and R is a ring of characteristic zero or coprime to $n!$.
Output: The values $f(0), f(1), \ldots, f(n)$, where $f = \sum_{0 \leq i \leq n} f_i x^{\underline{i}}$.

1. **for** $0 \leq i \leq n$ **do** $g_i \longleftarrow 1/i!$

2. $\displaystyle \sum_{0 \leq i \leq 2n} h_i x^i \longleftarrow \left(\sum_{0 \leq i \leq n} f_i x^i \right) \cdot \left(\sum_{0 \leq i \leq n} g_i x^i \right)$

3. **return** $h_0, h_1, 2h_2, \ldots, n! \cdot h_n$

Theorem 4.24. *Algorithm 4.23 works correctly as specified and takes $O(\mathsf{M}(n))$ additions and multiplications plus $O(n)$ divisions by integers in R.*

Proof. Let $0 \leq k \leq n$. Then

$$f(k) = \sum_{0 \leq i \leq n} f_i k^{\underline{i}} = \sum_{0 \leq i \leq k} f_i \frac{k!}{(k-i)!} = k! \sum_{0 \leq i \leq k} f_i g_{k-i} = k! \cdot h_k , \qquad (4.11)$$

and the correctness follows. The cost for steps 1 and 3 is $O(n)$ divisions and multiplications, respectively, and step 2 takes $O(\mathsf{M}(n))$ additions and multiplications. $\square$

Algorithm 4.25 (Interpolation in the falling factorial basis).
Input: $u_0, \ldots, u_n \in R$, where $n \in \mathbb{N}$ and R is a field of characteristic zero or coprime to $n!$.
Output: The falling factorial coefficients $f_0, \ldots, f_n \in R$ of the interpolating polynomial $f = \sum_{0 \leq i \leq n} f_i x^{\underline{i}}$ such that $u_i = f(i)$ for $0 \leq i \leq n$.

1. **for** $0 \leq i \leq n$ **do** $h_i \longleftarrow u_i/i!, \quad v_i \longleftarrow (-1)^i/i!$

2. $\displaystyle\sum_{0 \leq i \leq 2n} w_i x^i \longleftarrow \left(\sum_{0 \leq i \leq n} h_i x^i \right) \cdot \left(\sum_{0 \leq i \leq n} v_i x^i \right)$

3. **return** $w_0, w_1, \ldots, w_n$

Theorem 4.26. *Algorithm 4.25 works correctly as specified and takes $O(\mathsf{M}(n))$ additions and multiplications plus $O(n)$ divisions by integers in R.*

Proof. Let $g = \sum_{0 \leq i \leq n} x^i/i!$ and $v = \sum_{0 \leq i \leq n} (-1)^i x^i/i!$. For $1 \leq k \leq n$, the coefficient of x^k in gv is

$$\sum_{0 \leq i \leq k} \frac{(-1)^{k-i}}{i!\,(k-i)!} = \frac{1}{k!} \sum_{0 \leq i \leq k} (-1)^{k-i} \binom{k}{i} = \frac{(1+(-1))^k}{k!} = 0 \,,$$

and hence $gv \equiv 1 \bmod x^{n+1}$. (This also follows from the fact that g and v are initial segments of the formal exponential series and its inverse, respectively.) If we now let $h = \sum_{0 \leq i \leq n} h_i x^i$ and $w = \sum_{0 \leq i \leq n} w_i x^i$, then (4.11) shows that

$$\sum_{0 \leq k \leq n} \frac{w(k)}{k!} x^k \equiv gw \equiv gvh \equiv h = \sum_{0 \leq k \leq n} \frac{u_k}{k!} x^k \bmod x^{n+1} \,,$$

and the correctness follows from the uniqueness of the interpolating polynomial. The cost for step 1 is $O(n)$ multiplications and divisions, and step 2 takes $O(\mathsf{M}(n))$ additions and multiplications. □

Theorem 4.27. *Let $f_0, \ldots, f_n, g_0, \ldots, g_n \in R$, where $n \in \mathbb{N}$ and R is a ring of characteristic zero or coprime to $(2n)!$. Then we can compute $h_0, \ldots, h_{2n} \in R$ such that*

$$\sum_{0 \leq i \leq 2n} h_i x^{\underline{i}} = \left(\sum_{0 \leq i \leq n} f_i x^{\underline{i}} \right) \left(\sum_{0 \leq i \leq n} g_i x^{\underline{i}} \right)$$

using $O(\mathsf{M}(n))$ additions and multiplications plus $O(n)$ divisions by integers in R.

Proof. We use Algorithm 4.23 to compute $f(i)$ and $g(i)$ for $0 \leq i \leq 2n$ and then call Algorithm 4.25 to compute $h \in R[x]$ of degree at most $2n$ such that $h(i) = f(i)g(i)$ for $0 \leq i \leq 2n$. By Theorems 4.24 and 4.26, the cost for the two evaluations and the interpolation is $O(\mathsf{M}(n))$ additions and multiplications plus $O(n)$ divisions, and this dominates the additional cost for the $2n+1$ pointwise multiplications to compute the values $h(i)$. □

Our last algorithm in this section is for Taylor shift in the falling factorial basis. It is a straightforward adaption of the convolution method (method F in Sect. 4.1). More generally, this method works for any polynomial basis satisfying a binomial identity.

Theorem 4.28. *Let $f_0, \ldots, f_n, b \in R$, where $n \in \mathbb{N}$ and R is a ring of characteristic zero or coprime to $n!$. Then we can compute $g_0, \ldots, g_n \in R$ such that*

$$\sum_{0 \le k \le n} g_k x^k = \sum_{0 \le i \le n} f_i \cdot (x+b)^i \tag{4.12}$$

using $O(\mathsf{M}(n))$ additions and multiplications plus $O(n)$ divisions by integers in R.

Proof. We apply Vandermonde's convolution (see, e.g., Exercise 23.9 in von zur Gathen & Gerhard 1999)

$$(x+b)^i = i! \binom{x+b}{i} = i! \sum_{0 \le k \le i} \binom{x}{k}\binom{b}{i-k} = i! \sum_{0 \le k \le i} \binom{b}{i-k} \frac{x^k}{k!}$$

to (4.12) and find

$$g_k = \sum_{k \le i \le n} \frac{i! \, f_i}{k!} \binom{b}{i-k}$$

for $0 \le k \le n$. If we now let $u = \sum_{0 \le i \le n} i! f_i x^{n-i}$ and $v = \sum_{0 \le j \le n} \binom{b}{j} x^j$, then $k! \, g_k$ is the coefficient of x^{n-k} in the product polynomial uv. Using the recursion formula

$$\binom{b}{j+1} = \frac{b-j}{j+1}\binom{b}{j}$$

for $j \in \mathbb{N}$, computing the coefficients of u and v takes $O(n)$ additions, multiplications, and divisions in R. The same estimate is valid for recovering the g_k from the coefficients of the product polynomial uv, and the multiplication itself takes $O(\mathsf{M}(n))$ additions and multiplications. $\square$

5. Modular Squarefree and Greatest Factorial Factorization

5.1 Squarefree Factorization

The presentation in this section follows Gerhard (2001). Let Z be a UFD and $f \in Z[x]$ nonzero primitive of degree n. Then f has a decomposition

$$f = \prod_{1 \leq i \leq n} f_i^i \tag{5.1}$$

into nonzero primitive polynomials $f_1, \ldots, f_n \in Z[x]$ such that each f_i is square-free and $\gcd(f_i, f_j) = 1$ if $i \neq j$. We call this a *primitive squarefree decomposition* of f, and $f_1 \cdots f_n$ is a *primitive squarefree part* of f. Both are unique up to multiplication by units. If $Z = \mathbb{Z}$, then we can make them unique by requiring f and $f_1, \ldots, f_n$ to have positive leading coefficients, and call them the *normalized squarefree decomposition* and the *normalized squarefree part*, respectively. If Z is a field and $f, f_1, \ldots, f_n$ are all monic, then the decomposition (5.1) is unique as well, and we call it the *monic squarefree decomposition* and $f_1 \cdots f_n$ the *monic squarefree part* of f.

Problem 5.1 (Squarefree factorization). *Let Z be a UFD. Given a nonzero primitive polynomial $f \in Z[x]$ of degree n, compute nonzero primitive squarefree and pairwise coprime polynomials $f_1, \ldots, f_n \in Z[x]$ satisfying*

$$f = f_1 f_2^2 \cdots f_n^n .$$

If Z has characteristic zero or prime characteristic greater than n and $f_1, \ldots, f_n$ is a primitive (or monic) squarefree decomposition of f, then we have the following well-known equality

$$\gcd(f, f') = f_2 f_3^2 \cdots f_n^{n-1} \tag{5.2}$$

for the primitive (or monic) gcd of f and its formal derivative f', which consequently has the primitive (or monic) squarefree decomposition $f_2, \ldots, f_{1+\deg g}$. This immediately leads to a simple algorithm for computing the squarefree decomposition. A more efficient algorithm has been given by Yun (1976, 1977a). We restate it for completeness and later reference.

Algorithm 5.2 (Yun's squarefree factorization).
Input: A nonzero monic polynomial $f \in F[x]$ of degree n, where F is a field of characteristic zero or greater than n.
Output: The squarefree decomposition of f.

J. Gerhard: Modular Algorithms, LNCS 3218, pp. 61-77, 2004.

1. $g \longleftarrow \gcd(f, f'), \quad u_1 \longleftarrow \dfrac{f}{g}, \quad v_1 \longleftarrow \dfrac{f'}{g}$
2. **for** $1 \leq i \leq n$ **do**
3. $\qquad h_i \longleftarrow \gcd(u_i, v_i - u_i'), \quad u_{i+1} \longleftarrow \dfrac{u_i}{h_i}, \quad v_{i+1} \longleftarrow \dfrac{v_i - u_i'}{h_i}$
4. **return** $h_1, \ldots, h_n$

Proposition 5.3. *Let* $f_1, \ldots, f_n \in F[x]$ *be the squarefree decomposition of* f. *Then the invariants*

$$u_i = \prod_{i \leq j \leq n} f_j, \quad v_i = \sum_{i \leq j \leq n} (j - i + 1)\frac{u_i}{f_j}f_j', \quad h_i = f_i$$

hold for $1 \leq i \leq n$ *and imply that Algorithm 5.2 works correctly. It takes* $O(n^2)$ *and* $O(\mathsf{M}(n) \log n)$ *arithmetic operations in* F *with classical and fast arithmetic, respectively.*

For a proof, see Yun (1976, 1977a) or von zur Gathen & Gerhard (1999), Theorem 14.23, for fast arithmetic. A similar reasoning also yields the estimate for classical arithmetic.

The following lemma, which is the key tool for our modular algorithm, says when the modular image of a squarefree decomposition is again a squarefree decomposition.

Lemma 5.4. *Let* $n \in \mathbb{N}$, Z *be a UFD of characteristic zero or prime characteristic greater than* n, $f \in Z[x]$ *nonzero primitive of degree* n, *and* $f_1, \ldots, f_n \in Z[x]$ *a primitive squarefree decomposition of* f. *Moreover, let* $I \subseteq Z$ *be a maximal ideal not containing* $\mathrm{lc}(f)$, *denote reduction modulo* I *by a bar, and assume that* $\overline{Z} = Z/I$ *has characteristic zero or greater than* n *as well. Finally, assume that* $g_1, \ldots, g_n \in \overline{Z}[x]$ *is the monic squarefree decomposition of* $\overline{f}/\mathrm{lc}(\overline{f})$. *Then*

$$\deg \gcd(f, f') \leq \deg \gcd(\overline{f}, \overline{f}') \text{ and } \deg(g_1 \cdots g_n) \leq \deg(f_1 \cdots f_n), \quad (5.3)$$

and for each of the two inequalities, we have equality if and only if $\overline{\mathrm{lc}(f_i)}g_i = \overline{f_i}$ *for* $1 \leq i \leq n$.

Proof. Since Z has characteristic zero or prime characteristic greater than n, we have

$$\gcd(f, f') \sim f_2 f_3^2 \cdots f_n^{n-1}, \quad \frac{f}{\gcd(f, f')} \sim f_1 \cdots f_n,$$
$$\gcd(f', f_1 \cdots f_n) \sim \gcd(\gcd(f, f'), f_1 \cdots f_n) \sim f_2 \cdots f_n, \quad (5.4)$$

where $\sim$ denotes equality up to associates. Similarly,

$$\gcd(\overline{f}, \overline{f}') \sim g_2 g_3^2 \cdots g_n^{n-1}, \quad \frac{\overline{f}}{\gcd(\overline{f}, \overline{f}')} \sim g_1 \cdots g_n,$$
$$\gcd(\overline{f}', g_1 \cdots g_n) \sim \gcd(\gcd(\overline{f}, \overline{f}'), g_1 \cdots g_n) \sim g_2 \cdots g_n. \quad (5.5)$$

Now $\overline{\gcd(f, f')}$ is a common divisor of $\overline{f}$ and $\overline{f'}$, and hence

$$\overline{\gcd(f, f')} \mid \gcd(\overline{f}, \overline{f}') \text{ and } g_1 \cdots g_n \mid \overline{f_1 \cdots f_n} . \tag{5.6}$$

This proves the first assertion.

The "if" direction of the second claim is clear, and we prove the "only if" part by induction on n. There is nothing to prove if $n = 0$. Otherwise, $f_2, f_3, \ldots$ is a primitive squarefree decomposition of $\gcd(f, f')$ and $g_2, g_3, \ldots$ is the monic squarefree decomposition of $\gcd(\overline{f}, \overline{f}')$. If equality holds in (5.3), then (5.6) implies that

$$g_2 g_3^2 \cdots g_n^{n-1} \sim \gcd(\overline{f}, \overline{f}') \sim \overline{\gcd(f, f')} \sim \overline{f_2 f_3^2 \cdots f_n^{n-1}}$$

and $g_1 \cdots g_n \sim \overline{f_1 \cdots f_n}$. Applying (5.6) to $\gcd(f, f')$ instead of f, we find that $g_2 \cdots g_n$ divides $\overline{f_2 \cdots f_n}$. On the other hand, we have

$$\overline{f_2 \cdots f_n} \sim \overline{\gcd(f', f_1 \cdots f_n)} \mid \gcd(\overline{f}', \overline{f_1 \cdots f_n}) \sim \gcd(\overline{f}', g_1 \cdots g_n) \sim g_2 \cdots g_n ,$$

by (5.4) and (5.5). Thus $g_1 \sim \overline{f_1}$ and $g_2 \cdots g_n \sim \overline{f_2 \cdots f_n}$, and the induction hypothesis, applied to $\gcd(f, f')$, shows that $g_i \sim \overline{f_i}$ for $2 \leq i \leq n$. $\square$

Corollary 5.5. *With the assumptions of Lemma 5.4, let $\delta = \deg \gcd(f, f')$, and let $\sigma \in Z$ be the δth subresultant of f and f'. Then the following are equivalent.*

(i) $\overline{\sigma} \neq 0$,
(ii) $\overline{\mathrm{lc}(f)} \neq 0$ and $\deg \gcd(\overline{f}, \overline{f}') = \delta$,
(iii) $\overline{\mathrm{lc}(f)} \neq 0$ and $\overline{\mathrm{lc}(f_i)} g_i = \overline{f_i}$ for $1 \leq i \leq n$.

Proof. Since $\mathrm{lc}(f)$ divides the first row of the matrix in Figure 3.1 for $g = f'$, it divides σ. The claims now follow from Corollary 3.12 and Lemma 5.4. $\square$

We now present a new modular algorithm for squarefree factorization, based on the Chinese Remainder Theorem. We recall that ω is the word size of our processor. We will assume that our polynomials have degree at most $2^{\omega-1}$, which is not a serious restriction in practice for $\omega \geq 32$. The cardinality of a finite set S is denoted by $\#S$.

Algorithm 5.6 (Small primes modular squarefree factorization). ▬▬▬▬▬▬▬
Input: A normalized polynomial $f \in \mathbb{Z}[x]$ of degree $1 \leq n \leq 2^{\omega-1}$ and max-norm $\|f\|_\infty < 2^\lambda$.
Output: The normalized squarefree decomposition of f, or "FAIL".

  1. $b \longleftarrow \mathrm{lc}(f)$, $\quad B \longleftarrow \lfloor (n+1)^{1/2} 2^{n+\lambda} \rfloor$, $\quad s \longleftarrow \lceil (\log_2 2bB)/(\omega - 1) \rceil$
  2. choose a set S_0 of $2s$ single precision primes
    $S_1 \longleftarrow \{p \in S_0 : p \nmid b\}$

3. **for** all $p \in S_1$ **do**
4.   **call** Yun's algorithm 5.2 to compute the monic squarefree decomposition

$$f \equiv b \prod_{1 \le i \le n} g_{p,i}^i \bmod p$$

of f/b modulo p, with monic polynomials $g_{p,1}, \ldots, g_{p,n} \in \mathbb{Z}[x]$ of max-norm less than $p/2$ that are squarefree and pairwise coprime modulo p
5. /* remove unlucky primes from S_1 */
   $d \longleftarrow \max\{\deg g_{p,1} + \cdots + \deg g_{p,n} : p \in S_1\}$
   $S_2 \longleftarrow \{p \in S_1 : \deg g_{p,1} + \cdots + \deg g_{p,n} = d\}$
   if $\#S_2 \ge s$ **then** remove $\#S_2 - s$ primes from S_2 **else return** "FAIL"
6. **for** $1 \le i \le n$ **do**
7.   use the Chinese Remainder Algorithm to compute $f_i^* \in \mathbb{Z}[x]$ of max-norm less than $(\prod_{p \in S_2} p)/2$ such that $f_i^* \equiv bg_{p,i} \bmod p$ for all $p \in S_2$
8. **if** $\prod_{1 \le i \le n} \|\mathrm{normal}(f_i^*)\|_1^i > B$ **then return** "FAIL"
9. **if** $\prod_{1 \le i \le n} \mathrm{lc}(\mathrm{normal}(f_i^*))^i = \mathrm{lc}(f)$
   then return $\mathrm{normal}(f_1^*), \ldots, \mathrm{normal}(f_n^*)$ **else return** "FAIL"

Definition 5.7. *Let* $Z = \mathbb{Z}$ *and* $f \in \mathbb{Z}[x]$ *nonzero primitive. We say that a prime* $p \in \mathbb{N}$ *is* lucky *with respect to the squarefree factorization problem 5.1 if* $p \nmid \mathrm{lc}(f)$ *and* $\deg \gcd(f, f') = \deg \gcd(f \bmod p, f' \bmod p)$.

Thus if p is a prime not dividing $\mathrm{lc}(f)$ and greater than $\deg f$, then p is an unlucky prime with respect to squarefree factorization if and only if the image modulo p of the squarefree decomposition of f is not a squarefree decomposition, by Lemma 5.4.

Theorem 5.8. *Algorithm 5.6 succeeds if and only if at most* $s \in \Theta(n+\lambda)$ *of the initial primes are unlucky with respect to squarefree factorization, and then it correctly returns the normalized squarefree decomposition of* f.

Proof. Let $f_1, \ldots, f_n \in \mathbb{Z}[x]$ be the normalized squarefree decomposition of f. Then $f_1 \cdots f_n$ is its normalized squarefree part. We first show that unless all $2s$ initial primes in S_0 are unlucky, a prime $p \in S_1$ is lucky if and only if $\deg(g_{p,1} \cdots g_{p,n}) = d$ in step 5. Lemma 5.4 implies that $\deg(g_{p,1} \cdots g_{p,n}) \le \deg(f_1 \cdots f_n)$, with equality if and only if p is lucky. Thus if at least one prime is lucky, we have $\deg(f_1 \cdots f_n) = d$, and the claim follows.

Now assume that at least s of the initial primes in S_0 are lucky. Then S_2 contains s primes after step 5, all lucky. Let $m = \prod_{p \in S_2} p$ and $1 \le i \le n$. Since f_i divides f, also $\mathrm{lc}(f_i)$ divides b, and hence $bf_i/\mathrm{lc}(f_i) \equiv f_i^* \bmod m$, by Lemma 5.4. Mignotte's bound (Fact 3.3) implies that $\|f_i\|_\infty \le B$, and hence

$$\|bf_i/\mathrm{lc}(f_i)\|_\infty \le bB \le 2^{(\omega-1)s-1} < \frac{m}{2} .$$

Thus the coefficients of $bf_i/\mathrm{lc}(f_i)$ and of f_i^* are at most $m/2$ in absolute value, so that both polynomials are equal. Since f_i is normalized, it follows that $f_i =$

normal(f_i^*), and again by Mignotte's bound, we find that the condition in step 8 is false, the condition in step 9 is true, and the algorithm returns the correct result.

Conversely, suppose that the condition in step 8 is false. Then

$$
\left\| \prod_{1 \le i \le n} \text{normal}(f_i^*)^i \right\|_\infty \le \left\| \prod_{1 \le i \le n} \text{normal}(f_i^*)^i \right\|_1 \le \prod_{1 \le i \le n} \| \text{normal}(f_i^*) \|_1^i
$$

$$
\le \ B < \frac{m}{2} \ .
$$

By construction, $\prod_{1 \le i \le n} \text{normal}(f_i^*)^i$ and f agree modulo m up to a multiplicative constant. If the condition in step 9 is true, then $\prod_{1 \le i \le n} \text{normal}(f_i^*)^i \equiv f \bmod m$, and since the coefficients of both polynomials are absolutely bounded by $m/2$, they are equal. The normal(f_i^*) are squarefree and pairwise coprime in $\mathbb{Z}[x]$, since they are so modulo each prime in S_2 after step 5, and since they are normalized, we find that $f_i = \text{normal}(f_i^*) \equiv \text{lc}(f_i) g_{p,i} \bmod p$ for all $i \le n$ and $p \in S_2$, by the uniqueness of the normalized squarefree decomposition. Hence all s primes in S_2 are lucky after step 5, by Lemma 5.4. □

Theorem 5.9. *Let* $\delta = \deg \gcd(f, f')$ *and* $r \in \mathbb{Z}$ *be the* δ*th subresultant of* f *and* f'*. Then* r *is a nonzero integer with* $|r| \le D = (n^2 + n)^n 2^{2n\lambda}$*, and the number of single precision primes that are unlucky with respect to squarefree factorization is at most* $\lfloor (\log_2 D)/(\omega - 1) \rfloor \in \Theta(n(\lambda + \log n))$*. If* $n \ge 2$ *and* $\lambda \ge 1$*, the number of single precision primes exceeds* $2\lfloor (\log_2 D)/(\omega - 1) \rfloor$*, and the set* S_0 *in step 2 is chosen uniformly at random from among all subsets of cardinality* $2s$ *of the single precision primes, then Algorithm 5.6 returns "FAIL" with probability at most* $1/2$*.*

Proof. The first claim follows from Lemma 3.9 (i). We have $\|f'\|_2 \le n\|f\|_2$, and the subresultant bound (Corollary 3.2) shows that $|r| \le D$. Corollary 5.5 implies that the unlucky primes greater than n are precisely those dividing r. We have $2Bb \le D$ under the assumptions of the theorem, so that it is possible to find $2s$ single precision primes in step 2. By assumption, at least half of the single precision primes are lucky. Therefore the probability that at least half of the primes of a randomly chosen set of $2s$ single precision primes are lucky is at least $1/2$, by Lemma 3.24, and the claim follows from Theorem 5.8. □

Theorem 5.10. *If we neglect the cost for choosing primes in step 2, then Algorithm 5.6 takes* $O(n^3 + n\lambda^2)$ *word operations when using classical arithmetic, and* $O(n \, \mathsf{M}(n + \lambda) \log(n + \lambda) + \lambda \, \mathsf{M}(n) \log n)$ *or* $O^\sim(n^2 + n\lambda)$ *word operations when using fast arithmetic, where the* $O^\sim$*-notation suppresses logarithmic factors.*

Proof. Reducing a coefficient of f, of word length $O(\lambda)$, modulo at most $2s$ single precision primes takes $O(\lambda s)$ word operations with classical arithmetic and $O(\mathsf{M}(s) \log s)$ with fast arithmetic. There are at most $n + 1$ coefficients, and hence the overall cost for this is $O(n\lambda s)$ and $O(n \, \mathsf{M}(s) \log s)$ word operations, respectively.

Step 4 takes $O(n^2)$ arithmetic operations in $\mathbb{F}_p$ with classical arithmetic and $O(\mathsf{M}(n)\log n)$ with fast arithmetic, by Proposition 5.3. Since p is single precision, one arithmetic operation in $\mathbb{F}_p$ takes $O(1)$ word operations. There are at most $2s$ primes in S_1, and hence the overall cost of steps 3 and 4 is $O(sn(n+\lambda))$ or $O(s\,\mathsf{M}(n)\log n + n\,\mathsf{M}(s)\log s)$ word operations when using classical or fast arithmetic, respectively.

Let $m = \prod_{p\in S_2} p < 2^{\omega s}$. Then the cost for computing one coefficient of some f_i^* in step 7 by Chinese remaindering is $O(s^2)$ word operations with classical arithmetic and $O(\mathsf{M}(s)\log s)$ with fast arithmetic. Since $n = \sum_{1\leq i\leq n} i\deg f_i^*$, the total number of non-leading coefficients of all nonconstant f_i^* is at most n, and the overall cost of steps 6 and 7 is $O(ns^2)$ and $O(n\,\mathsf{M}(s)\log s)$ word operations, respectively. (In the unlikely event that all primes in S_2 are unlucky, it may happen that not all degree sequences $\deg g_{p,1},\ldots,\deg g_{p,n}$ for $p \in S_2$ are equal, but we can detect this during the Chinese remaindering stage in step 7 and then report "FAIL".)

The cost for normalizing all f_i^* in step 8 is $O(n)$ gcd's and divisions of integers of word length at most s, taking $O(ns^2)$ word operations with classical arithmetic and $O(n\,\mathsf{M}(s)\log s)$ with fast arithmetic. Computing the products in steps 8 and 9 takes $O(s^2)$ word operations with classical arithmetic and $O(\mathsf{M}(s)\log n)$ with fast arithmetic.

The cost for all other steps is negligible, the claims follow by adding costs and using $s \in O(n+\lambda)$, and taking $\mathsf{M}(n) = n\log n\,\mathrm{loglog}\,n$ gives the $O^\sim$-estimate. □

Remark 5.11. • *When $\mathsf{M}(n) \in \Omega(n^{1+\varepsilon})$ for some positive ε, then we can drop the logarithmic factors in the cost estimate for Algorithm 5.6.*

• *When using fast arithmetic, Algorithm 5.6 is – up to logarithmic factors – asymptotically optimal in the diagonal case where $n \approx \lambda$, since its running time is essentially linear in the input size, which is about n^2 words.*

• *Yun (1976) states an estimate of $O(k^4(\nu^2\delta + \nu\delta^2))$ word operations for his algorithm for univariate polynomials over $\mathbb{Z}$, employing modular gcd techniques. Here $k \leq n$ is the largest index such that f_k is nonconstant, and δ and ν are upper bounds on the degrees and the bit size of the coefficients of all f_i, respectively. This result cannot be directly compared with ours, since it is expressed in terms of a different set of parameters. For example, in the case when k is small, we have the approximate correspondences $n \approx \delta$ and $\lambda \approx \nu$, and then his estimate agrees with our estimate for classical arithmetic, and our result for fast arithmetic is – up to logarithmic factors – better by one order of magnitude.*

• *The number $2s$ of initial primes in step 2 of Algorithm 5.6 is often much too large in practice, for two reasons. On the one hand, the estimate for the failure probability in Theorem 5.9 is too pessimistic when there are considerably more single precision primes than required by the theorem. On the other hand, the coefficients of the g_i are often significantly smaller than guaranteed by Mignotte's bound. One solution to overcome this is to work in an adaptive fashion: start with a reasonable number of primes (say about d/ω or even fewer), check whether $\prod_{1\leq i\leq n} \mathrm{normal}(g_i^*)^i = g$, and add more new primes in case of failure. In this way, the algorithm will never use more primes than needed, at the expense of* *an additional check after each new prime. Heuristically, most of the checks can even be avoided by first checking whether the equation holds for $x = 0$. This strategy has been successfully implemented by Shoup (1995) for computing* gcd*'s of integer polynomials in his software package* NTL.

The method above can be adapted to bivariate polynomials over a field when we count coefficient operations. Both the algorithm and its analysis are then much easier due to the absence of carries, and in particular when the field is large enough such that all moduli may be chosen linear. If the degree bounds in x and y for the input polynomials are n and m, respectively, then the cost for this algorithm is $O(n^2 m)$ arithmetic operations with classical arithmetic and $O^\sim(nm)$ with fast arithmetic.

A similar approach yields also fast modular algorithms (with the cost estimate from Theorem 5.10) for *factor refinement*: given one (or several) nontrivial partial factorizations $f = g_1 \cdots g_t$ in $\mathbb{Z}[x]$, compute the finest partial factorization $f = \prod_i f_i^{e_i}$, with positive integers e_i and $f_i \in \mathbb{Z}[x]$ pairwise coprime, but not necessarily squarefree, that you can obtain from the given factorization by gcd computations. See Bach, Driscoll & Shallit (1993) for algorithms in the integer case.

Yun (1976) gives a modular algorithm for the squarefree factorization of multivariate polynomials based on Hensel lifting, but does not state a time estimate. We now analyze a variant of Yun's algorithm for univariate integer polynomials (see also Exercise 15.27 in von zur Gathen & Gerhard 1999).

Algorithm 5.12 (Prime power modular squarefree factorization).
Input: A normalized polynomial $f \in \mathbb{Z}[x]$ of degree $1 \leq n \leq 2^{\omega-1}$ and max-norm $\|f\|_\infty < 2^\lambda$.
Output: The normalized squarefree decomposition of f, or otherwise "FAIL".

1. $g \longleftarrow \mathrm{normal}(\gcd(f, f'))$, $u \longleftarrow \dfrac{f}{g}$, $v \longleftarrow \dfrac{f'}{g}$

2. **repeat**
      choose a single precision prime $2^{\omega-1} < p < 2^\omega$
   until $p \nmid \mathrm{lc}(f)$ and $\deg g = \deg \gcd(f \bmod p, f' \bmod p)$

3. $b \longleftarrow \mathrm{lc}(u)$, $B \longleftarrow \lfloor (n+1)^{1/2} 2^{\lambda+\deg u} \rfloor$, $s \longleftarrow \lceil (\log_2 2bB)/(\log_2 p) \rceil$

4. **call** steps 2 and 3 of Yun's algorithm 5.2 to compute the monic squarefree decomposition

$$f \equiv \mathrm{lc}(f) \prod_{1 \leq i \leq n} g_i^i \bmod p$$

   of $f/\mathrm{lc}(f)$ modulo p, with monic polynomials $g_1, \ldots, g_n \in \mathbb{Z}[x]$ of max-norm less than $p/2$ that are squarefree and pairwise coprime modulo p

5. use Hensel lifting to compute a factorization

$$b^{n-1} u \equiv \prod_{1 \leq i \leq n} h_i \bmod p^s$$

   with polynomials $h_1, \ldots, h_n \in \mathbb{Z}[x]$ of max-norm less than $p^s/2$ such that $h_i \equiv b g_i \bmod p$ for all i

6. **return** $\mathrm{normal}(h_1), \ldots, \mathrm{normal}(h_n)$

Theorem 5.13. *Algorithm 5.12 correctly computes the normalized squarefree decomposition of f.*

Proof. Let $f_1, \ldots, f_n \in \mathbb{Z}[x]$ be the normalized squarefree decomposition of f. Step 2 ensures that p is a lucky prime, and Lemma 5.4 implies $f_i \equiv \mathrm{lc}(f_i)g_i \bmod p$ for all i and

$$b^{n-1}u = \frac{bf_1}{\mathrm{lc}(f_1)} \cdots \frac{bf_n}{\mathrm{lc}(f_n)} \equiv (bg_1) \cdots (bg_n) \bmod p \ .$$

Now the g_i are pairwise coprime modulo p and $bf_i/\mathrm{lc}(f_i) \equiv bg_i \equiv h_i \bmod p$, and the uniqueness of Hensel lifting (Fact 3.20) yields $bf_i/\mathrm{lc}(f_i) \equiv h_i \bmod p^s$ for all i. Let $1 \leq i \leq n$. Mignotte's bound 3.3 shows that $\|f_i\|_\infty \leq B$ and $\|bf_i/\mathrm{lc}(f_i)\|_\infty \leq bB < p^s/2$. Thus both $bf_i/\mathrm{lc}(f_i)$ and h_i have max-norm less than $p^s/2$, whence they are equal. Since f_i is normalized, it follows that $f_i = \mathrm{normal}(h_i)$. $\square$

Theorem 5.14. *If the number of single precision primes exceeds*

$$2\lfloor \log_2((n^2 + n)^n 2^{2n\lambda})/(\omega - 1)\rfloor \in \Theta(n(\lambda + \log n)) \ ,$$

then the expected number of iterations of step 2 is at most two. If we neglect the cost for step 2, then Algorithm 5.12 takes $O(n^4 + n^2\lambda^2))$ word operations with classical arithmetic and $O((\mathsf{M}(n)\log n + n\log\lambda)\mathsf{M}(n + \lambda))$ or $O^\sim(n^2 + n\lambda)$ with fast arithmetic.

Proof. As in the proof of Theorem 5.9, at most $\lfloor \log_2((n^2 + n)^n 2^{2n\lambda})/(\omega - 1)\rfloor$ single precision primes are unlucky, and since there are at least twice as many of them, the expected number of iterations to find a lucky prime is at most two.

Using a modular gcd algorithm based on Chinese remaindering, like Algorithm 8.6 below with $b = 0$, the (expected) cost for step 1 is $O(n^3 + n\lambda^2)$ word operations with classical arithmetic and $O(\lambda\,\mathsf{M}(n)\log n + n\,\mathsf{M}(n + \lambda)\log(n + \lambda))$ with fast arithmetic, by Theorem 8.10. (Yun's original algorithm employs Hensel lifting for the gcd computation as well.) In step 4, we first reduce u and v modulo p, taking $O(n\log B)$ word operations. The cost for Yun's algorithm is $O(n^2)$ word operations with classical arithmetic and $O(\mathsf{M}(n)\log n)$ with fast arithmetic, by Proposition 5.3. The cost for the Hensel lifting is $O(n^2 s^2)$ and $O(\mathsf{M}(n)\log n\,(\mathsf{M}(s) + \log n))$ with classical and fast arithmetic, respectively, by Fact 3.20. Finally, the cost for the normalization in step 6 is $O(ns^2)$ and $O(n\,\mathsf{M}(s)\log s)$ word operations with classical and fast arithmetic, respectively. The claims now follow from $s, \log B \in O(n + \lambda)$. $\square$

The estimate above for classical arithmetic is slower by a factor of n than the corresponding estimate for Algorithm 5.6, and the time bounds for fast arithmetic agree up to logarithmic factors.

5.2 Greatest Factorial Factorization

In this section, we adapt Yun's (1976) algorithm (Algorithm 5.12) for squarefree factorization to its discrete analog, the *greatest factorial factorization*, which was introduced by Paule (1995). We also discuss a new modular algorithm for computing this factorization. It is quite analogous to the modular algorithm 5.6 for computing the squarefree factorization. The presentation follows closely Gerhard (2000). In the following, E denotes the *shift operator*, which acts by $(Ef)(x) = f(x+1)$ on a polynomial f in the indeterminate x.

Definition 5.15. *Let F be a field and $f \in F[x]$ nonzero monic of degree n. A greatest factorial factorization (or gff) of f is a sequence of monic polynomials $f_1, \dots, f_n \in F[x]$ with the following properties.*

(i) $f = \prod_{1 \leq i \leq n} f_i^i,$

(ii) $\gcd(f_i^i, Ef_j) = \gcd(f_i^i, E^{-j}f_j) = 1$ *for* $1 \leq i \leq j \leq n$.

The intuition is that f_i collects all maximal falling factorials of length i. We note that (ii) is not equivalent to $\gcd(f_i^i, f_j^j) = 1$ for all i, j. For example, the sequence $x, x+1, 1$ is a gff of $x^3 + x^2$ but does not satisfy the latter condition.

Paule (1995) has shown that every nonzero polynomial has a unique gff when F has characteristic zero (see also Lemma 23.10 in von zur Gathen & Gerhard 1999). This is false in positive characteristic: for example, the polynomial $x^p - x = x^p \in \mathbb{F}_p[x]$ has no gff. Existence and uniqueness of the gff still hold if we require in addition that the degree of the polynomial be less than the characteristic. An alternative would be to modify the definition, namely to require (ii) only for $i < j$ and to add the property $\gcd(f_i, Ef_i) = 1$ for $1 \leq i \leq n$, in analogy to the squarefree decomposition. Then the gff is still not unique: for example, $(x+1)^p$ is also a gff of $x^p - x$. However, since we do not need the concept of gff when the degree exceeds the characteristic, we will stick to the original definition. The gff of the constant polynomial 1 is the empty sequence.

There are strong analogies between the squarefree factorization and the gff. Bauer & Petkovšek (1999) discuss a common generalization of both types of factorizations. A different analog and generalization of the squarefree decomposition, the shiftless decomposition, is discussed in Gerhard, Giesbrecht, Storjohann & Zima (2003)

The following lemma, which we quote without proof, is due to Paule (1995); see also Theorem 23.12 in von zur Gathen & Gerhard (1999). It is the analog of the well-known property (5.2) of the squarefree decomposition.

Lemma 5.16 (Fundamental Lemma). *Let F be a field of characteristic zero or greater than $n \in \mathbb{N}$ and $f \in F[x]$ nonzero monic of degree n. If $f_1, \dots, f_n$ is the gff of f and $g = \gcd(f, Ef)$, then*

$$g = f_2 f_3^2 \cdots f_n^{n-1}, \quad \frac{f}{g} = f_1(E^{-1}f_2) \cdots (E^{-n+1}f_n), \quad \frac{Ef}{g} = E(f_1 \cdots f_n) .$$

Moreover, $f_2, \dots, f_{1+\deg g}$ is the gff of g.

The following technical lemma about the gff will be used later.

Lemma 5.17. *Let F be a field, $f \in F[x]$ a nonzero monic polynomial of degree n, and $f_1, \ldots, f_n \in F[x]$ the gff of f. Moreover, let $h \in F[x]$ be an irreducible divisor of f.*

(i) *Let $i, j, k, l \in \mathbb{N}$ be such that $0 \le k < i \le j$ and $0 \le l < j \le n$. If $h \mid E^{-k} f_i$ and $h \mid E^{-l} f_j$, then $0 \le l - k \le j - i$.*

(ii) *Let $1 \le e, i \le n$ be such that $h^e \mid f_i^i$. Then there is a unique $k \in \{0, \ldots, i-1\}$ such that $h^e \mid E^{-k} f_i$.*

(iii) *Assume that $e \ge 1$ is such that $h^e \mid (f/f_1) = f_2^{\overline{2}} \cdots f_n^{\overline{n}}$. Then $Eh^e \mid (f/f_1)$ or $E^{-1} h^e \mid (f/f_1)$. If in addition $h \mid f_1$ and $h^{e+1} \nmid (f/f_1)$, then $Eh^{e+1} \nmid (f/f_1)$ and $E^{-1} h^{e+1} \nmid (f/f_1)$.*

Proof.   (i) If $0 > l - k$, then $E^{l+1} h \mid E^{l-k+1} f_i \mid f_i^i$ and $E^{l+1} h \mid E f_j$. Thus $E^{l+1} h \mid \gcd(f_i^i, E f_j)$, contradicting Definition 5.15 (ii). Similarly, if $l - k > j - i$, then $E^{l-j} h \mid E^{l-k-j} f_i \mid f_i^i$ and $E^{l-j} h \mid E^{-j} f_j$. Thus $E^{l-j} h \mid \gcd(f_i^i, E^{-j} f_j)$, again contradicting Definition 5.15 (ii).

(ii) This follows from (i) with $i = j$.

(iii) Let $e_j \in \mathbb{N}$ be the multiplicity of h in f_j^j, for $1 \le j \le n$. Then $e \le e_2 + \cdots + e_n$. Moreover, let $i \in \mathbb{N}$ be minimal with $i \ge 2$ and $e_i > 0$. By (ii), there is a unique $k \in \{0, \ldots, i-1\}$ with $h^{e_i} \mid E^{-k} f_i$. Assume first that $k = 0$. Then $i \ge 2$ implies that $E^{-1} h^{e_i} \mid E^{-1} f_i \mid f_i^i$. If $e \le e_i$, then we are done. Otherwise, let $j > i$ be such that $e_j \ge 1$. Again by (ii), there is a unique $l \in \{0, \ldots, j-1\}$ such that $h^{e_j} \mid E^{-l} f_j$. Then (i) shows that $l \le j - i < j - 1$, and hence $E^{-1} h^{e_j} \mid E^{-l-1} f_j \mid f_j^j$. Since this holds for all $j > i$, we conclude that $E^{-1} h^{e_j} \mid f_j^j$ for $2 \le j \le n$, and the first claim follows. Similar arguments show that $Eh^{e_j} \mid f_j^j$ for $2 \le j \le n$ if $k > 0$.

For the remaining claim, we first note that the additional assumptions imply $e = e_2 + \cdots + e_n$. Suppose that the claim is wrong and $Eh^{e+1} \mid (f/f_1)$, and let $i \in \{2, \ldots, n\}$ be such that $Eh^{e_i+1} \mid f_i^i$. Then (ii) yields $k \in \{0, \ldots, i-1\}$ satisfying $Eh^{e_i+1} \mid E^{-k} f_i$. If $k < i - 1$, then $h^{e_i+1} \mid E^{-k+1} f_i \mid f_i^i$, a contradiction to the definition of e_i. Thus $k = i-1$ and $Eh^{e_i+1} \mid f_i$, and hence $h \mid E^{-i} f_i$. By assumption, $h \mid f_1 = f_1^1$, and therefore $h \mid \gcd(f_1^1, E^{-i} f_i)$, which contradicts Definition 5.15 (ii). We conclude that $Eh^{e+1} \nmid (f/f_1)$, and the other remaining claim follows by similar arguments. $\square$

We now present two algorithms for computing the gff. The first one is due to Paule (1995). The second one seems to be new and is analogous to Yun's (1976) algorithm for computing the squarefree decomposition.

Algorithm 5.18 (Gff computation).

Input: A nonzero monic polynomial $f \in F[x]$ of degree n, where F is a field of characteristic zero or greater than n.
Output: The gff of f.

0. **if** $f = 1$ **then return** the empty sequence
1. $g \longleftarrow \gcd(f, Ef), \quad m \longleftarrow \deg g$
2. **call** the algorithm recursively to compute the gff $g_1, \ldots, g_m$ of g

3. $h \longleftarrow \dfrac{f}{g \cdot (E^{-1} g_1)(E^{-2} g_2) \cdots (E^{-m} g_m)}$

4. **return** $h, g_1, \ldots, g_m, 1, \ldots, 1$

Theorem 5.19. *Algorithm 5.18 works correctly and takes $O(n^3)$ field operations with classical arithmetic and $O(n \, \mathsf{M}(n) \log n)$ with fast arithmetic.*

Proof. Let $f_1, f_2, \ldots, f_n$ be the gff of f. Then $g = f_2 f_3^2 \cdots f_n^{n-1}$, by the Fundamental Lemma 5.16, and by induction, we have $f_i = g_{i-1}$ for $2 \le i \le m + 1$ and $f_i = 1$ for $m + 2 \le i \le n$. Finally, $f/g = f_1 \cdot (E^{-1} f_2) \cdots (E^{-n+1} f_n)$, again by the Fundamental Lemma, so that $h = f_1$, and the correctness is proved.

Let $T(n)$ denote the cost of the algorithm. Including the cost for the Taylor shift to compute $(Ef)(x) = f(x + 1)$ (Theorems 4.3 and 4.5), step 1 takes $O(n^2)$ field operations with classical arithmetic and $O(\mathsf{M}(n) \log n)$ with fast arithmetic. The cost for step 2 is $T(m)$. Step 3 takes $O(n^2)$ operations with classical arithmetic, by Lemma 3.15 and Theorem 4.3. By Theorem 4.5, we can compute $E^{-i} g_i$ with $O(\mathsf{M}(\deg g_i))$ operations, for each i, together $O(\mathsf{M}(m))$, by the sub-additivity of M. Computing the product in the denominator takes $O(\mathsf{M}(m) \log m)$, and the division with remainder takes $O(\mathsf{M}(n))$, by Lemma 3.15 and Fact 3.13, respectively. Thus the cost for step 3 with fast arithmetic is $O(\mathsf{M}(n) + \mathsf{M}(m) \log m)$. The claims now follow from $m \le n - 1$ and unraveling the recursion for T. $\square$

The cost estimate for fast arithmetic appears also in Theorem 23.14 of von zur Gathen & Gerhard (1999). The upper bounds of the theorem are achieved for $f = x^n$, up to a factor of $\log n$ in the case of fast arithmetic. The following method improves this by an order of magnitude.

Algorithm 5.20 (Gff computation à la Yun).
Input: A nonzero monic polynomial $f \in F[x]$ of degree n, where F is a field of
 characteristic zero or greater than n.
Output: The gff of f.

0. **if** $f = 1$ **then return** the empty sequence
1. $g \longleftarrow \gcd(f, Ef), \quad u_1 \longleftarrow \dfrac{f}{E^{-1} g}, \quad v_1 \longleftarrow \dfrac{f}{g}$
2. **for** $1 \le i \le n$ **do**
3. $\qquad h_i \longleftarrow \gcd(u_i, v_i), \quad u_{i+1} \longleftarrow \dfrac{u_i}{h_i}, \quad v_{i+1} \longleftarrow E \dfrac{v_i}{h_i}$
4. **return** $h_1, \ldots, h_n$

Clearly the algorithm may be aborted as soon as $u_i = 1 \, (= v_i)$.

Theorem 5.21. *Algorithm 5.20 works correctly and takes $O(n^2)$ field operations with classical arithmetic and $O(\mathsf{M}(n) \log n)$ with fast arithmetic.*

Proof. Let $1 \leq i < j, k \leq n$ and $f_1, \ldots, f_n \in F[x]$ be the gff of f. We have $E^{i-j+1} f_j \mid f_j^j$, and the properties of the gff imply that $\gcd(f_k, E^{i-j} f_j) = E^{-1} \gcd(E f_k, E^{i-j+1} f_j) = 1$ if $j \leq k$. Similarly, we have $E^{-i} f_k \mid f_k^k$ and $\gcd(f_k, E^{i-j} f_j) = E^i \gcd(E^{-i} f_k, E^{-j} f_j) = 1$ if $j \leq k$. From these observations, the invariants

$$u_i = \prod_{i \leq k \leq n} f_k, \quad v_i = \prod_{i \leq j \leq n} E^{i-j} f_j, \quad h_i = f_i$$

for $1 \leq i \leq n$ follow easily by induction, and the correctness is proved.

The cost for step 1 is $O(n^2)$ field operations with classical arithmetic and $O(\mathsf{M}(n) \log n)$ with fast arithmetic. Similarly, one execution of step 3 takes $O((\deg u_i)^2)$ and $O(\mathsf{M}(\deg u_i) \log(\deg u_i))$ field operations with classical and fast arithmetic, respectively. If $n_j = \deg f_j$ for all j, then $\deg u_i = \sum_{i \leq j \leq n} n_j$ for all i and $\sum_{1 \leq i \leq n} \deg u_i = n$, and the cost estimates follow by the sub-additivity of M. $\square$

In positive characteristic p, Algorithms 5.18 and 5.20 may get into an infinite loop or return the wrong result if $p \leq n$. For example, when $f = x^p - x = x^p$, then $\gcd(f, Ef) = f$, and Algorithm 5.18 does not terminate, while Algorithm 5.20 incorrectly returns $h_1 = f$ and $h_2 = \cdots = h_n = 1$.

We now discuss a modular gff algorithm. Let Z be a UFD with field of fractions F. We say that a sequence $f_1, \ldots, f_n \in Z[x]$ of primitive polynomials is a *primitive gff* if $f = \prod_i f_i^i$ and $f_1/\mathrm{lc}(f_1), \ldots, f_n/\mathrm{lc}(f_n)$ is the (monic) gff of $f/\mathrm{lc}(f)$ in $F[x]$. The primitive gff is unique up to multiplication by units in $Z^\times$. For $Z = \mathbb{Z}$, we can make the primitive gff unique by requiring $f, f_1, \ldots, f_n$ to be normalized, and then call it the *normalized gff*.

Problem 5.22 (Greatest factorial factorization). *Let Z be a UFD. Given a nonzero primitive polynomial $f \in Z[x]$ of degree n, compute nonzero primitive polynomials $f_1, \ldots, f_n \in Z[x]$ satisfying*

$$f = f_1 f_2^2 \cdots f_n^n,$$

$$\gcd(f_i^i, E f_j) = \gcd(f_i^i, E^{-j} f_j) = 1 \text{ for } 1 \leq i \leq j \leq n.$$

We first investigate when the modular image of a gff is again a gff. Interestingly, the straightforward analog of Lemma 5.4 for the gff, where f' is replaced by Ef, does not hold. For example, let $I = 7\mathbb{Z}$ and $f = (x+8)(x+7)x(x-1)(x-2) \in \mathbb{Z}[x]$. The primitive and monic gff of f is $1, x+8, x, 1, 1$ and $\gcd(f, Ef) = (x+8)x(x-1)$. Now $\gcd(\overline{f}, E\overline{f}) = (x+1)x(x-1) \bmod 7\mathbb{Z} = \overline{\gcd(f, Ef)}$, but the monic gff of $\overline{f}$ is $x, 1, 1, x+1, 1$, and this is not the modular image of the gff of f. The problem here is that the two factors $x+7$ and x are coprime in $\mathbb{Z}[x]$, but not modulo I. Under the additional assumption that any two distinct irreducible factors of f remain coprime modulo I, we can prove the following analog of Lemma 5.4.

Lemma 5.23. *Let Z be a UFD of characteristic zero or prime characteristic greater than n, $f \in Z[x]$ be a nonzero primitive polynomial of degree n, and $f_1, \ldots, f_n \in Z[x]$ be a primitive gff of f. Moreover, let $I \subseteq Z$ be a maximal ideal not containing $\mathrm{lc}(f)$, denote the residue class map modulo I by a bar, and assume that $\overline{Z} = Z/I$ has characteristic zero or greater than n as well. Finally, assume that $g_1, \ldots, g_n \in \overline{Z}[x]$ is the monic gff of $\overline{f}/\mathrm{lc}(f)$. Then*

$$\deg \gcd(f, Ef) \le \deg \gcd(\overline{f}, E\overline{f}) \text{ and } \deg(g_1 \cdots g_n) \le \deg(f_1 \cdots f_n) . \quad (5.7)$$

If in addition

$$\deg \gcd(f, f') = \deg \gcd(\overline{f}, \overline{f'}) , \quad (5.8)$$

then for each of the two inequalities, equality holds if and only if $\overline{\mathrm{lc}(f_i)}g_i = \overline{f_i}$ for $1 \le i \le n$.

We note that the condition (5.8) alone is not is not sufficient to ensure that the modular image of the gff is again a gff, as the example $f = x^2(x + 8) \in \mathbb{Z}[x]$ and $I = 7\mathbb{Z}$ shows. On the other hand, the, example $f = x(x + 7) \in \mathbb{Z}[x]$ and $I = 7\mathbb{Z}$ illustrates that (5.8) is not necessary.

Proof. Let F be the field of fractions of Z. We first note that

$$\gcd(f, Ef) \sim f_2 f_3^2 \cdots f_n^{n-1}, \quad E(f_1 \cdots f_n) \sim \frac{Ef}{\gcd(f, Ef)} ,$$

by the Fundamental Lemma 5.16, where $\sim$ denotes equality up to associates in $F[x]$, and similarly

$$\gcd(\overline{f}, E\overline{f}) \sim g_2 g_3^2 \cdots g_n^{n-1}, \quad E(g_1 \cdots g_n) \sim \frac{E\overline{f}}{\gcd(\overline{f}, E\overline{f})} . \quad (5.9)$$

Now $\overline{\gcd(f, Ef)}$ is a common divisor of $\overline{f}$ and $E\overline{f}$, which implies that

$$\overline{\gcd(f, Ef)} \mid \gcd(\overline{f}, E\overline{f}) \text{ and } g_1 \cdots g_n \mid \overline{f_1 \cdots f_n} , \quad (5.10)$$

and (5.9) shows that either both inequalities in (5.7) are strict or both are equalities. This proves the first claim.

The "if" direction of the second claim is true without the additional assumptions. We prove the "only if" part by induction on n. There is nothing to prove if $n = 0$. If $n > 0$, then $f_2, f_3, \ldots$ is a gff of $\gcd(f, Ef)$ and $g_2, g_3, \ldots$ is a gff of $\gcd(\overline{f}, E\overline{f})$, by the Fundamental Lemma 5.16. If equality holds in (5.7), then (5.10) yields

$$g_2 g_3^2 \cdots g_n^{n-1} \sim \gcd(\overline{f}, E\overline{f}) \sim \overline{\gcd(f, Ef)} \sim \overline{f_2 f_3^2 \cdots f_n^{n-1}}$$

and $g_1 \cdots g_n \sim \overline{f_1 \cdots f_n}$. Applying (5.10) to $\gcd(f, Ef)$ instead of f, we find that $g_2 \cdots g_n$ divides $\overline{f_2 \cdots f_n}$. We show below that also $g_1 \mid \overline{f_1}$. Thus $g_1 \sim \overline{f_1}$ and $g_2 \cdots g_n \sim \overline{f_2 \cdots f_n}$. Lemma 5.4 together with the property (5.2) implies that the condition (5.8) holds for $\gcd(f, Ef)$ as well, and the induction hypothesis, applied to $\gcd(f, Ef)$, yields $g_i \sim \overline{f}_i$ for $2 \leq i \leq n$.

To prove that $g_1 \mid \overline{f}_1$, let $t \in \overline{Z}[x]$ be an irreducible factor of g_1 of multiplicity $i > 0$, and let $j \geq 0$ be its multiplicity in $\overline{f}/g_1$. Moreover, let $h \in Z[x]$ be the unique primitive irreducible factor of f such that $t \mid \overline{h}$, and let $e \geq 1$ be its multiplicity in f. Equation (5.8) and Lemma 5.4 imply that the distinct irreducible factors of f in $Z[x]$ remain squarefree and pairwise coprime modulo I. Thus $t^2 \nmid \overline{h}$ and $t \nmid (\overline{f}/\overline{h}^e)$, and hence $e = i + j$. Now assume that $h^i \nmid f_1$. Then $h^{j+1} \mid (f/f_1) = f_2^2 \cdots f_n^n$, and Lemma 5.17 implies that $Eh^{j+1} \mid (f/f_1)$ or $E^{-1}h^{j+1} \mid (f/f_1)$. For simplicity, we assume that $Eh^{j+1} \mid (f/f_1)$; the arguments for the other case are analogous. Then $Et^{j+1} \mid \overline{f}$, but $Et \nmid g_1$, by Definition 5.15 (ii), such that $Et^{j+1} \mid (\overline{f}/g_1)$. Now Lemma 5.17 (iii) yields $t^{j+1} \mid (\overline{f}/g_1)$. This contradiction to the definition of j proves that our assumption is wrong, $h^i \mid f_1$, and $t^i \mid \overline{f}_1$. Since this holds for any irreducible factor of g_1, we conclude that $g_1 \mid \overline{f}_1$. $\square$

Lemma 5.23 is the main tool for the following modular gff algorithm, which is based on the Chinese Remainder Theorem. We assume that ω is the word size of our processor. For a nonzero polynomial $f \in \mathbb{Z}[x]$, $\mathrm{normal}(f)$ denotes the unique normalized polynomial that divides f and has the same degree.

Algorithm 5.24 (Small primes modular gff computation). ▬▬▬
Input: A normalized polynomial $f \in \mathbb{Z}[x]$ of degree $1 \leq n \leq 2^{\omega-1}$ and max-norm $\|f\|_\infty < 2^\lambda$.
Output: The normalized gff of f, or otherwise "FAIL".

1. $b \longleftarrow \mathrm{lc}(f)$, $B \longleftarrow \lfloor (n+1)^{1/2} 2^{n+\lambda} \rfloor$, $s \longleftarrow \lceil \log_2(2 \cdot 3^{n/3} bB)/(\omega - 1) \rceil$
2. choose a set S_0 of $2s$ single precision primes
   $S_1 \longleftarrow \{p \in S_0 : p \nmid b\}$
3. **for** all $p \in S_1$ **do**
4.   **call** Algorithm 5.20 to compute the monic gff

$$f \equiv b \prod_{1 \leq i \leq n} g_{p,i}^i \bmod p$$

   of f/b modulo p, with monic polynomials $g_{p,1}, \ldots, g_{p,n} \in \mathbb{Z}[x]$ of max-norm less than $p/2$
   $e_p \longleftarrow \deg \gcd(f \bmod p, f' \bmod p)$
5. /∗ remove unlucky primes from S_1 ∗/
   $d \longleftarrow \max\{\deg g_{p,1} + \cdots + \deg g_{p,n} : p \in S_1\}$
   $e \longleftarrow \min\{e_p : p \in S_1\}$
   $S_2 \longleftarrow \{p \in S_1 : \deg g_{p,1} + \cdots + \deg g_{p,n} = d \text{ and } e_p = e\}$
   if $\#S_2 \geq s$ **then** remove $\#S_2 - s$ primes from S_2 **else return** "FAIL"
6. **for** $0 \leq j < i \leq n$ **do**
7.   use the Chinese Remainder Algorithm to compute $f_{ij}^* \in \mathbb{Z}[x]$ of max-norm less than $(\prod_{p \in S_2} p)/2$ such that $f_{ij}^* \equiv bg_{p,i}(x - j) \bmod p$ for all $p \in S_2$

8. **if** $\prod_{0 \leq j < i \leq n} \|\mathrm{normal}(f_{ij}^*)\|_1 > B$ **then return** "FAIL"
9. **if** $\prod_{1 \leq i \leq n} \mathrm{lc}(\mathrm{normal}(f_{i0}^*))^i = \mathrm{lc}(f)$
    then return $\mathrm{normal}(f_{10}^*), \ldots, \mathrm{normal}(f_{n0}^*)$ **else return** "FAIL"

Definition 5.25. *Let* $Z = \mathbb{Z}$ *and* $f \in \mathbb{Z}[x]$ *nonzero primitive. A prime* $p \in \mathbb{N}$ *is* lucky *with respect to the greatest factorial factorization problem 5.22 if* $p \nmid \mathrm{lc}(f)$ *and* $\deg \gcd(f, f') = \deg \gcd(f \bmod p, f' \bmod p)$ *(i.e.,* p *is lucky with respect to the squarefree factorization problem 5.1) and* $\deg \gcd(f, Ef) = \deg \gcd(f \bmod p, Ef \bmod p)$.

For a prime $p > \deg f$, Lemma 5.23 implies that the image modulo p of a greatest factorial factorization of f is again a greatest factorial factorization if p is a lucky prime with respect to greatest factorial factorization, but not vice versa.

Theorem 5.26. *Algorithm 5.24 succeeds if at least* $s \in \Theta(n + \lambda)$ *of the initial primes in* S_0 *are lucky. If the algorithm does not return "FAIL", then it correctly returns the normalized gff of* f.

Proof. Let $f_1, \ldots, f_n \in \mathbb{Z}[x]$ be the normalized gff of f, let $h = \gcd(f, f')$, and let $p \in S_1$. Lemmas 5.23 and 5.4 imply that $\deg(g_{p,1} \cdots g_{p,n}) \leq \deg(f_1 \cdots f_n)$ and $e_p \geq \deg h$, respectively. By definition, we have equality in both cases if and only if p is a lucky prime. Thus, if at least one prime in S_0 is lucky, then we have $d = \deg(f_1 \cdots f_n)$ and $e = \deg h$ in step 5, and all primes in S_2 are lucky.

If at least s of the primes in S_0 are lucky, then S_2 contains s lucky primes after step 5. Let $m = \prod_{p \in S_2} p$ and $0 \leq j < i \leq n$. Then Lemma 5.23 implies that $b f_i(x - j)/\mathrm{lc}(f_i) \equiv f_{ij}^* \bmod m$. Since $f_i(x - j)$ divides f, Mignotte's bound (Fact 3.3) implies that $\|f_i(x - j)\|_\infty \leq B$ and $\|b f_i(x - j)/\mathrm{lc}(f_i)\|_\infty < m/2$. Thus the coefficients of $b f_i(x - j)/\mathrm{lc}(f_i)$ and f_{ij}^* are both absolutely less than $m/2$, and hence they are equal. Since f_i is normalized, so is $f_i(x - j)$, and we have $f_i(x - j) = \mathrm{normal}(f_{ij}^*)$. Again by Mignotte's bound, the condition in step 8 is false, the condition in step 9 is true, and the algorithm does not return "FAIL". This proves the first assertion.

To prove the second claim, suppose that the condition in step 8 is true. Then

$$\left\| \prod_{0 \leq j < i \leq n} \mathrm{normal}(f_{ij}^*) \right\|_\infty \leq \left\| \prod_{0 \leq j < i \leq n} \mathrm{normal}(f_{ij}^*) \right\|_1$$
$$\leq \prod_{0 \leq j < i \leq n} \|\mathrm{normal}(f_{ij}^*)\|_1 \leq B < \frac{m}{2}.$$

Let $0 \leq j < i \leq n$. Then

$$\|f_{ij}^*\|_\infty \leq \|f_{ij}^*\|_1 \leq b \|\mathrm{normal}(f_{ij}^*)\|_1 \leq bB < \frac{m}{2},$$
$$\|f_{i0}^*(x - j)\|_\infty \leq (j + 1)^{\deg f_{i0}^*} \|f_{i0}^*\|_1 \leq i^{n/i} b \|\mathrm{normal}(f_{i0}^*)\|_1 \leq i^{n/i} bB < \frac{m}{2},$$

where we have used Lemma 4.2 and the fact that $i^{1/i} \leq 3^{1/3}$ for all $i \in \mathbb{N}_{\geq 1}$. Thus the congruence $f_{ij}^* \equiv f_{i0}^*(x - j) \bmod m$ is in fact an equality and $\mathrm{normal}(f_{ij}^*) = E^{-j}\mathrm{normal}(f_{i0}^*)$. The polynomials $\prod_{1 \leq i \leq n} \mathrm{normal}(f_{i0}^*)^i$ and f agree modulo m up to a multiplicative constant. If the condition in step 9 is true, then

$$\prod_{1 \leq i \leq n} \mathrm{normal}(f_{i0}^*)^i = \prod_{0 \leq j < i \leq n} \mathrm{normal}(f_{ij}^*) \equiv f \bmod m ,$$

and since both polynomials have max-norms less than $m/2$, they are equal. Now the $\mathrm{normal}(f_{i0}^*)$ are a gff of f, since the required gcd conditions hold modulo each prime in S_2. Since these polynomials are normalized, the uniqueness of the normalized gff implies that $f_i = \mathrm{normal}(f_{i0}^*)$ for all i, and the algorithm returns the correct result. $\square$

We recall from Sect. 3 that for two polynomials $f, g \in R[x]$ and $0 \leq \delta \leq \min\{\deg f, \deg g\}$, $\sigma_\delta(f, g) \in R$ denotes the δth subresultant of f and g.

Theorem 5.27. Let $\delta = \deg \gcd(f, Ef)$, $\delta^* = \deg \gcd(f, f')$, let $r = \sigma_\delta(f, Ef)$, and let $r^* = \sigma_{\delta^*}(f, f')$. Then $0 < |rr^*| \leq D = n^{3n/2}(n+1)^n 2^{n(4\lambda+n+1)}$, and the number of single precision primes that are unlucky with respect to greatest factorial factorization is at most $\lfloor (\log_2 D)/(\omega - 1) \rfloor \in \Theta(n^2 + n\lambda)$. If the number of single precision primes exceeds $2\lfloor (\log_2 D)/(\omega - 1) \rfloor$ and the set S_0 in step 2 is chosen uniformly at random from among all subsets of cardinality $2s$ of the single precision primes, then Algorithm 5.24 returns "FAIL" with probability at most $1/2$.

Proof. The claim that r and r^* are nonzero is clear from Lemma 3.9 (i). We have $\|f\|_2 \leq (n+1)^{1/2}\|f\|_\infty$, $\|Ef\|_2 < 2^{n+1}\|f\|_\infty$, by Lemma 4.2, and

$$\|f'\|_2 \leq n^{1/2}\|f'\|_\infty \leq n^{3/2}\|f\|_\infty ,$$

and Corollary 3.2 implies that

$$|r| \leq \|f\|_2^{n-\delta}\|Ef\|_2^{n-\delta} \leq (n+1)^{n/2}2^{n^2+n}\|f\|_\infty^{2n} ,$$
$$|r^*| \leq \|f\|_2^{n-\delta^*-1}\|f'\|_2^{n-\delta^*} \leq (n+1)^{n/2}n^{3n/2}\|f\|_\infty^{2n} .$$

We conclude that $|rr^*| < D$. Now $2bB \leq D$, so that under the assumptions of the theorem, we can indeed choose $2s$ primes in step 2. A prime is unlucky if and only if it divides rr^*, by Corollary 3.12, and hence there are at most $\lfloor (\log_2 D)/(\omega - 1) \rfloor$ unlucky single precision primes. By assumption, there are at least twice as many single precision primes. Therefore the probability that at least half of the primes of a randomly chosen set of $2s$ single precision primes are lucky is at least $1/2$ (Lemma 3.24), and the claim follows from Theorem 5.26. $\square$

Theorem 5.28. If we neglect the cost for choosing primes in step 2, then Algorithm 5.24 takes $O(n^3 + n\lambda^2)$ word operations when using classical arithmetic, and $O(n\,\mathsf{M}(n + \lambda)\log(n + \lambda) + \lambda\,\mathsf{M}(n)\log n)$ or $O^\sim(n^2 + n\lambda)$ word operations when using fast arithmetic, where the $O^\sim$-notation suppresses logarithmic factors.

Proof. We first reduce $\mathrm{lc}(f)$ and all other coefficients of f modulo all primes in S_0 and S_1, respectively. This takes $O(n\lambda s)$ and $O(n\,\mathsf{M}(s)\log s)$ word operations with classical and fast arithmetic, respectively. The cost for step 4 is $O(n^2)$ and $O(\mathsf{M}(n)\log n)$ word operations for each of the $O(s)$ primes with classical and fast arithmetic, respectively, by Theorem 5.21 and Fact 3.13. Computing a coefficient of some f_{ij}^* in step 7 takes $O(s^2)$ word operations with classical arithmetic and $O(\mathsf{M}(s)\log s)$ with fast arithmetic. There are n non-leading coefficients in total, and hence the cost for steps 6 and 7 is $O(ns^2)$ and $O(n\,\mathsf{M}(s)\log s)$, respectively. The cost for normalizing all f_{ij}^* in step 8 is $O(n)$ gcd's and divisions of integers of word length at most s, taking $O(ns^2)$ word operations with classical arithmetic and $O(n\,\mathsf{M}(s)\log s)$ with fast arithmetic. Computing the products in steps 8 and 9 takes $O(s^2)$ word operations with classical arithmetic and $O(\mathsf{M}(s)\log n)$ with fast arithmetic, by Lemma 3.15. The claims now follow from $s \in O(n + \lambda)$ and using $\mathsf{M}(n) = n \log n \log\log n$ by adding costs. $\square$

We could also think of a modular algorithm for computing the gff $f_1, \ldots, f_n$ of f based on Hensel lifting. A direct adaption of Yun's algorithm 5.12 is not obvious since the f_i are not necessarily coprime and Hensel lifting cannot be applied. However, we could first compute the squarefree decomposition $g_1, \ldots, g_n$ of f, next compute the gff of $g_i g_{i+1} \cdots g_n$ for all i, and finally put things together. As in the case of squarefree factorization, this would lead to similar or even worse time estimates than those of Theorem 5.28, and we do not analyze this algorithm here.

Remark 5.29. *(i) When $\mathsf{M}(n) \in \Omega(n^{1+\varepsilon})$ for some positive ε, then we can drop the factor $\log(n+d)$ in the cost estimate of Theorem 5.28.*

(ii) When using fast arithmetic, then Algorithm 5.24 is – up to logarithmic factors – asymptotically optimal in the diagonal case where $n \approx \lambda$, since its running time is essentially linear in the input size, which is about n^2 words.

(iii) The number $2s$ of initial primes in step 2 of Algorithm 5.24 is often much too large in practice; see Remark 5.11 for a discussion.

(iv) See Remark 3.23 for a discussion how to find random single precision primes in step 2 of Algorithm 5.24 and what this costs.

6. Modular Hermite Integration

The problem of rational function integration that we discuss in this chapter is, given two nonzero polynomials $f, g \in \mathbb{Z}[x]$, to compute $\int (f/g)$. Most undergraduate calculus textbooks contain a solution by factoring the denominator g into linear polynomials over the complex numbers (or at most quadratic polynomials over the real numbers) and performing a partial fraction decomposition. For rational functions with only simple poles, this algorithm first appears in Johann Bernoulli (1703). For symbolic computation, this approach is inefficient since it involves polynomial factorization and computation with algebraic numbers, and the algorithms implemented in most computer algebra systems pursue a different approach due to Hermite (1872).

The idea is to extract the "largest possible rational part" from the integral $\int (f/g)$, i.e., to find polynomials $a, b, c, d \in \mathbb{Q}[x]$ such that

$$\int \frac{f}{g} = \frac{c}{d} + \int \frac{a}{b} \, , \tag{6.1}$$

where $\deg a < \deg b$ and the numerator b is of minimal degree. It turns out that this condition is equivalent to b being squarefree and coprime to a, and that the solution is unique if we stipulate that in addition b and d are monic, c and d are coprime, and $c(0) = 0$. Hermite's algorithm only requires the squarefree decomposition of g, which is much easier to compute than the irreducible factorization, and the partial fraction decomposition along this squarefree decomposition, and it involves only rational arithmetic. Algebraic numbers may solely occur in the subsequent computation of $\int (a/b)$, and the methods of Rothstein (1976, 1977) and Trager (1976) and of Lazard, Rioboo and Trager do this in an efficient way with as small an algebraic extension as possible. The latter step will be discussed in Chap. 7 below.

We present two new modular algorithms for computing the decomposition (6.1). The idea is either to choose several single precision primes, execute the Hermite algorithm modulo each prime, and then obtain the result in $\mathbb{Q}[x]$ by Chinese remaindering and rational number reconstruction, or to choose one single precision prime and use Hensel lifting. The main contributions are estimates of the number of "unlucky" primes, for which the modular computation might not work, and of the number of "lucky" primes required to reconstruct the result, and a cost analysis in terms of processor operations on machine words. More precisely, if $0 < \deg g = n \le m$, $\deg f < m$, and the coefficients of f and g are absolutely less

J. Gerhard: Modular Algorithms, LNCS 3218, pp. 79–95, 2004.
© Springer-Verlag Berlin Heidelberg 2004

than 2^λ, then we show that our modular algorithms for Hermite integration of f/g take $O(m^3(n^2 + \log^2 m + \lambda^2))$ word operations with classical arithmetic, and the small primes modular algorithm takes $O^\sim(m^2(n + \lambda))$ with fast arithmetic, where we neglect the cost for finding our prime moduli. The presentation follows Gerhard (2001).

6.1 Small Primes Modular Algorithm

The direct way of implementing Hermite integration is to perform all computations in $\mathbb{Z}[x]$ or $\mathbb{Q}[x]$ and use modular algorithms only for the gcd computations involved. We believe that our algorithms are superior to this approach, already for medium size inputs, for several reasons. The all-modular approach avoids the well-known phenomenon of intermediate expression swell. Moreover, it delays arithmetic on rational numbers, which is expensive in practice, until the very last step, and it has the overhead for modular reduction and Chinese remaindering only once at the beginning and at the end of the algorithm, respectively.

Let $f, g \in \mathbb{Z}[x] \setminus \{0\}$ be polynomials with $0 < n = \deg g \le m$ and $\deg f < m$. *Hermite reduction* finds polynomials $a, b, c, d, h \in \mathbb{Q}[x]$ with $\deg a < \deg b$, $\deg c < \deg d$, $\deg b + \deg d \le n$, $\deg h \le m - n$, $h(0) = 0$, and b squarefree, such that

$$\int \frac{f}{g} = h + \frac{c}{d} + \int \frac{a}{b} \,,$$

or equivalently,

$$\frac{f}{g} = h' + \left(\frac{c}{d}\right)' + \frac{a}{b} \,. \tag{6.2}$$

The polynomial h is usually called the *polynomial part* of the integral $\int(f/g)$, c/d is the *rational part*, and $\int(a/b)$ is the *logarithmic part*. It is well-known (Ostrogradsky 1845; Horowitz 1971) that the denominators in the representation (6.2) may be chosen as $b = g_*$, the squarefree part of g, and $d = g/g_*$, and then the polynomials a, c, and h are uniquely determined.

We may assume that f and g are primitive and coprime and that g is normalized, and let $g_1, \ldots, g_n \in \mathbb{Z}[x]$ be the normalized squarefree decomposition of g, $b = g_* = g_1 \cdots g_n$ its (normalized) squarefree part, and $d = g/g_*$. Performing a partial fraction decomposition, we may rewrite (6.2) as

$$\frac{f}{g} = h' + \sum_{1 \le j < i \le n} \left(\frac{c_{ij}}{g_i^j}\right)' + \sum_{1 \le i \le n} \frac{a_i}{g_i} \,, \tag{6.3}$$

with unique polynomials $c_{ij}, a_i \in \mathbb{Q}[x]$ of degree less than $\deg g_i$ for all i, j.

Problem 6.1 (Hermite integration). *Given nonzero primitive and coprime polynomials $f, g \in \mathbb{Z}[x]$ with $0 < n = \deg g \le m$ and $\deg f < m$ such that g is normalized, plus the normalized squarefree decomposition $g_1, \ldots, g_n \in \mathbb{Z}[x]$*

of g, compute $h \in \mathbb{Q}[x]$ of degree at most $m - n$ with $h(0) = 0$ and polynomials $c_{ij}, a_i \in \mathbb{Q}[x]$ with $\deg c_{ij}, \deg a_i < \deg g_i$ for all i, j satisfying

$$\frac{f}{g} = h' + \sum_{1 \leq j < i \leq n} \left(\frac{c_{ij}}{g_i^j}\right)' + \sum_{1 \leq i \leq n} \frac{a_i}{g_i} .$$

For completeness, we briefly recall how Hermite reduction works (see von zur Gathen & Gerhard 1999, Sect. 22.2). After splitting off the derivative of the polynomial part by a division with remainder, the partial fraction decomposition

$$\frac{f}{g} = h' + \sum_{1 \leq j \leq i \leq n} \frac{\gamma_{ij}}{g_i^j} \tag{6.4}$$

is computed, with $\gamma_{ij} \in \mathbb{Q}[x]$ of degree less than $\deg g_i$ for all i, j. Then for each $i \geq 2$ independently, the following reduction step is performed for $j = i, \ldots, 2$. First we compute $\sigma, \tau \in \mathbb{Q}[x]$ of degree less than $\deg g_i$ such that $\sigma g_i + \tau g_i' = \gamma_{ij}$, by means of the Extended Euclidean Algorithm. Then we set

$$c_{i,j-1} \longleftarrow \frac{-\tau}{j-1}, \qquad \gamma_{i,j-1} \longleftarrow \gamma_{i,j-1} + \sigma + \frac{\tau'}{j-1} . \tag{6.5}$$

The final γ_{i1} is then equal to a_i, for all i. Both the cost for computing the partial fraction decomposition and the total cost for all reduction steps is $O(n^2)$ arithmetic operations with classical arithmetic and $O(\mathsf{M}(n) \log n)$ with fast arithmetic (see Fact 3.16 for the partial fraction decomposition and, for example, von zur Gathen & Gerhard (1999), Theorem 22.7, for the reduction steps with fast arithmetic; the corresponding result with classical arithmetic follows by a similar reasoning).

Lemma 6.2. *Let F be a field of characteristic zero or greater than $m \in \mathbb{N}$, $f, g \in F[x]$ nonzero polynomials with $0 < \deg g = n \leq m$ and $\deg f < m$, and $g = \prod_{1 \leq i \leq n} g_i^i$ a squarefree decomposition of g, with nonzero and pairwise coprime $g_1, \ldots, g_n \in F[x]$. Then there exist unique polynomials $h \in F[x]$ of degree at most $m - n$ with $h(0) = 0$ and c_{ij}, a_i in $F[x]$ of degree less than $\deg g_i$ for all i, j, satisfying (6.3), and they can be computed by Hermite reduction as described above.*

Proof. For the existence, we note that Hermite reduction works literally as described above in positive characteristic, since the degrees of all occurring polynomials are less than the characteristic, so that taking derivatives does not interfere with the characteristic, and also all integers occurring in denominators in (6.5) are units in F. The uniqueness follows from the uniqueness of the decomposition (6.2), the uniqueness of quotient and remainder in polynomial division, and the uniqueness of partial fraction decomposition (Fact 3.16). $\square$

Back to equation (6.3), we have

$$\left(\frac{c_{ij}}{g_i^j}\right)' = \frac{c'_{ij}g_i - jc_{ij}g'_i}{g_i^{j+1}},$$

and after multiplying both sides in (6.3) by g, we obtain the equivalent equation

$$f = gh' + \sum_{1 \le j < i \le n} \left(\frac{g}{g_i^j}c'_{ij} - jg'_i \frac{g}{g_i^{j+1}}c_{ij}\right) + \sum_{1 \le i \le n} \frac{g}{g_i}a_i. \qquad (6.6)$$

Both taking derivatives and multiplying by a fixed polynomial are linear operators on polynomials, and by comparing coefficients, we see that (6.6) is in turn equivalent to a system of m linear equations in m unknowns, namely the coefficients of h (without the constant coefficient $h(0)$), the c_{ij}, and the a_i.

The idea for a modular algorithm now is as follows. Using Cramer's rule and Hadamard's inequality, we obtain a bound on the numerators and the denominators of the coefficients of h, the c_{ij}, and the a_i. Then we choose sufficiently many single precision primes, so that their product exceeds this bound and we can uniquely reconstruct the coefficients from all the modular images, use Hermite's method modulo each of the primes, and obtain the result by rational number reconstruction. In practice, it may be advantageous to do this in an adaptive fashion, so that only as many primes as needed to recover the result are used; see Remark 6.8 below.

So let $A \in \mathbb{Z}^{m \times m}$ be the coefficient matrix of the linear system equivalent to (6.6) (a similar matrix for the linear system equivalent to (6.2) has been given explicitly by Horowitz 1971), and identify the polynomial $f \in \mathbb{Z}[x]$ with its coefficient vector in $\mathbb{Z}^m$ (this is the "right hand side" of the linear system). Thus, if $y \in \mathbb{Q}^m$ contains the coefficients of h and the c_{ij} and a_i, then (6.6) is equivalent to $Ay = f$. Since (6.6) has a unique solution for a fixed g and arbitrary f of degree less than m, the matrix A, depending only on g but not on f, is nonsingular.

A variant of the following theorem for the linear system equivalent to (6.2) appears in Horowitz (1971).

Theorem 6.3. *Assume that $\|f\|_\infty, \|g\|_\infty < 2^\lambda$, and let $B = \lfloor (n+1)^{1/2}2^{n+\lambda} \rfloor$. Let A^* be the matrix resulting from A by replacing some column by f. Then $\det A$ and $\det A^*$ are absolutely less than $(mB)^m$, and so are the (coprime) numerators and denominators of the unique solutions h, c_{ij}, and a_i of (6.6).*

Proof. For $0 \le k < m - n$, the entries of the column of the matrix A corresponding to the coefficient of x^{k+1} in h are coefficients of g multiplied by $k + 1 \le m$. Now let $0 \le k < n_i = \deg g_i$. The entries of the column of A corresponding to the coefficient of x^k in a_i are coefficients of the polynomial g/g_i, for all i. For $1 \le j < i \le n$, the entries of the column corresponding to the coefficient of x^k in c_{ij} are of the form $(k+1)\alpha - j\beta$ if $0 \le k < n_i - 1$ and $-j\beta$ if $k = n_i - 1$, where α is some coefficient of g/g_i^j and β is some coefficient of $g'_i g/g_i^{j+1}$, by (6.6).

Now Mignotte's bound (Fact 3.3) implies that the one-norm of any g/g_i^j is at most B, and also that

$$\|g_i'g/g_i^{j+1}\|_2 \le \|g_i'g/g_i^{j+1}\|_1 \le \|g_i'\|_1 \|g/g_i^{j+1}\|_1 \le n_i \|g_i\|_1 \|g/g_i^{j+1}\|_1 \le n_i B .$$

Thus $\|(k+1)g/g_i^j\|_2 \le n_i B$ and $\|jg_i'g/g_i^{j+1}\|_2 \le jn_i B \le (i-1)n_i B$. Now $in_i \le n \le m$ and $\|f\|_2 \le m^{1/2} B$, and hence the Euclidean norm of any column of A or A^* is at most mB. Hadamard's inequality (Lemma 3.1) implies that $|\det A| \le (mB)^m$ and $|\det A^*| \le (mB)^m$, and the last claim follows from Cramer's rule which implies that a coefficient of h, c_{ij}, or a_i is of the form $(\det A^*)/\det A$. $\square$

Here is the modular algorithm for Hermite reduction. We recall that ω is the word size of our processor. We will assume that our polynomials have degree at most $2^{\omega-1}$, which is not a serious restriction in practice for $\omega \ge 32$.

Algorithm 6.4 (Small primes modular Hermite integration). ▰▰▰▰▰
Input: Primitive nonzero and coprime polynomials $f, g \in \mathbb{Z}[x]$ of max-norm less
   than 2^λ such that g is normalized, $0 < n = \deg g \le m$ and $\deg f < m < 2^{\omega-1}$,
   and the normalized squarefree decomposition $g_1, \dots, g_n \in \mathbb{Z}[x]$ of g.
Output: The unique polynomials $h, c_{ij}, a_i \in \mathbb{Q}[x]$ solving (6.3) with $h(0) = 0$,
   $\deg h \le m - n$, and $\deg c_{ij}, \deg a_i < \deg g_i$ for all i, j.

1. $B \longleftarrow \lfloor (n+1)^{1/2} 2^{n+\lambda} \rfloor, \quad C \longleftarrow (mB)^m,$
   $t \longleftarrow \left\lceil \dfrac{n \log_2((n^2 + n)4^\lambda)}{\omega - 1} \right\rceil, \quad s^* \longleftarrow \left\lceil \dfrac{\log_2(2C^2)}{\omega - 1} \right\rceil$
2. choose a set S_0 of $s = s^* + t$ single precision primes
   $S_1 \longleftarrow \{p \in S : p \nmid \mathrm{lc}(g) \text{ and } \deg \gcd(g, g') = \deg \gcd(g \bmod p, g' \bmod p)\}$
3. **for each** $p \in S_1$ **do**
4.   compute the partial fraction decomposition (6.4) modulo p
5.   use the Hermite reduction steps (6.5) to solve (6.3) modulo p
6. compute the coefficients of h and all c_{ij} and a_i from their images modulo all
   primes in S_1 by rational number reconstruction ▰▰▰▰▰▰

Theorem 6.5. *If there are at least s single precision primes, then Algorithm 6.4 solves the Hermite integration problem 6.1 correctly as specified.*

Proof. Let h and c_{ij}, a_i in $\mathbb{Q}[x]$ be the unique solution of (6.3). By definition, each prime $p \in S_1$ is lucky with respect to the squarefree factorization of g, and Lemma 5.4 implies that $g_1 \bmod p, \dots, g_n \bmod p$ are a squarefree decomposition of $g \bmod p$. Now $h \bmod p$ and the $c_{ij} \bmod p$ and $a_i \bmod p$ form a solution of (6.3) in $\mathbb{F}_p[x]$, and Lemma 6.2 implies that these are computed in steps 3–5 for all primes $p \in S_1$. In particular, the uniqueness statement of the lemma implies that none of the primes in S_1 divides the determinant of the coefficient matrix A of the linear system equivalent to (6.3), and by Cramer's rule, the denominators of h and the c_{ij} and a_i are not divisible by any prime in S_1.

As in the proof of Theorem 5.9, we find that all primes in $S_0 \setminus S_1$ are divisors of a certain subresultant r of g and g' with

$$0 < |r| \le (n^2 + n)^n 2^{2n\lambda} \le 2^{(\omega-1)t} ,$$

so that S_1 contains at least $s^* = s - t$ primes. Now $\prod_{p \in S_1} p > 2^{(\omega-1)s^*} \geq 2C^2$, the numerators and denominators of the coefficients of h and the c_{ij} and a_i are absolutely bounded by C, by Theorem 6.3, and hence we can apply Fact 3.18 to reconstruct them. □

Theorem 6.6. *If there are at least* $s \in \Theta(m(n + \log m + \lambda))$ *single precision primes and we neglect the cost for finding them, then the cost for Algorithm 6.4 is* $O(m^3(n^2 + \log^2 m + \lambda^2))$ *word operations with classical arithmetic, and* $O(m\,\mathsf{M}(m(n + \log m + \lambda)) \log(m(n + \lambda)))$ *or* $O^{\sim}(m^2(n + \lambda))$ *with fast arithmetic.*

Proof. We can evaluate f, g, and all g_i modulo all primes in S_1 and S_0, respectively, using $O(ms^2)$ word operations with classical arithmetic and $O(m\,\mathsf{M}(s) \cdot \log s)$ with fast arithmetic. The cost for computing $\gcd(g \bmod p, g' \bmod p)$ for all $p \in S_0$ is $O(n^2 s)$ or $O(\mathsf{M}(n) \log n \cdot s)$, respectively. The cost for computing the polynomial part modulo a prime $p \in S_1$ is essentially one division with remainder, taking $O(mn)$ word operations with classical arithmetic and $O(\mathsf{M}(m))$ with fast arithmetic. By the discussion preceding Lemma 6.2, the cost for steps 4 and 5 is $O(mn)$ word operations per prime p, in total $O(mns)$ word operations with classical arithmetic, and $O((\mathsf{M}(m) + \mathsf{M}(n) \log n)s)$ with fast arithmetic. Finally, by Fact 3.18 the cost for rational number reconstruction in step 6 is $O(s^2)$ word operations per coefficient, or $O(ms^2)$ word operations in total, with classical arithmetic, and $O(m\,\mathsf{M}(s) \log s)$ with fast arithmetic. Now the claims follow from $s \in O(m(n + \log m + \lambda))$ by adding up costs, and taking $\mathsf{M}(n) = n \log n \log\log n$ gives the $O^{\sim}$-estimate. □

Since the cost for the modular Hermite reduction dominates the cost for computing the squarefree decomposition (Theorem 5.10), we have the following corollary.

Corollary 6.7. *Suppose that* $f, g \in \mathbb{Z}[x]$ *are two primitive nonzero and coprime polynomials with max-norms less than* 2^λ *such that* g *is normalized,* $0 < n = \deg g \leq m$, *and* $\deg f < m < 2^{\omega-1}$. *If there are at least* s *single precision primes, with* s *as in Algorithm 6.4, then we can compute the decomposition (6.3) using* $O(m^3(n^2 + \log^2 m + \lambda^2))$ *word operations with classical arithmetic, and* $O(m\,\mathsf{M}(m(n + \log m + \lambda)) \log(m(n + \lambda)))$ *or* $O^{\sim}(m^2(n + \lambda))$ *with fast arithmetic.*

Remark 6.8. • *When* $\mathsf{M}(n) \in \Omega(n^{1+\varepsilon})$ *for some positive* ε, *then we can drop the factor* $\log(m(n + \lambda))$ *in the cost estimates of Theorem 6.6 and Corollary 6.7.*
• *For those inputs where the upper bound of* $O(m^2(n + \log m + \lambda))$ *on the output size provided by Theorem 6.3 is achieved, Algorithm 6.4 is – up to logarithmic factors – asymptotically optimal when using fast arithmetic.*
• *Horowitz (1971) has analyzed two variants of Hermite reduction for univariate polynomials over* $\mathbb{Z}$ *in the case* $m = n$: *a non-modular variant of Algorithm 6.4 and an algorithm solving (6.2) by linear algebra methods. In Theorems 4.1 and 4.3, he gives cost estimates of* $O(k^3 n^5 \log^2(n\gamma))$ *and* $O(n^5 \log^2(n\gamma))$ *word*

operations, respectively, where $k \leq n$ is the largest index such that g_k is noncon-stant and γ corresponds essentially to our B, such that $\log(n\gamma) \in O(n+\lambda)$. Even for classical arithmetic, the time bound of Theorem 6.6 is better by two orders of magnitude.

- *In practice, the number s of initial primes in step 2 of Algorithm 6.4 may be too large. A reason for this is that the Mignotte bound B in Theorem 6.3 is often too large. One solution to overcome this is to work in an adaptive fashion, as described in Remark 5.11.*

6.2 Prime Power Modular Algorithm

We now present a modular algorithm for Hermite integration using Hensel lifting, by adapting an algorithm for computing the solution of a nonsingular square system of linear equations (Moenck & Carter 1979; Dixon 1982; see also Mulders & Storjohann 1999).

Algorithm 6.9 (Prime power modular Hermite integration). ▬▬▬▬▬▬
Input: Primitive nonzero and coprime polynomials $f, g \in \mathbb{Z}[x]$ of max-norm less than 2^λ such that g is normalized, $0 < n = \deg g \leq m$ and $\deg f < m < 2^{\omega-1}$, and the normalized squarefree decomposition $g_1, \ldots, g_n \in \mathbb{Z}[x]$ of g.
Output: The unique polynomials $h, c_{ij}, a_i \in \mathbb{Q}[x]$ solving (6.3) with $h(0) = 0$, $\deg h \leq m - n$, and $\deg c_{ij}, \deg a_i < \deg g_i$ for all i, j.

1. **repeat**
      choose a single precision prime $2^{\omega-1} < p < 2^\omega$
    until $p \nmid \mathrm{lc}(g)$ and $\deg \gcd(g, g') = \deg \gcd(g \bmod p, g' \bmod p)$
2. $B \longleftarrow \lfloor (n+1)^{1/2}2^{n+\lambda} \rfloor, \quad C \longleftarrow (mB)^m, \quad s \longleftarrow \left\lceil \dfrac{\log_2(2C^2)}{\log_2 p} \right\rceil$
3. let $A \in \mathbb{Z}^{m \times m}$ denote the coefficient matrix of the linear system equivalent to (6.6)
    $e_0 \longleftarrow f$
4. **for** $0 \leq i < s$ **do**
5.   compute the partial fraction decomposition (6.4), with f replaced by e_i, modulo p
6.   use the Hermite reduction steps (6.5) to solve (6.3), with f replaced by e_i, modulo p
        let $z_i \in \{0, \ldots, p-1\}^m$ be the vector with the coefficients of the solution h, c_{ij}, and a_i
7. $e_{i+1} \longleftarrow (e_i - Az_i)/p$
8. $z \longleftarrow \displaystyle\sum_{0 \leq i < s} z_i p^i$
9. compute the coefficients of h and all c_{ij} and a_i from their images in z modulo p^s by rational number reconstruction ▬▬▬▬▬▬

Theorem 6.10. *Algorithm 6.9 solves the Hermite integration problem 6.1 correctly as specified. If at least $2 \log_2((n^2 + n)^n 2^{2n\lambda})/(\omega - 1) \in \Theta(n(\lambda + \log n))$ single precision primes exist, then the expected number of iterations of step 1 is at most two. The cost for steps 2 through 9 is $O(m^3(n^2 + \log^2 m + \lambda^2))$ word operations with classical arithmetic and*

$$O(m(n + \log m + \lambda)(\mathsf{M}(m(n + \log m + \lambda)) + \mathsf{M}(n^2 + n\lambda) \log n)))$$

or $O^\sim(m^2(n^2 + \lambda^2)))$ with fast arithmetic.

Proof. We follow closely Mulders & Storjohann (1999). Let h and c_{ij}, a_i in $\mathbb{Q}[x]$ be the unique solution of (6.3). The invariant

$$f - A\left(\sum_{0 \le j < i} z_j p^j\right) = e_i p^i \tag{6.7}$$

is easily proved by induction on i. In particular, for $i = s$ we find $Az \equiv f \bmod p^s$. As in the proof of Theorem 6.5, the uniqueness statement of Lemma 6.2 implies that A is invertible modulo p, and hence the vector z contains the images of h, the c_{ij}, and the a_i modulo p^s. By construction, we have $p^s \ge 2C^2$, and as in the proof of Theorem 6.5, we can use Fact 3.18 for the rational reconstruction.

As in the proof of Theorem 5.9, there are at most $\log_2((n^2 + n)^n 2^{2n\lambda})/(\omega - 1)$ single precision primes that are unlucky with respect to the squarefree factorization of g, and since there are at least twice as many single precision primes, the expected number of iterations of step 1 to find a lucky prime is at most two.

As in the proof of Theorem 6.3, the entries of A are at most mB in absolute value. If we let $y_i = \sum_{0 \le j < i} z_j p^j$ and denote the maximal absolute value of an entry of A by $\|A\|_\infty$, then (6.6) and (6.7) imply that

$$
\begin{aligned}
p^i\|e_i\|_\infty &= \|e_i p^i\|_\infty = \|f - Ay_i\|_\infty \le \|f\|_\infty + \|Ay_i\|_\infty \\
&< 2^\lambda + m\|A\|_\infty\|y_i\|_\infty \le 2^\lambda + m^2 B(p^i - 1) \\
&\le m^2 B p^i,
\end{aligned}
$$

and hence $\|e_i\|_\infty < m^2 B$ for all i. With classical arithmetic, we compute the entries of A in step 3 from equation (6.6), at a cost of $O(n^2 \log^2(mB))$ word operations. With fast arithmetic, we do not compute the entries of A at all. In step 4, we first reduce the coefficients of g and the g_i, of max-norm at most B, by Mignotte's bound 3.3, modulo p. This takes $O(n \log B)$ word operations. In step 5, we first reduce the coefficients of e_i modulo p, taking $O(m \log(mB))$ word operations. The cost for the partial fraction decomposition and step 6, including the initial division with remainder, is $O(mn)$ and $O(\mathsf{M}(m) + \mathsf{M}(n) \log n)$ with classical and fast arithmetic, respectively, by the discussion preceding Lemma 6.2. Using the at most $O(mn)$ precomputed nonzero entries of A, computing Az_i in step 7 with classical arithmetic takes $O(mn)$ additions and about the same number of multiplications by single precision integers. All intermediate results are of word size $O(\log(mB))$, and hence the cost with classical arithmetic is $O(mn \log(mB))$.

With fast arithmetic, we perform step 7 via the formula (6.6). Computing gh' takes $O(\mathsf{M}(m \log(mB)))$ word operations, by Remark 3.14. Computing the second sum in (6.6) takes $O(\mathsf{M}(n \log(nB)) \log n)$ word operations, by Lemma 3.17 (i). To compute the first sum, we first compute for each i separately

$$C_{i,1} = \sum_{1 \leq j < i} c'_{ij} g^{i-1-j} \text{ and } C_{i,2} = \sum_{1 < j \leq i} (j-1)c_{i,j-1} g^{i-j} \; .$$

If we let $n_i = \deg g_i$, then this takes $O(\mathsf{M}(in_i \log(in_iB)) \log i)$ word operations, and

$$\|C_{i,1}\|_1, \|C_{i,2}\|_1 \leq in_i(p-1)B \; ,$$

by Lemma 3.17 (ii). Next, we compute $C_i = g_i C_{i,1} - g'_i C_{i,2}$ and g_i^i, at a cost of $O(\mathsf{M}(in_i \log(in_iB)))$ word operations, by Remark 3.14, and note that $\|C_i\|_1 \leq in_i^2(p-1)B^2$. Finally, we obtain the first sum in (6.6) as $\sum_{1 \leq i \leq n} C_i g/g_i^i$, taking $O(\mathsf{M}(n \log(nB)) \log n)$ word operations, again by Lemma 3.17 (i). Using the fact that $\sum_{1 \leq i \leq n} in_i = n$, we obtain a total cost of $O(\mathsf{M}(m \log(mB)) + \mathsf{M}(n \log(nB)) \log n)$ for step 7 with fast arithmetic.

Thus the overall cost for steps 4 through 7 is $O(mns \log(mB))$ word operations with classical and $O(s(\mathsf{M}(m \log(mB)) + \mathsf{M}(n \log(nB)) \log n))$ with fast arithmetic. Step 8 takes $O(ms^2)$ and $O(m \mathsf{M}(s) \log s)$ word operations with classical and fast arithmetic, respectively, by Theorem 9.15 in von zur Gathen & Gerhard (1999), and the same bound is valid for step 9. The claims now follow from $\log B \in O(n + \lambda)$ and $s \in O(m(n + \log m + \lambda))$. $\square$

The classical time bound for Algorithm 6.9 agrees with the corresponding one for Algorithm 6.4, and the estimate for fast arithmetic is slower by a factor of about $n + \log m + \lambda$. In both algorithms, the dominant cost with classical arithmetic is for the rational reconstruction in the last step.

6.3 Implementation

We implemented variants of both the small primes algorithm 6.4 and the prime power algorithm 6.9 in C++, using Victor Shoup's library NTL version 5.0c, which provides efficient and convenient data types for univariate polynomials over the integers and over finite fields and has elaborate support for homomorphic imaging.

We ran a series of experiments with our implementation. The software was compiled with the GNU compiler version 2.8.1 with optimization level 2. The timings were taken on a Linux PC with a 800 MHz Pentium III CPU.

We used NTL's built-in routine for computing the squarefree decomposition of the denominator. In our experiments, the running time for this computation turned out to be negligible in comparison to the total running time of both Algorithm 6.4 and Algorithm 6.9.

Our implementation of the small primes algorithm differs from Algorithm 6.4. It chooses the primes deterministically and in an adaptive fashion, starting with a small number of primes. Then it reconstructs only some of the rational coefficients of h, the c_{ij}, and the a_i, and checks whether the constant coefficients on both sides of equation (6.6) coincide, i.e., whether the equation holds for $x = 0$. If this check fails, then some new primes are chosen, and so on. Only if the constant coefficient check succeeds, then all rational coefficients are reconstructed, and finally the implementation checks that the equation (6.6) holds true. We note that in none of our experiments, we ever encountered an unlucky prime in step 2 of Algorithm 6.4.

The implementation does not perform a constant coefficient check after each new prime, but only after about $c, c^2, c^3, c^4, \ldots$ primes, for some suitably chosen real constant $c > 1$. We have chosen $c = 11/10$ in our experiments. This ensures that the number of primes used never exceeds the minimal number of primes needed by more than 10%. At the same time, it keeps the overall cost for the constant coefficient check reasonably small, as can be seen from the timings in Tables 6.1 through 6.3 below.

Our implementation of the prime power algorithm 6.9 has similar modifications. It works in an adaptive fashion as well, increasing the lifting exponent by one in each step, performing a constant coefficient check after about $c, c^2, c^3, \ldots$ steps, with $c = 11/10$ as above. Only if the constant coefficient check is successful, our implementation reconstructs all rational coefficients of the solution and checks that the equation (6.6) holds true. The time for deterministically finding a suitable prime in step 1 of Algorithm 6.9 was less than 10 milliseconds in all our experiments, and we never encountered an unlucky prime.

The constant coefficient check was extremely useful in our experiments: in about 94% of all cases, the rational reconstruction of all coefficients and the check of equation (6.6) was performed only once in the end of the computation, and in all but one case two trials were sufficient.

There are timings for three series of experiments, where the input is a quotient of two pseudorandomly chosen polynomials with integer coefficients. Since such a polynomial is irreducible with very high probability and any algorithm for Hermite integration becomes trivial when the denominator is irreducible, we chose our denominators so as to have a nontrivial squarefree decomposition.

In the first series, we pseudorandomly and independently chose two polynomials h_1 and h_2 of degrees $n/2$ and $n/4$, respectively, and with nonnegative coefficients less than 2^{15} for h_1 and less than 2^7 for h_2, for various values of n that are divisible by 4. Then we put $g = h_1 h_2^2$ as the denominator polynomial. This is a polynomial of degree n with about $30 + \log_2 n$-bit coefficients. In fact, the squarefree decomposition of g is equal to $h_1, h_2, 1, \ldots, 1$ in all our experiments.

In the second series of experiments, we pseudorandomly chose a polynomial h_1 of degree $n/2$ with nonnegative coefficients less than 2^{15}, and put $g = h_1 \cdot (x^2 + 1)^{n/4}$ as the denominator polynomial, for various values of n that are divisible by 4. Then g is a polynomial of degree n with coefficients of bit size about $15 + n$. In all our experiments, the squarefree decomposition of g is $g_1 = h_1, g_{n/4} = x^2 + 1$, and all other factors are trivial.

Finally, in the third series, we pseudorandomly and independently chose polynomials $h_1, h_2, \ldots, h_k \in \mathbb{Z}[x]$ of degree 5 and with nonnegative coefficients less than 4, and put $g = h_1 h_2^2 \cdots h_k^k$ as the denominator polynomial, for various values of k. Here, g is a polynomial of degree $n = 5k(k+1)/2$ with coefficients of bit size about $3k(k+1)/2 = 3n/5$.

The numerator polynomial f was chosen pseudorandomly with coefficients of about the same bit size as the denominator polynomial g. For the first and the second series mentioned above, we performed one experiment with $\deg f < \deg g$, such that the input f/g is a proper rational function and the polynomial part h of its integral is zero, and another experiment with $\deg f < 2 \deg g$. In the latter case, the degree of the polynomial part of the integral is about $\deg g$.

We implemented two variants of rational reconstruction. The first variant reconstructs each coefficient of the output polynomials independently. This always yields the coefficients reduced to lowest terms. The second variant proceeds as follows. We first reconstruct the numerator a and the denominator b of the first coefficient. Since all denominators divide the determinant of the matrix A from Theorem 6.3, we can expect, heuristically, that the gcd of any two denominators is "large". Thus, when reconstructing the second coefficient, we only look for denominators that are multiples of b. This works by reconstructing b times the second coefficient instead of the coefficient itself. We proceed in this fashion. If b is indeed very close to the real denominator, then this is considerably faster than reconstructing the corresponding coefficient independently. This approach leads to rational coefficients that are not necessarily in lowest terms.

However, our experiments show that this is not always a good strategy: the least common denominator of all coefficients may be considerably larger than the individual denominators. In this case, the second variant of rational reconstruction may require a substantially larger number of primes, and the running times may be even slower than for the straightforward implementation.

Tables 6.1, 6.2, and 6.3 give timings for the three series of experiments described above with our implementation of the small primes modular Hermite algorithm 6.4. In the first column, n denotes the degree of the denominator polynomial g. The other columns contain the averages of the following data over 10 pseudorandomly chosen inputs for the same value of n:

- $\#p$ is the number of single precision primes that the algorithm uses. By default, a single precision prime in NTL lies between 2^{29} and 2^{30}.
- $\#$ bits is the bit size of the least common denominator of all rational coefficients of the result.
- time is the total time.
- red is the time used for reducing the coefficients of the input polynomials modulo single precision primes.
- mod is the time for the modular computation that the algorithm spends in steps 4 and 5.

Table 6.1. Timings for the first series with the small primes modular Hermite integration algorithm, with a designed squarefree decomposition of the denominator of type (degree $n/2$) · (degree $n/4$)^2. The coefficients of the numerator and the denominator are of bit size about $30 + \log_2 n$. The denominator has degree n, and the numerator has degree about n in the upper table and about $2n$ in the lower table.

n	#p	# bits	time	red	mod	cra	chk0	ratrec	chk
20	24.0	338.3	0.02	0.00	0.00	0.00	0.00	0.01	0.00
	24.0	338.3	0.02	0.00	0.00	0.00	0.00	0.00	0.00
40	51.0	720.4	0.12	0.01	0.03	0.02	0.02	0.03	0.01
	51.0	720.4	0.08	0.02	0.04	0.01	0.00	0.01	0.00
80	106.2	1509.3	0.75	0.09	0.29	0.09	0.04	0.21	0.02
	106.2	1509.3	0.50	0.10	0.24	0.09	0.02	0.03	0.02
160	227.0	3160.4	5.58	0.79	2.10	0.79	0.19	1.61	0.10
	227.0	3160.4	4.00	0.75	2.03	0.78	0.09	0.24	0.10
320	482.0	6649.6	43.27	6.05	14.85	7.37	0.77	12.84	1.38
	473.0	6649.6	31.22	5.99	14.64	7.06	0.29	1.82	1.39
640	963.0	14040.4	299.39	39.25	83.08	58.44	2.81	100.96	14.80
	963.0	14040.4	210.69	39.20	82.94	58.32	1.23	14.28	14.69

n	#p	# bits	time	red	mod	cra	chk0	ratrec	chk
20	39.8	685.6	0.06	0.00	0.01	0.00	0.01	0.01	0.00
	49.1	685.6	0.05	0.00	0.01	0.01	0.00	0.01	0.00
40	79.6	1375.4	0.28	0.02	0.07	0.05	0.03	0.09	0.01
	98.8	1375.4	0.28	0.04	0.08	0.09	0.03	0.03	0.01
80	163.8	3047.7	1.94	0.17	0.54	0.41	0.12	0.65	0.06
	216.7	3047.7	1.97	0.20	0.74	0.68	0.08	0.21	0.06
160	327.8	6063.3	14.29	1.16	4.05	3.42	0.37	4.57	0.70
	429.6	6063.3	14.97	1.45	5.28	5.75	0.28	1.50	0.71
320	675.8	12441.3	108.30	8.75	27.20	29.01	1.41	35.63	6.28
	877.4	12441.3	113.56	11.42	35.30	48.78	1.04	10.67	6.31

- cra is the time that the implementation spends for Chinese remaindering, i.e., for computing a solution modulo the product of all primes from the solutions modulo the single precision primes.
- chk0 is the time spent for constant coefficient checks.
- ratrec is the time used for the rational reconstruction of all coefficients after a successful constant coefficient check.
- chk is the time used for the final check of equation (6.6).

All times are in CPU seconds. The numerator f is a pseudorandom polynomial of degree less than n in the table above the caption, and less than $2n$ in the table below the caption. There are two rows for each value of n. The entries in the upper row are for the straightforward variant of rational reconstruction, and the entries in the lower row are for the second variant. The inputs for the upper and the lower row are identical.

Tables 6.1 and 6.2 show that the second variant of rational reconstruction is considerably faster than the straightforward implementation for the first and the second series of experiments, respectively. However, in the case where deg $f \approx 2n$

Table 6.2. Timings for the second series with the small primes modular Hermite integration algorithm, with a designed squarefree decomposition of the denominator of type (degree $n/2$) $\cdot$ (degree $2)^{n/4}$. The coefficients of the numerator and the denominator are of bit size about $15 + n$. The denominator has degree n, and the numerator has degree about n in the upper table and about $2n$ in the lower table.

n	#p	# bits	time	red	mod	cra	chk0	ratrec	chk
20	23.4	291.8	0.03	0.00	0.00	0.00	0.01	0.01	0.00
	23.4	291.8	0.02	0.00	0.01	0.00	0.00	0.00	0.00
40	46.0	599.3	0.11	0.01	0.04	0.01	0.02	0.02	0.00
	46.0	599.3	0.07	0.01	0.03	0.01	0.01	0.00	0.01
80	92.2	1247.1	0.61	0.07	0.21	0.08	0.07	0.14	0.02
	92.2	1247.1	0.44	0.07	0.19	0.10	0.03	0.03	0.02
160	192.7	2579.3	4.49	0.53	1.55	0.73	0.48	1.02	0.18
	188.9	2579.3	3.23	0.51	1.52	0.69	0.15	0.18	0.18
320	397.6	5185.0	34.11	4.11	10.72	6.23	3.41	7.45	2.17
	382.8	5185.0	24.45	3.95	10.44	5.86	0.80	1.23	2.14
640	788.4	10313.8	228.71	25.00	57.62	49.13	19.68	56.28	20.97
	780.4	10313.8	165.49	24.86	56.98	48.31	5.16	9.11	21.06

n	#p	# bits	time	red	mod	cra	chk0	ratrec	chk
20	65.1	889.5	0.10	0.00	0.02	0.02	0.02	0.02	0.00
	61.8	889.5	0.08	0.01	0.02	0.02	0.01	0.01	0.00
40	125.8	1728.6	0.52	0.03	0.11	0.14	0.08	0.16	0.01
	124.8	1728.6	0.35	0.02	0.11	0.13	0.03	0.04	0.01
80	255.0	3484.8	3.61	0.24	0.77	1.13	0.32	1.07	0.08
	250.4	3484.8	2.57	0.18	0.76	1.10	0.18	0.26	0.08
160	506.0	7216.0	27.33	1.47	5.77	8.97	2.12	7.84	1.13
	511.0	7216.0	20.82	1.47	5.88	9.29	1.05	1.97	1.14
320	1011.5	14269.3	198.46	10.86	36.60	72.56	9.54	59.12	9.74
	1001.8	14269.3	145.41	10.80	36.27	70.96	5.36	12.25	9.74

in the first series, this does not speed up the computation, since significantly more primes are needed. The effect is even more drastic in Table 6.3, where the second variant of rational reconstruction slows down the whole computation considerably.

Another difference between the first two series and the third series is that the time for the final check of equation (6.6) is reasonably small for the first and the second series, while it is the dominating factor in the third series. This can be attributed to the fact that the coefficient sizes in the third series is

The bit size of the coefficients of the output can be read off the second and the third columns in Tables 6.1 through 6.3. The number of primes, multiplied by NTL's bit size 30 for a single precision prime, corresponds approximately to twice the maximal bit size of a single numerator or denominator. The entry in the third column is the bit size of the least common denominator of all coefficients. The latter number is in general larger than the maximal bit size of an individual denominator. The tables indicate that using the second variant of rational reconstruction is beneficial whenever the two sizes are close together.

Table 6.3. Timings for the third series with the small primes modular Hermite integration algorithm, with a designed squarefree decomposition of the denominator of type (degree 5) · (degree 5)^2 · (degree 5)^3 ···. The coefficients of the numerator and the denominator are of bit size about $3n/5$. The denominator has degree n, and the numerator has degree about n.

n	$\#p$	# bits	time	red	mod	cra	chk0	ratrec	chk
30	7.6	113.8	0.01	0.00	0.01	0.00	0.00	0.00	0.00
	8.3	113.8	0.01	0.00	0.01	0.00	0.00	0.00	0.00
50	16.7	301.1	0.06	0.01	0.03	0.00	0.01	0.01	0.00
	21.4	301.1	0.06	0.01	0.03	0.00	0.00	0.00	0.01
75	26.7	593.4	0.19	0.01	0.09	0.01	0.02	0.02	0.03
	42.8	593.4	0.23	0.03	0.14	0.01	0.01	0.01	0.02
105	38.3	960.4	0.58	0.02	0.36	0.02	0.05	0.04	0.10
	68.1	960.4	0.78	0.08	0.50	0.07	0.02	0.02	0.07
140	48.1	1291.9	1.31	0.06	0.84	0.05	0.09	0.07	0.21
	93.2	1291.9	2.15	0.14	1.53	0.16	0.06	0.05	0.20
180	60.9	1902.2	2.79	0.11	1.79	0.09	0.14	0.14	0.53
	139.5	1902.2	5.41	0.25	3.88	0.48	0.15	0.11	0.53
225	85.3	2828.0	5.83	0.23	3.43	0.23	0.33	0.29	1.32
	200.9	2828.0	11.50	0.54	7.87	1.20	0.31	0.27	1.30
275	102.6	3444.2	10.25	0.35	6.15	0.41	0.47	0.45	2.42
	245.4	3444.2	21.73	0.91	15.08	2.20	0.53	0.48	2.52
330	134.2	4896.9	21.76	0.64	11.88	0.82	1.01	0.87	6.54
	349.7	4896.9	47.06	1.71	31.28	5.21	1.16	1.04	6.65
390	146.3	5957.7	35.31	0.88	16.79	1.13	1.29	1.38	13.82
	424.2	5957.7	76.83	2.58	49.19	9.07	1.91	1.75	12.31
455	177.5	7328.2	54.38	1.44	25.97	1.92	2.27	2.15	20.62
	514.7	7328.2	122.82	4.30	76.19	15.41	3.18	2.94	20.77
525	206.8	8912.1	86.89	1.89	39.83	2.98	3.20	3.16	35.82
	627.1	8912.1	198.76	5.84	120.99	26.24	5.13	4.82	35.72
600	237.0	9817.3	142.21	3.11	62.05	4.55	4.78	4.11	63.59
	689.2	9817.3	304.77	8.95	181.08	37.25	7.20	6.60	63.66

Tables 6.4, 6.5, and 6.6 give timings for the same three series of experiments as described above, but now with our implementation of the prime power modular Hermite algorithm 6.9. The inputs are the same as for the experiments with the small primes modular algorithm. In the first column, n denotes the degree of the denominator polynomial g. The other columns contain the averages of the following data over 10 pseudorandomly chosen inputs for the same value of n:

- s is the maximal lifting exponent. It corresponds to the number of single precision primes that the small primes algorithm uses. The prime power algorithm chooses a suitable single precision prime p, which is between 2^{29} and 2^{30} in NTL, computes the solution modulo p, and then lifts it modulo p^s.
- # bits is the bit size of the least common denominator of all rational coefficients of the result.
- time is the total time.
- red is the time used for reducing e_i modulo p in step 5; the reduction of all other data modulo p is done only once at the beginning. The time for this preprocessing

Table 6.4. Timings for the first series with the prime power modular Hermite integration algorithm, with a designed squarefree decomposition of the denominator of type (degree $n/2$) · (degree $n/4$)^2. The coefficients of the numerator and the denominator are of bit size about $30 + \log_2 n$. The denominator has degree n, and the numerator has degree about n in the upper table and about $2n$ in the lower table.

n	s	# bits	time	red	mod	upd	chk0	ratrec	chk
20	24.0	338.3	0.02	0.00	0.00	0.00	0.00	0.01	0.00
	24.0	338.3	0.02	0.00	0.00	0.01	0.00	0.01	0.00
40	51.0	720.4	0.13	0.00	0.04	0.03	0.02	0.03	0.00
	51.0	720.4	0.09	0.00	0.05	0.03	0.01	0.01	0.00
80	106.2	1509.3	0.79	0.00	0.32	0.18	0.05	0.21	0.02
	106.2	1509.3	0.53	0.00	0.25	0.19	0.03	0.03	0.02
160	227.0	3160.4	5.60	0.01	2.09	1.58	0.19	1.61	0.10
	227.0	3160.4	4.04	0.01	2.04	1.54	0.09	0.25	0.10
320	486.5	6649.6	40.49	0.04	15.06	10.86	0.78	12.32	1.39
	486.5	6649.6	29.53	0.04	15.10	10.80	0.32	1.84	1.39
640	963.0	14040.4	267.96	0.17	82.88	66.01	3.44	100.67	14.67
	963.0	14040.4	180.04	0.15	83.00	66.19	1.28	14.41	14.89

n	s	# bits	time	red	mod	upd	chk0	ratrec	chk
20	39.8	685.6	0.06	0.00	0.01	0.02	0.01	0.01	0.00
	49.1	685.6	0.05	0.00	0.01	0.02	0.01	0.01	0.00
40	79.6	1375.4	0.30	0.00	0.08	0.08	0.04	0.08	0.01
	98.8	1375.4	0.30	0.00	0.10	0.13	0.02	0.03	0.01
80	163.8	3047.7	2.01	0.01	0.55	0.65	0.10	0.65	0.05
	218.6	3047.7	2.04	0.01	0.73	0.91	0.09	0.22	0.06
160	327.8	6063.3	14.24	0.02	4.07	4.61	0.35	4.47	0.70
	429.6	6063.3	14.46	0.02	5.33	6.70	0.26	1.41	0.72
320	675.8	12441.3	100.82	0.11	26.98	30.34	1.67	35.38	6.30
	894.2	12441.3	101.59	0.15	35.78	46.63	1.11	11.63	6.25

was less than 10 milliseconds in all our experiments and does therefore not appear in the tables.

- mod is the time for the modular computation that the algorithm spends in steps 5 and 6.
- upd is the time that the implementation spends in steps 7 and 8. This is comparable to the time for the Chinese remaindering in the small primes algorithm.
- chk0 is the time spent for constant coefficient checks.
- ratrec is the time used for the rational reconstruction of all coefficients after a successful constant coefficient check.
- chk is the time used for the final check of equation (6.6).

The comparison of the various experiments and variants leads to the same observations as for the small primes modular algorithm. We now compare the experiments with the small primes algorithm to those with the prime power algorithm.

We first note that in all experiments, the number of single precision primes in the small primes implementation agrees roughly with the lifting exponent in the prime power algorithm. Moreover, we have approximate coincidence for the bit size of the

Table 6.5. Timings for the second series with the prime power modular Hermite integration algorithm, with a designed squarefree decomposition of the denominator of type (degree $n/2$) · (degree 2)$^{n/4}$. The coefficients of the numerator and the denominator are of bit size about $15 + n$. The denominator has degree n, and the numerator has degree about n in the upper table and about $2n$ in the lower table.

n	s	# bits	time	red	mod	upd	chk0	ratrec	chk
20	23.4	291.8	0.03	0.00	0.01	0.01	0.01	0.00	0.00
	23.4	291.8	0.03	0.00	0.01	0.01	0.00	0.00	0.00
40	46.0	599.3	0.12	0.00	0.04	0.04	0.02	0.02	0.00
	46.0	599.3	0.10	0.00	0.04	0.04	0.01	0.00	0.01
80	93.1	1247.1	0.73	0.00	0.22	0.26	0.07	0.14	0.02
	93.1	1247.1	0.58	0.00	0.22	0.27	0.03	0.03	0.02
160	192.7	2579.3	5.44	0.01	1.56	2.20	0.44	1.02	0.18
	192.7	2579.3	4.28	0.02	1.55	2.21	0.14	0.17	0.17
320	397.6	5185.0	44.67	0.04	10.80	20.53	3.63	7.43	2.17
	382.8	5185.0	34.52	0.05	10.45	19.75	0.79	1.23	2.17
640	788.4	10313.8	355.58	0.36	58.07	199.60	20.09	56.23	20.90
	780.4	10313.8	289.63	0.35	57.12	196.73	5.25	9.00	20.85

n	s	# bits	time	red	mod	upd	chk0	ratrec	chk
20	67.2	889.5	0.11	0.00	0.02	0.04	0.03	0.02	0.00
	61.8	889.5	0.08	0.00	0.01	0.04	0.01	0.01	0.00
40	127.1	1728.6	0.56	0.00	0.11	0.19	0.10	0.15	0.01
	124.8	1728.6	0.39	0.00	0.11	0.19	0.04	0.04	0.01
80	255.0	3484.8	3.64	0.01	0.78	1.37	0.36	1.03	0.08
	250.4	3484.8	2.67	0.01	0.77	1.39	0.17	0.23	0.08
160	516.0	7216.0	29.54	0.04	6.05	12.05	2.28	7.95	1.14
	521.0	7216.0	22.53	0.06	6.05	12.06	1.18	2.03	1.13
320	1011.5	14269.3	215.47	0.32	36.71	94.93	14.57	59.12	9.73
	1001.8	14269.3	157.80	0.32	36.31	93.78	5.37	12.15	9.78

common denominator of the output, the time for the modular Hermite reductions, the time for the constant coefficient checks, the time for the rational reconstruction, and the time for the final check of equation (6.6).

There are two main differences between the small primes implementation and the prime power implementation. The time for the modular reduction is much smaller for the prime power variant, since we need to reduce most of the data only once at the beginning of the algorithm. This is the main reason why the prime power implementation is faster in the first series of experiments. However, the cost for the "updating" in our implementation of the prime power algorithm is greater than the corresponding cost for the Chinese remaindering in our implementation of the small primes algorithm, and this makes the prime power variant slower in the second and the third series of experiments.

Table 6.6. Timings for the third series with the prime power modular Hermite integration algorithm, with a designed squarefree decomposition of the denominator of type (degree 5) · (degree 5)^2 · (degree 5)^3 · · ·. The coefficients of the numerator and the denominator are of bit size about $3n/5$. The denominator has degree n, and the numerator has degree about n.

n	s	# bits	time	red	mod	upd	chk0	ratrec	chk
30	7.7	113.8	0.02	0.00	0.01	0.00	0.00	0.00	0.00
	8.3	113.8	0.02	0.00	0.01	0.00	0.00	0.00	0.00
50	16.7	301.1	0.07	0.00	0.02	0.03	0.01	0.01	0.01
	21.6	301.1	0.10	0.00	0.05	0.03	0.01	0.00	0.01
75	26.7	593.4	0.23	0.00	0.07	0.10	0.02	0.02	0.02
	42.8	593.4	0.34	0.00	0.14	0.14	0.01	0.01	0.02
105	38.3	960.4	0.81	0.00	0.33	0.29	0.05	0.05	0.08
	68.8	960.4	1.25	0.00	0.57	0.54	0.03	0.02	0.07
140	49.2	1291.9	1.98	0.00	0.81	0.78	0.08	0.07	0.20
	93.2	1291.9	3.42	0.00	1.54	1.52	0.07	0.04	0.21
180	61.5	1902.2	4.48	0.00	1.69	2.01	0.12	0.13	0.50
	139.5	1902.2	9.56	0.01	3.93	4.78	0.15	0.11	0.55
225	86.0	2828.0	11.28	0.01	3.37	5.91	0.32	0.29	1.32
	200.9	2828.0	24.28	0.03	8.05	14.24	0.31	0.26	1.31
275	102.6	3444.2	21.54	0.02	6.29	11.64	0.48	0.48	2.53
	245.4	3444.2	47.85	0.04	15.35	28.85	0.53	0.45	2.51
330	134.2	4896.9	46.25	0.05	11.85	25.80	1.01	0.86	6.54
	349.7	4896.9	109.53	0.09	31.36	69.04	1.18	1.03	6.68
390	146.3	5957.7	75.73	0.06	16.92	43.93	1.23	1.26	12.10
	424.2	5957.7	195.58	0.14	49.45	129.97	1.90	1.74	12.16
455	177.5	7328.2	131.16	0.08	25.83	79.80	2.30	2.13	20.71
	531.2	7328.2	349.60	0.24	78.22	243.58	3.48	3.02	20.75
525	208.7	8912.1	225.32	0.15	40.07	142.51	3.20	3.16	35.82
	632.6	8912.1	614.37	0.40	123.70	443.43	5.26	4.93	36.19
600	237.0	9817.3	353.69	0.17	62.29	218.02	4.77	4.00	63.82
	689.2	9817.3	920.01	0.51	183.99	656.09	7.30	6.73	64.74

7. Computing All Integral Roots of the Resultant

In this chapter, we discuss modular algorithms for the following problem. Given two bivariate polynomials $f, g \in \mathbb{Z}[x, y]$, compute all integral roots of the resultant $r = \mathrm{res}_x(f, g) \in \mathbb{Z}[y]$ with respect to y. The straightforward solution is to compute the coefficients of r, either by linear algebra or by a variant of the Euclidean Algorithm, and then to compute all linear factors of r in $\mathbb{Z}[y]$. In one of our main applications, namely Gosper's (1978) algorithm for hypergeometric summation, the resultant is of the form $r = \mathrm{res}_x(f(x), g(x + y))$ for two univariate polynomials $f, g \in \mathbb{Z}[x]$. In this case, the straightforward algorithm is inefficient, since the coefficients of the linear factors of r are much smaller than the coefficients of r itself. For that reason, we pursue a different approach. We choose a suitable single precision prime p and compute the coefficients of r modulo p^k, were $k \in \mathbb{N}$ is an integer such that p^k bounds the absolute value of all integral roots of r. Then we find all roots of r modulo p^k via Hensel lifting. To compute r modulo p^k, we employ the monic Extended Euclidean Algorithm 3.4 over the ring $\mathbb{Z}/p^k\mathbb{Z}$. Since this ring has zero divisors when $k \geq 1$, the analysis of our algorithm is somewhat intricate. It is easy to modify our algorithm so as to compute all rational roots of r. We also discuss an alternative approach for the hypergeometric case, which does not compute the resultant r at all, but instead computes the irreducible factorizations of f and g modulo p^k. This approach turns out to be faster than the resultant-based approach.

Besides hypergeometric summation, we also discuss applications of our algorithm to the continuous analog, namely hyperexponential integration. Moreover, we give a complete cost analysis of a modular variant of the algorithm by Lazard, Rioboo and Trager for computing the logarithmic part of the integral of a rational function in $\mathbb{Q}(x)$. This algorithms requires not only the computation of the resultant, but also of certain subresultants. It turns out that in the worst case, the cost estimate dominates the cost for the Hermite integration discussed in the previous chapter.

The idea of our modular algorithm is to choose a single precision prime p and a sufficiently large set U of integers, to compute $r(u) \bmod p^k$ for all $u \in U$, where k is as described above, and then to recover $r \bmod p^k$ by interpolation. The following definition formalizes when a prime p and an evaluation point u are suitable.

Definition 7.1. *Let* $f, g \in \mathbb{Z}[x, y]$ *be nonzero polynomials, and for* $0 \leq d < \min\{\deg_x f, \deg_x g\}$, *let* $\sigma_d \in \mathbb{Z}[y]$ *be the dth subresultant of f and g with respect to x.*

(i) *A prime $p \in \mathbb{N}$ is lucky (for f, g) if p neither divides $\mathrm{lc}_x(fg)$ nor σ_d, for all integers d with $0 \leq d < \min\{\deg_x f, \deg_x g\}$ and $\sigma_d \neq 0$.*

(ii) *A point $u \in \mathbb{Z}$ is lucky (for f, g) with respect to a prime p if $\mathrm{lc}_x(fg) \not\equiv 0 \bmod \langle p, y - u \rangle$ and $\sigma_d \not\equiv 0 \bmod \langle p, y - u \rangle$ holds for all integers d with $0 \leq d < \min\{\deg_x f, \deg_x g\}$ and $\sigma_d \neq 0$.*

Obviously every $u \in \mathbb{Z}$ is unlucky with respect to p if p itself is unlucky.

Here is the modular algorithm for computing the resultant of two bivariate polynomials modulo a prime power. In fact, it computes all intermediate results of the Extended Euclidean Algorithm. It works by computing the intermediate results of the EEA for $f(x, u) \bmod p^k$ and $g(x, u) \bmod p^k$ for sufficiently many points u and reconstructing the intermediate results of $f(x, y) \bmod p^k$ and $g(x, y) \bmod p^k$ by interpolation with respect to y. For $k = 1$, the algorithm boils down to Algorithm 6.36 in von zur Gathen & Gerhard (1999).

Algorithm 7.2 (Modular bivariate EEA).

Input: Nonzero polynomials $f, g \in \mathbb{Z}[x, y]$ with $\deg_x g = m \leq \deg_x f = n$ and $\deg_y g, \deg_y f \leq \nu$, where $n\nu \geq 2$, a prime $p \geq 5(n+1)^2\nu$ not dividing $\mathrm{lc}_x(fg)$, and $k \in \mathbb{N}_{>0}$.

Output: The Euclidean length ℓ of f, g, considered as polynomials in the main variable x, their degree sequence $d_1 = m > \cdots > d_\ell \geq 0$, and the images modulo p^k of the subresultants $\sigma_{d_i} \in \mathbb{Z}[y]$ and the polynomials $\sigma_{d_i} r_{d_i}, \sigma_{d_i} s_{d_i}, \sigma_{d_i} t_{d_i}$ in $\mathbb{Z}[x, y]$, where $r_{d_i}, s_{d_i}, t_{d_i}$ is the row of degree d_i in the monic EEA of f, g in $\mathbb{Q}(y)[x]$, for $2 \leq i \leq \ell$, or "FAIL".

1. let $U \subseteq \{0, 1, \ldots, 5(n+1)^2\nu - 1\}$ be a set of $4n\nu$ evaluation points remove those u from U with

$$\deg_x f(x, u) \bmod p < \deg_x f(x, y) \text{ or } \deg_x g(x, u) \bmod p < \deg_x g(x, y)$$

2. **for** $u \in U$ **do**
3. $f_u \longleftarrow f(x, u) \bmod p^k, \quad g_u \longleftarrow g(x, u) \bmod p^k$
4.   **call** the monic Extended Euclidean Algorithm 3.4 over $\mathbb{Z}_{p^k}$ with input f_u and g_u
        if it returns regularly
        then let $\ell_u, d_{u,i} \in \mathbb{N}, \gamma_{u,d_i} \in \mathbb{Z}_{p^k}$ and $r_{u,d_i}, s_{u,d_i}, t_{u,d_i} \in \mathbb{Z}_{p^k}[x]$ for $2 \leq i \leq \ell_u$ be its results
        else remove u from U and **continue** the loop 2
5. **if** $\#U \leq 2n\nu$ **then return** "FAIL"
6. let $d_1 = \deg_x g > d_2 > \cdots > d_{\ell^*}$ be all degrees that occur in the degree sequence of f_u, g_u for some $u \in U$
7. **for** $2 \leq i \leq \ell^*$ **do**
8.   let $U_{d_i} = \{u \in U : d_i \text{ occurs in the degree sequence of } f_u, g_u\}$
        if $\#U_{d_i} \leq 2(n - d_i)\nu$ **then return** "FAIL"

9.   remove $\#U_{d_i} - 2(n - d_i)\nu - 1$ points from U_{d_i} and compute by inter-
     polation $\gamma_{d_i} \in \mathbb{Z}_{p^k}[y]$ and $r_{d_i}^*, s_{d_i}^*, t_{d_i}^* \in \mathbb{Z}_{p^k}[x, y]$ of degrees at most
     $2(n - d_i)\nu$ in y such that the following holds for all $u \in U_{d_i}$:

$$\begin{array}{rclcrcl}
\gamma_{d_i}(u) & = & \gamma_{u,d_i}, & \quad & r_{d_i}^*(x, u) & = & \gamma_{u,d_i}\, r_{u,d_i}, \\
s_{d_i}^*(x, u) & = & \gamma_{u,d_i}\, s_{u,d_i}, & \quad & t_{d_i}^*(x, u) & = & \gamma_{u,d_i}\, t_{u,d_i}.
\end{array}$$

10. **return** $\gamma_{d_i}, r_{d_i}^*, s_{d_i}^*, t_{d_i}^*$ for $2 \leq i \leq \ell^*$

Theorem 7.3 (Correctness of Algorithm 7.2). *Let f, g be as in Algorithm 7.2, and
for $0 \leq d < m$, let $\sigma_d \in \mathbb{Z}[y]$ be the dth subresultant of f, g with respect to x. If
Algorithm 7.2 returns "FAIL", then at most $2n\nu$ of the initial points in U are lucky
for f, g with respect to p. Otherwise, for each $d < m$, the degree d occurs in the
degree sequence $d_2, \ldots, d_{\ell^*}$ of the output if and only if $p \nmid \sigma_d$, and in that case we
have*

$$\begin{array}{rclcrcl}
\gamma_d & = & \sigma_d \bmod p^k, & \quad & r_d^* & = & \sigma_d r_d \bmod p^k, \\
s_d^* & = & \sigma_d s_d \bmod p^k, & \quad & t_d^* & = & \sigma_d t_d \bmod p^k.
\end{array}$$

If p is lucky for f, g, then $\ell = \ell^*$ and the algorithm returns the correct results.

Proof. By Lemma 3.11 (ii) with $I = \langle p^k, y - u \rangle$, we have $\sigma_d(u) \equiv 0 \bmod p$ for
some nonzero σ_d if the monic EEA for $\overline{f}, \overline{g}$ does not return regularly in step 4. Thus
a point u that is removed from U either in step 1 or in step 4 is unlucky with respect
to p, and at most $2n\nu$ of the initial points in U are lucky if the algorithm returns
"FAIL" in step 5. Let $2 \leq i \leq \ell^*$. Since there exists some $v \in U$ such that d_i occurs
in the degree sequence of f_v and g_v, Lemma 3.11 (i) implies that $\sigma_{d_i} \neq 0$. If now
d_i does not occur in the degree sequence of f_u and g_u for some other point $u \in U$,
then the same Lemma implies that u is unlucky with respect to p. Therefore at most
$2n\nu$ of the initial points in U are lucky if the algorithm returns "FAIL" in step 8.

We now assume that the algorithm does not return "FAIL", and let r_d, s_d, t_d in
$\mathbb{Q}(y)[x]$ be a row in the monic EEA of f, g over $\mathbb{Q}(y)$, such that $sf + tg = r$ is
monic of degree $d < m$ in x. Then $\sigma_d \neq 0$, by Lemma 3.9 (i). If $p \mid \sigma_d$, then
all points in U are roots of σ_d modulo p, and Lemma 3.11 (i) implies that d does
not occur in the degree sequence of f_u, g_u for any $u \in U$. Thus $d \notin \{d_2, \ldots, d_{\ell^*}\}$.
Otherwise, if $p \nmid \sigma_d$, then $\deg_y \sigma_d \leq 2(n-d)\nu$ implies that at most $2(n-d)\nu$ points
in U are roots of σ_d modulo p, and the degree d occurs in the degree sequence of
f_u, g_u for at least $\#U - 2(n-d)\nu > 0$ points in U after step 5, by Lemma 3.11 (i).
Thus $d \in \{d_2, \ldots, d_{\ell^*}\}$. In particular, if p is lucky, then $p \mid \sigma_d \iff \sigma_d = 0$, so
that d occurs in the degree sequence of f, g over $\mathbb{Q}(y)[x]$ if and only if it occurs in
the degree sequence of the output of Algorithm 7.2, and hence $\ell = \ell^*$.

Now let $2 \leq i \leq \ell$ be such that $d = d_i$ in step 6, and let $u \in U_{d_i}$ in step 9 for
that value of i. Then $\sigma_d(u) \bmod p^k$ is the dth subresultant of f_u, g_u, and Proposition
3.7 (v) yields $\sigma_d(u) \bmod p^k = \gamma_{u,d}$. By Corollary 6.49 (ii) in von zur Gathen &
Gerhard (1999) and Cramer's rule, the denominators of r_d, s_d, t_d are divisible by σ_d,
so that $\sigma_d r_d, \sigma_d s_d, \sigma_d t_d$ are in $\mathbb{Z}[x, y]$, and the degrees in y of the latter polynomials
are less than $2(n - d)\nu$. By Lemma 3.11 (i), we have $r_d(x, u) \bmod p^k = r_{u,d}$,
and similarly for s_d, t_d. Since we have assumed that the algorithm does not return

"FAIL" in step 8, there are $\#U_d = 2(n-d)\nu + 1$ interpolation points. These points are distinct modulo p, and hence we can indeed reconstruct $\sigma_d, \sigma_d r_d, \sigma_d s_d, \sigma_d t_d$ modulo p^k by interpolation. $\square$

Theorem 7.4. *Let $n, \nu \in \mathbb{N}$ with $n\nu \geq 2$, $f, g \in \mathbb{Z}[x, y]$ with $\deg_x f, \deg_x g \leq n$ and $\deg_y f, \deg_y g \leq \nu$, and $p \in \mathbb{N}$ a prime that is lucky for f, g. If $p \geq 5(n+1)^2\nu$ and U is a uniformly randomly chosen subset of $S = \{0, 1, \ldots, 5(n+1)^2\nu - 1\}$ of cardinality $\#U = 4n\nu$, then with probability at least $3/4$, more than $2n\nu$ of the points $u \in U$ are lucky for f, g with respect to p.*

Proof. Let $m = \min\{\deg_x f, \deg_x g\}$ and $\sigma_d \in \mathbb{Z}[y]$ be the dth subresultant of f, g with respect to x for $0 \leq d < m$, and let $\sigma = \mathrm{lc}_x(fg) \prod_{\sigma_d \neq 0} \sigma_d \in \mathbb{Z}[y]$. By Corollary 3.2, we have $\deg \sigma_d \leq 2(n-d)\nu$ for $0 \leq d < m$. Since p is lucky for f, g, we find that $\sigma \bmod p \in \mathbb{F}_p[y]$ is a nonzero polynomial of degree at most

$$2\nu + 2\nu \sum_{0 \leq d < m} (n-d) \leq 2\nu + 2\nu \sum_{1 \leq d \leq n} d = (n^2 + n + 2)\nu \leq (n+1)^2\nu .$$

Thus $\sigma \bmod p$ has at most $(n+1)^2\nu$ roots in $\mathbb{F}_p$, and at least $4/5$ of the points in S are lucky for f, g with respect to p. Now Lemma 3.24 (ii) with $b = (n+1)^2\nu$, $w = 4b$, and $k = 4n\nu$ shows that for a randomly chosen subset $U \subseteq S$ of cardinality $\#U = k$, the probability that at most half of the points $u \in U$ are unlucky is at most $1/4$. $\square$

Theorem 7.5 (Success probability of Algorithm 7.2). *Let f, g be as in Algorithm 7.2. If p is lucky for f, g and U in step 1 is chosen uniformly at random from among all subsets of $\{0, 1, \ldots, 5(n+1)^2\nu - 1\}$ of cardinality $\#U = 4n\nu$, then Algorithm 7.2 returns "FAIL" with probability at most $1/4$.*

Proof. This follows immediately from Theorems 7.3 and 7.4. $\square$

Remark 7.6. *Algorithm 7.2 can be turned into a deterministic algorithm if $k = 1$, as follows (see also Sect. 6.7 and 11.2 in von zur Gathen & Gerhard 1999). We choose a set U of $(4n + 2)\nu$ arbitrary distinct evaluation points in step 1. By assumption, $\mathrm{lc}_x(fg) \bmod p$ is a nonzero polynomial in $\mathbb{F}_p[y]$ of degree at most 2ν. Then U has at least $4n\nu$ points after step 1. The Euclidean Algorithm always terminates regularly over the field $\mathbb{F}_p$, and hence $\#U \geq 4n\nu$ and Algorithm 7.2 does not return "FAIL" in step 5. (Thus step 5 may be omitted.) Now let $2 \leq i \leq \ell^*$. Since d_i occurs in the degree sequence of the EEA of f_u and g_u for some $u \in U$, the d_ith subresultant σ_{d_i} of $f \bmod p$ and $g \bmod p$ is a nonzero polynomial in $\mathbb{F}_p[y]$, by Lemma 3.11 (i), which has degree at most $2(n - d_i)\nu$, by Corollary 3.2. Thus the d_ith subresultant of f_u and g_u, which equals $\sigma_{d_i}(u)$, vanishes for at most $2(n-d_i)\nu$ values $u \in U$, and hence d_i occurs in the degree sequence of the EEA of f_u and g_u for at least $2(n+d_i)\nu$ points from U. Thus $\#U_{d_i} \geq 2n\nu$ for all i, and the algorithm does not return "FAIL" in step 8. (This means that the "if" statement in step 8 may be omitted as well.)*

Theorem 7.7 (Cost of Algorithm 7.2). *Let $f, g \in \mathbb{Z}[x, y]$ be as in Algorithm 7.2, and assume that the integer coefficients of f and g are absolutely bounded by p^k.*

(i) *Algorithm 7.2 takes $O(n^4 \nu^2)$ arithmetic operations in $\mathbb{Z}_{p^k}$ with classical polynomial arithmetic and $O(n^2 \mathsf{M}(n\nu) \log(n\nu))$ or $O^\sim(n^3 \nu)$ operations with fast arithmetic.*

(ii) *If only one row in the EEA is required, then the cost is $O(n^3 \nu^2)$ with classical and $O(n \mathsf{M}(n\nu) \log(n\nu))$ or $O^\sim(n^2 \nu)$ with fast arithmetic.*

(iii) *The cost for computing only the resultant is $O(n^2 \nu (n + \nu))$ with classical arithmetic and*

$$O(n^2 \mathsf{M}(\nu) \log \nu + n\nu \, \mathsf{M}(n) \log n + \mathsf{M}(n\nu) \log(n\nu))$$

or $O^\sim(n^2 \nu)$ with fast arithmetic.

Proof. In steps 1 and 3, we evaluate the at most $2n$ coefficients in $\mathbb{Z}_{p^k}[y]$ of $f \bmod p^k$ and $g \bmod p^k$, each of degree at most ν, at at most $4n\nu$ points. This takes $O(n^2 \nu^2)$ operations with classical arithmetic and $O(n^2 \mathsf{M}(\nu) \log \nu)$ with fast arithmetic. The cost for one execution of the monic EEA in step 4 is $O(n^2)$ operations, both with classical and fast arithmetic, in total $O(n^3 \nu)$. Interpolating one coefficient in $\mathbb{Z}_{p^k}[y]$, of degree at most $2n\nu$, in step 9 takes $O(n^2 \nu^2)$ operations with classical arithmetic and $O(\mathsf{M}(n\nu) \log(n\nu))$ with fast arithmetic, by Fact 3.13. There are $O(n^2)$ coefficients in total, and hence the overall cost for the loop in step 7 is $O(n^4 \nu^2)$ with classical arithmetic and $O(n^2 \mathsf{M}(n\nu) \log(n\nu))$ with fast arithmetic. This dominates the cost for the other steps.

If only one row in the EEA is required, then the overall cost for step 4 with fast arithmetic drops to $O(n\nu \, \mathsf{M}(n) \log n)$, by employing the fast Euclidean Algorithm, and the cost for the interpolation is $O(n^3 \nu^2)$ with classical and $O(n \mathsf{M}(n\nu) \log(n\nu))$ with fast arithmetic. Again, the cost for the interpolation dominates the cost for the other steps.

If only the resultant σ_0 is required, then the cost for the interpolation is only $O(n^2 \nu^2)$ with classical and $O(\mathsf{M}(n\nu) \log(n\nu))$ with fast arithmetic. $\square$

We note that in the generic case, where $\ell = n$ and $d_i = n - i$ for all i and the degrees in y of the numerators and denominators in the ith row of the monic EEA are close to the upper bound $2(n - d_i)\nu = 2i\nu$, the output size is $\Theta(n^2 \nu)$ for one row and $\Theta(n^3 \nu)$ for all rows, so that the algorithm is – up to logarithmic factors – asymptotically optimal when using fast arithmetic.

The following corollary applies Algorithm 7.2 in the case of hypergeometric summation.

Corollary 7.8. *Let $n \in \mathbb{N}_{\geq 2}$, $f, g \in \mathbb{Z}[x]$ be nonconstant of degrees at most n, $\ell \in \mathbb{N}$ the Euclidean length and $d_1 = \deg g > \cdots > d_\ell$ the degree sequence of $f(x)$ and $g(x + y)$ with respect to x, and let $\sigma_{d_i} \in \mathbb{Z}[y]$ and $r_{d_i}, s_{d_i}, t_{d_i} \in \mathbb{Z}[x, y]$ for $2 \leq i \leq \ell$ be the subresultants and the results of the monic EEA in $\mathbb{Q}(x)[y]$ of $f(x)$ and $g(x + y)$, respectively. Moreover, let $p \geq 5n(n + 1)^2$ be a prime that is lucky for $f(x), g(x + y)$.*

(i) *There is a probabilistic Las Vegas algorithm which for a given $k \geq 1$ computes ℓ, the d_i, and the images of $\sigma_{d_i}, \sigma_{d_i} r_{d_i}, \sigma_{d_i} s_{d_i}, \sigma_{d_i} t_{d_i}$ modulo p^k for $2 \leq i \leq \ell$. This algorithm takes $O(n^6)$ arithmetic operations in $\mathbb{Z}_{p^k}$ with classical polynomial arithmetic and $O(n^2 \mathsf{M}(n^2) \log n)$ or $O^\sim(n^4)$ with fast arithmetic, and it succeeds with probability at least $3/4$.*

(ii) *If only one row in the EEA is required, then the cost is $O(n^5)$ with classical and $O(n \, \mathsf{M}(n^2) \log n)$ or $O^\sim(n^3)$ with fast arithmetic.*

(iii) *If only the resultant is required, then the cost is $O(n^4)$ with classical and $O((n^2 \, \mathsf{M}(n) + \mathsf{M}(n^2)) \log n)$ or $O^\sim(n^3)$ with fast arithmetic.*

Proof. Correctness and the bound on the success probability follow from Theorems 7.3 and 7.5. We do not need the coefficients of the polynomial $g(x + y)$ explicitly, since we only want to evaluate it at several points $y = u$, and hence the coefficients of $g(x)$ are sufficient. For the cost analysis, we thus have to estimate the cost for computing $g(x + u) \bmod p^k$ for all $u \in U$ in step 3. Per point u, this takes $O(n^2)$ additions and multiplications modulo p^k with classical arithmetic, and $O(\mathsf{M}(n))$ with fast arithmetic, by Theorems 4.3 and 4.5. Thus the overall cost is $O(n^4)$ and $O(n^2 \mathsf{M}(n))$, respectively. This agrees with the cost estimate for step 3 from Theorem 7.7 with $\nu = n$, up to a factor $\log n$ in the case of fast arithmetic, and the claims follow from this theorem. $\square$

In the special case of Corollary 7.8, there are better bounds on the degrees of the intermediate results in the EEA. More precisely, Corollary 3.2 implies that $\deg_y \sigma_d \leq n(n - d)$ for all d, and the same bound is valid for the degree in y of $\sigma_d r_d, \sigma_d s_d, \sigma_d t_d$. Thus about half the number of evaluation points are sufficient, namely $\#U = 2n^2$ out of $5(n + 1)^2 n/2$ in step 1 of Algorithm 7.2.

The next corollary is the application of Algorithm 7.2 to the case of hyperexponential integration.

Corollary 7.9. *Let $n \in \mathbb{N}_{\geq 2}$, $f, g \in \mathbb{Z}[x]$ be nonconstant of degrees at most n, $\ell \in \mathbb{N}$ the Euclidean length and $d_1 > \cdots > d_\ell$ the degree sequence of g and $f - yg'$ with respect to x, and let $\sigma_{d_i} \in \mathbb{Z}[y]$ and $r_{d_i}, s_{d_i}, t_{d_i} \in \mathbb{Z}[x, y]$ for $2 \leq i \leq \ell$ be the subresultants and the results of the monic EEA in $\mathbb{Q}(x)[y]$ of g and $f - yg'$, respectively. Moreover, let $p \geq 5(n + 1)^2$ be a prime that is lucky for g and $f - yg'$.*

(i) *There is a probabilistic Las Vegas algorithm which for a given $k \geq 1$ computes ℓ, the d_i, and the images of $\sigma_{d_i}, \sigma_{d_i} r_{d_i}, \sigma_{d_i} s_{d_i}, \sigma_{d_i} t_{d_i}$ modulo p^k for $2 \leq i \leq \ell$. This algorithm takes $O(n^4)$ arithmetic operations in $\mathbb{Z}_{p^k}$ with classical polynomial arithmetic and $O(n^2 \mathsf{M}(n) \log n)$ or $O^\sim(n^3)$ with fast arithmetic, and it succeeds with probability at least $3/4$.*

(ii) *The cost for computing only one row is $O(n^3)$ operations with classical and $O(n \, \mathsf{M}(n) \log n)$ or $O^\sim(n^2)$ with fast arithmetic, and the same estimate is valid for computing the resultant.*

Proof. Correctness and the bound on the success probability follow from Theorems 7.3 and 7.5. The cost for computing g' and evaluating $f - yg'$ at $y = u$ for all $u \in U$ in step 3 is $O(n^2)$ additions and multiplications in $\mathbb{Z}_{p^k}$. This agrees with the cost estimate for step 3 from Theorem 7.7 with $\nu = 1$, and the cost estimates follow from the latter theorem. □

By a similar argument as after Corollary 7.8, we may reduce the number of evaluation points to $\#U = 2n$ out of $5(n + 1)^2/2$ in step 1 of Algorithm 7.2 in this special case.

7.1 Application to Hypergeometric Summation

We now return to hypergeometric summation and discuss a new modular algorithm for computing all integral roots of the resultant $r = \mathrm{res}_x(f(x), g(x + y)) \in \mathbb{Z}[y]$ for two given nonzero polynomials $f, g \in \mathbb{Z}[x]$.

Problem 7.10 (Integral root distances). *Given nonzero polynomials $f, g \in \mathbb{Z}[x]$ of degree at most n, compute at most n^2 integers containing all $z \in \mathbb{Z}$ such that $\gcd(f(x), g(x + z))$ is nonconstant.*

The following factorization is due to Euler; see Notes 6.8 and Exercise 6.12 in von zur Gathen & Gerhard (1999).

Fact 7.11. *Let F be a field and $f = \mathrm{lc}(f) \prod\limits_{1 \leq i \leq n} (x - \alpha_i)$, $g = \mathrm{lc}(g) \prod\limits_{1 \leq j \leq m} (x - \beta_j)$ be nonzero polynomials in $F[x]$.*

(i) $\mathrm{res}_x(f, g) = \mathrm{lc}(f)^m \prod\limits_{1 \leq i \leq n} g(\alpha_i) = \mathrm{lc}(f)^m \mathrm{lc}(g)^n \prod\limits_{\substack{1 \leq i \leq n \\ 1 \leq j \leq m}} (\alpha_i - \beta_j).$

(ii) $\mathrm{res}_x(f(x), g(x + y)) = \mathrm{lc}(f)^m \mathrm{lc}(g)^n \prod\limits_{\substack{1 \leq i \leq n \\ 1 \leq j \leq m}} (y + \alpha_i - \beta_j).$

(iii) If f and g are integral polynomials with coefficients absolutely less than 2^λ, then any $z \in \mathbb{Z}$ with $\mathrm{res}_x(f(x), g(x + z)) = 0$ satisfies $|z| < 2^{\lambda+2}$.

The third statement follows, for example, from Exercise 6.23 (ii) in von zur Gathen & Gerhard (1999).

This fact implies that each root of r is the difference of a root of g and a root of f. If f, g have integral coefficients, then (iii) says that the integral roots of r are "small", of about the same word size as the coefficients of f and g. On the other hand, Corollary 3.2 yields an upper bound of about $n + m$ times the size of the coefficients of f and g for the size of the coefficients of r. For that reason, we do not want to compute the coefficients of r exactly.

The following probabilistic algorithm, of Monte-Carlo type, computes all integral roots of r. First, it chooses a single precision prime p. Then it calls Algorithm 7.2 to compute r modulo p^k for a suitable k, determines all roots of r modulo p in $\mathbb{F}_p$, and simultaneously lifts them to roots of r modulo p^k. If p is lucky, then all integral roots of r are found in this way, and possibly also some non-roots of r. If p is unlucky, then there is no guarantee that all integral roots of r are found.

Algorithm 7.12 (Prime power modular integral root distances).
Input: Nonzero polynomials $f, g \in \mathbb{Z}[x]$ with $\deg g \leq \deg f \leq n$, where $n \geq 2$, and $\|f\|_\infty, \|g\|_\infty < 2^\lambda$.
Output: A list of at most $(\deg f)(\deg g)$ integers containing all $z \in \mathbb{Z}$ such that $\gcd(f(x), g(x + z))$ is nonconstant, or "FAIL".

1. choose a single precision prime $p > n^2$ not dividing $\operatorname{lc}_x(fg)$
   $k \longleftarrow \lceil (\lambda + 3)/\log_2 p \rceil$
2. **call** Algorithm 7.2 with input $f(x), g(x+y), p$, and k, to compute $\sigma \in \mathbb{Z}[y]$ with $\sigma \equiv \operatorname{res}_x(f(x), g(x + y)) \bmod p^k$, $\deg_y \sigma \leq (\deg f)(\deg g)$, and $\|\sigma\|_\infty < p^k/2$
   if the algorithm returns "FAIL" or 0 does not occur in the degree sequence of the output **then return** "FAIL"
3. use root finding over finite fields to compute the distinct roots $a_1, \ldots, a_t \in \{-(p-1)/2, \ldots, (p-1)/2\}$ of σ modulo p, with $0 \leq t \leq \deg \sigma$ and multiplicities $e_1, \ldots, e_t \in \mathbb{N}_{\geq 1}$, such that

$$\sigma \equiv (y - a_1)^{e_1} \cdots (y - a_t)^{e_t} v \bmod p,$$

   where $v \in \mathbb{Z}[y]$ has no roots modulo p
4. use Hensel lifting to lift the factorization from 3 to a factorization $\sigma \equiv w_1 \cdots w_t w \bmod p^k$, with $w_i \in \mathbb{Z}[x]$ monic of degree e_i and of max-norm less than $p^k/2$ and $w_i \equiv (y - a_i)^{e_i} \bmod p$, for all i
5. **for** $1 \leq i \leq t$ **do**
       let $z_i \in \{-(p^k - 1)/2, \ldots, (p^k - 1)/2\}$ be such that $-e_i z_i$ is congruent modulo p^k to the coefficient of $x^{e_i - 1}$ in w_i
       if $(y - z_i)^{e_i} \not\equiv w_i \bmod p^k$ **then return** "FAIL"
6. **return** $z_1, \ldots, z_t$

   The last check in step 5 is not necessary for the estimate of the algorithm's correctness probability below. However, in some cases it prevents that a wrong result is returned; see Remark 7.19 below.
   To prove correctness of the above algorithm, we need the following technical lemma about Hensel lifting.

Lemma 7.13. *Let $r \in \mathbb{Z}[y]$ be nonconstant and $p \in \mathbb{N}$ be a prime that is lucky with respect to the squarefree factorization of r. Moreover, let $s, u_1, v_1 \in \mathbb{Z}[y]$ and $e, j \in \mathbb{N}$ be such that s is an irreducible factor of r of multiplicity e, $u_1 \bmod p$ is an irreducible factor of $s \bmod p$ and coprime to $v_1 \bmod p$, and $r \equiv u_1^j v_1 \bmod p$. Then the following hold.*

(i) *$e = j$.*
(ii) *For every $k \in \mathbb{N}_{\geq 1}$, there exists a factorization $r \equiv u_k^e v_k \bmod p^k$ with polynomials $u_k, v_k \in \mathbb{Z}$ such that $u_k \equiv u_1 \bmod p$ and $v_k \equiv v_1 \bmod p$.*
(iii) *If $r \equiv UV \bmod p^k$, with $U, V \in \mathbb{Z}[y]$, is any factorization such that $U \equiv u_1^e \bmod p$ and $V \equiv v_1 \bmod p$, then $U \equiv u_k^e \bmod p^k$ and $V \equiv v_k \bmod p^k$.*

Proof. Let $t, w_1 \in \mathbb{Z}[y]$ be such that $r = s^e t$ and $s \equiv u_1 w_1 \bmod p$. Then s and t are coprime in $\mathbb{Q}[y]$ and $u_1^j v_1 \equiv r = s^e t \equiv u_1^e w_1^e t \bmod p$. Since p is lucky with respect to the squarefree factorization of r, Lemma 5.4 implies that $s \bmod p$ is squarefree and coprime to $t \bmod p$. Thus u_1 divides modulo p neither w_1 nor t, and hence $j \leq e$. On the other hand, u_1 and v_1 are coprime modulo p, which implies that $j = e$ and $v_1 \equiv w_1^e t \bmod p$. By Hensel's lemma (Theorem 15.12 in von zur Gathen & Gerhard 1999), we obtain a factorization $s \equiv u_k w_k \bmod p^k$, with $u_k, w_k \in \mathbb{Z}[y]$ such that $u_k \equiv u_1 \bmod p$ and $w_k \equiv w_1 \bmod p$. Letting $v_k = w_k^e t$, we arrive at the required factorization $r = s^e t \equiv u_k^e w_k^e t = u_k^e v_k \bmod p^k$. The third statement follows from the uniqueness of Hensel lifting (Fact 3.20). $\square$

Theorem 7.14 (Correctness of Algorithm 7.12). *Let f, g be as in Algorithm 7.12 and $r = \mathrm{res}_x(f(x), g(x+y)) \in \mathbb{Z}[y]$. If the algorithm returns "FAIL" in step 2, then p is unlucky for $f(x), g(x + y)$. If the algorithm does not return "FAIL" in step 2, then $\sigma \equiv r \bmod p^k$, $p \nmid \mathrm{lc}(r)$, and $t \leq \deg \sigma = \deg r = (\deg f)(\deg g) \leq n^2$ in step 3. Moreover, if p is lucky with respect to the squarefree factorization of r, then the algorithm does not return "FAIL" in step 5 and correctly solves the integral root distances problem 7.10.*

Proof. Fact 7.11 implies that r is not the zero polynomial. If Algorithm 7.2 returns "FAIL" or 0 does not occur in the degree sequence of its output, then Theorem 7.3 implies that p is unlucky for $f(x), g(x + y)$. Otherwise, and if Algorithm 7.2 does not return "FAIL", then the same theorem shows that $\sigma \equiv r \bmod p^k$. Fact 7.11 implies that $\deg r = (\deg f)(\deg g) \leq n^2$ and that $\mathrm{lc}(r) = \mathrm{lc}(f)^{\deg g} \mathrm{lc}(g)^{\deg f}$. Thus $\deg \sigma = \deg r$ and $p \nmid \mathrm{lc}(r)$.

Now we assume that the algorithm does not return "FAIL" in step 2. For $1 \leq i \leq t$, Lemma 7.13 for $U = w_i$ and $u_1 = y - a_i$, together with the assumption that p is lucky with respect to the squarefree factorization of r, implies that w_i is an e_ith power modulo p^k, and the algorithm does not return "FAIL" in step 5.

By Fact 7.11 (ii), each root of r is less than $2^{\lambda+2}$ in absolute value. Let $z \in \mathbb{Z}$ be a root of r of multiplicity $e \geq 1$. Then there is a unique $i \in \{1, \ldots, t\}$ such that $z \equiv a_i \bmod p$ and $(y - z)^{e_i} \equiv (y - a_i)^{e_i} \bmod p$. Lemma 7.13 with $s = y - z$, $u_1 = y - a_i$, and $U = (y - z)^e$ implies that $e = e_i$ and $(y - z)^{e_i} \equiv (y - z_i)^{e_i} \bmod p^k$. Comparing coefficients of $x^{e_i - 1}$, we find that $e_i z \equiv e_i z_i \bmod p^k$, and since $p > n^2 \geq e_i > 0$, we can cancel e_i and conclude that $z \equiv z_i \bmod p^k$. Both sides are less than $p^k/2$ in absolute value, and hence they are equal. Thus all integral roots of r occur among the z_i. $\square$

Note that $z \in \mathbb{Z}$ is a root of the resultant $r = \mathrm{res}_x(f(x), g(x + y))$ if and only if $\gcd(f(x), g(x + z))$ is nonconstant, by Corollary 3.12.

Lemma 7.15. *Let $f, g \in \mathbb{Z}[x]$ with $0 < \deg g \leq \deg f \leq n$ and max-norm less than 2^λ. For $0 \leq d < \deg g$, let $\sigma_d \in \mathbb{Z}[y]$ be the dth subresultant of $f(x)$ and $g(x + y)$ with respect to x. Then $\|\sigma_d\|_1 < (n2^{n+\lambda+3/2})^{2n-2d}$.*

Proof. We have

$$B = \|g(x+y)\|_1 < 2^\lambda \left\| \sum_{0 \le i \le n} (x+y)^i \right\|_1 = 2^\lambda \sum_{0 \le i \le n} 2^i < 2^{n+\lambda+1} ,$$

and this is also a bound on the maximal one-norm of a coefficient in $\mathbb{Z}[y]$ of $f(x)$ or $g(x+y)$. Then Corollary 3.2 implies that

$$\|\sigma_d\|_1 \le (2n-2d)! B^{2n-2d} \le 2^{n-d}(nB)^{2n-2d} < (n2^{n+\lambda+3/2})^{2n-2d} . \quad \square$$

Definition 7.16. *Let $f, g \in \mathbb{Z}[x]$ be nonzero. A prime $p \in \mathbb{N}$ is lucky with respect to the integral root distances problem 7.10 if it is lucky for $f(x), g(x+y)$ and lucky with respect to the squarefree factorization of $\mathrm{res}_x(f(x), g(x+y))$.*

Theorem 7.17 (Correctness probability of Algorithm 7.12). *Let n, λ be as in Algorithm 7.12, ω the precision of our processor, $C = (n2^{n+\lambda+3/2})^{4n^3+(n+1)^2}$, and $\gamma = \lfloor (\log_2 C)/(\omega-1) \rfloor \in \Theta(n^3(n+\lambda))$. The number of single precision primes that are unlucky with respect to the integral root distances problem 7.10 is at most γ. If p is chosen uniformly at random from among at least 4γ single precision primes between $2^{\omega-1}$ and 2^ω in step 1, then Algorithm 7.12 returns the correct result with probability at least $1/2$.*

Proof. For $0 \le d < \deg g$, let $\sigma_d \in \mathbb{Z}[y]$ be the dth subresultant of $f(x), g(x+y)$ with respect to x, let $r = \sigma_0 = \mathrm{res}_x(f(x), g(x+y)) \in \mathbb{Z}[y] \setminus \{0\}$, let $\delta = \deg \gcd(r, r') \le n^2$, and let $\tau \in \mathbb{Z} \setminus \{0\}$ be the δth subresultant of r and r'. A prime $p > n$ is lucky with respect to the integral root distances problem 7.10 if and only if it is lucky for $f(x), g(x+y)$ and does not divide τ, by Corollary 5.5. By Theorem 7.14, the algorithm returns the correct result if p is such a lucky prime. Let $\sigma = \mathrm{lc}(fg) \prod_{d < \deg g, \, \sigma_d \ne 0} \sigma_d$. Lemma 7.15 yields $|\mathrm{lc}(\sigma_d)| \le \|\sigma_d\|_\infty < (n2^{n+\lambda+3/2})^{2n-2d}$ for all d. If $h = \sum_{0 \le d < \deg g}(2n-2d) \le n^2 + n$ and B is as in the proof of Lemma 7.15, then

$$\begin{aligned}
|\mathrm{lc}(\sigma)| &= |\mathrm{lc}(fg)| \cdot \prod_{d < \deg g, \, \sigma_d \ne 0} |\mathrm{lc}(\sigma_d)| < 2^{2\lambda}(n2^{n+\lambda+3/2})^h \\
&\le (n2^{n+\lambda+3/2})^{(n+1)^2} , \\
\|r\|_2 &\le \|r\|_1 \le (2n)! \, B^{2n} \le n^{2n-1} 2^{(2n+2\lambda+3)n} , \\
|\tau| &\le \|r\|_2^{n^2} \|r'\|_2^{n^2} \le (n\|r\|_2)^{2n^2} \le (n2^{n+\lambda+3/2})^{4n^3} , \\
|\mathrm{lc}(\sigma)\tau| &< (n2^{n+\lambda+3/2})^{4n^3+(n+1)^2} = C ,
\end{aligned}$$

by Corollary 3.2. A single precision prime p is lucky for $f(x), g(x+y)$ if it does not divide $\mathrm{lc}(\sigma)$, and hence it is lucky for the integral root distances problem if it does not divide $\mathrm{lc}(\sigma)\tau$. The above estimate shows that at most γ single precision primes are unlucky with respect to the integral root distances problem, and since there are at least four times as many of them, the probability that a randomly chosen prime is lucky is at least $3/4$. By Theorem 7.5, the conditional probability that Algorithm 7.2 does not return "FAIL" if p is lucky is at least $3/4$ as well, and hence the probability that Algorithm 7.12 returns the correct result is at least $(3/4)^2 \geq 1/2$. $\square$

Theorem 7.18 (Cost of Algorithm 7.12). *If we neglect the cost for prime finding in step 1, then Algorithm 7.12 takes $O(n^4(\log n + \lambda^2))$ word operations with classical arithmetic and $O((n^2 \, \mathsf{M}(n) + \mathsf{M}(n^2)) \log n)(\log n)\mathsf{M}(\lambda) \log \lambda)$ or $O^\sim(n^3\lambda)$ with fast arithmetic.*

Proof. The cost for step 2 is $O(n^4)$ arithmetic operations modulo p^k or $O(n^4\lambda^2)$ word operations with classical arithmetic, and $O((n^2 \, \mathsf{M}(n) + \mathsf{M}(n^2)) \log n)$ arithmetic operations or $O((n^2 \, \mathsf{M}(n) + \mathsf{M}(n^2))(\log n)\mathsf{M}(\lambda) \log \lambda)$ word operations with fast arithmetic, by Corollary 7.8. In step 3, we perform a squarefree factorization to determine the multiplicities, taking $O(n^4)$ word operations with classical arithmetic and $O(\mathsf{M}(n^2) \log n)$ with fast arithmetic, by Proposition 5.3, and then compute all roots of all polynomials in the squarefree decomposition. Since the degrees of all these polynomials sum to at most n^2, the latter task takes an expected number of $O(n^4 \log n)$ word operations with classical arithmetic and $O(\mathsf{M}(n^2) \log^2 n)$ with fast arithmetic, by Fact 3.21 (i). The cost for step 4 is $O(n^4\lambda^2)$ word operations with classical arithmetic and $O(\mathsf{M}(n^2)(\log n)(\log n + \mathsf{M}(\lambda)))$ with fast arithmetic, by Fact 3.20. Computing $(y - z_i)^{e_i} \bmod p^k$ in step 5 by repeated squaring takes $O(e_i^2)$ additions and multiplications modulo p^k with classical arithmetic and $O(\mathsf{M}(e_i))$ with fast arithmetic. Since $\sum_{1 \leq i \leq t} e_i \leq \deg \sigma \leq n^2$, this amounts to a total of $O(n^4\lambda^2)$ word operations with classical arithmetic and $O(\mathsf{M}(n^2) \, \mathsf{M}(\lambda))$ with fast arithmetic. $\square$

Remark 7.19. *(i) Since the roots of r are the differences of the roots of f and g, it is sufficient to apply Algorithm 7.12 to the squarefree parts of f and g.*

(ii) We note that the algorithm is not Las Vegas if $k \geq 2$: For example, if $\sigma = y(y - 2p)$ and $k = 2$, then $\sigma \equiv y^2 \bmod p$ and $\sigma \equiv (y - p)^2 \bmod p^2$, and the algorithm incorrectly returns $z_1 = p$. In fact, there are exponentially many possible factorizations of σ modulo p^2: we have $\sigma \equiv (y - kp)(y - (2 + k)p) \bmod p^2$ for $0 \leq k < p$ (see also von zur Gathen & Hartlieb 1998). However, $\sigma \not\equiv (y - p)^2 \bmod p^3$, and hence the last check in step 5 of Algorithm 7.12 would prevent that an incorrect result is returned if $k \geq 3$.

This nuisance goes away if we choose $k = 1$ and a (possibly multiprecision) prime $p > 2^{\lambda+3}$ in step 1 and return $a_1, \ldots, a_t$ after step 3. Then the condition that p should not divide the δth subresultant of r and r' can be dropped, and the bit length of the bound corresponding to C in Theorem 7.17 is only $O(n^2(n + \lambda))$, which is smaller by a factor of about n. However, the cost for root finding in step 3 is an expected number of $O(n^4 \log n \cdot \lambda^3)$ word operations with classical arithmetic and $O(\mathsf{M}(n^2)(\lambda + \log n) \log n \cdot \mathsf{M}(\lambda) \log \lambda)$ or $O^\sim(n^2\lambda^2)$ with fast arithmetic, and this introduces an additional factor of λ in the running time estimate of Theorem 7.18.

Another possibility to make the algorithm Las Vegas would be to compute r exactly, not only modulo p^k, and then compute its roots (or the roots of its *squarefree part). Since the bit length of the bound on the integer coefficients of r is $\Theta(n^2 + n\lambda)$, as in the proof of Theorem 7.17, this would introduce an additional factor of at least n in the timing estimates.*

(iii) *Instead of computing the resultant modulo p^k, one might think about computing it modulo p and then lifting it somehow modulo p^k. Currently, there is apparently no algorithm known for lifting the resultant, but all we need is some factor of the resultant having the same roots, such as the least common multiple of the denominators of the Bézout coefficients $s, t \in \mathbb{Q}(y)[x]$ satisfying $s \cdot f(x) + t \cdot g(x + y) = 1$. One may imagine various ways of computing these s, t modulo p^k, by performing substitution of one point (or several points) u for y plus $(y - u)$-adic lifting (or interpolation), p-adic lifting, and rational reconstruction. in any order. (See Lauer (2000) for an analysis in the case $y = 0$.) Some of these tasks would have to be performed over rings with zero divisors, which makes the correctness proofs quite complicated, as we have already seen for Algorithm 7.2. We do not analyze any of these combinations, since we expect that none of them would yield an essential improvement on the estimate of Theorem 7.18, by the following heuristic argument. The number of coefficients in $\mathbb{Q}$ of s, t is of order n^3 in the worst case, so that we expect a running time which is at least cubic in n, even with fast arithmetic, and Theorem 7.18 already achieves this bound up to logarithmic factors.*

(iv) *We compare the cost estimate from Theorem 7.18 with an estimate for a modular algorithm that computes the resultant r exactly. By Lemma 7.15, the coefficients of r are of word length $O(n^2 + n\lambda)$. Thus we would call Algorithm 7.2 with $k = 1$ for $O(n^2 + n\lambda)$ distinct single precision primes and then reconstruct r via Chinese remaindering. The cost for the modular computation is then $O(n^6 + n^5\lambda)$ word operations with classical arithmetic and $O^\sim(n^5 + n^4\lambda)$ with fast arithmetic, by Corollary 7.8 (iii). The resultant r has $O(n^2)$ coefficients, and reconstructing all of them from their $O(n^2 + n\lambda)$ modular images takes $O(n^6 + n^4\lambda^2)$ word operations with classical arithmetic and $O^\sim(n^4 + n^3\lambda)$ with fast arithmetic. The cost for all other steps is negligible, and hence the overall cost is $O(n^6 + n^4\lambda^2)$ with classical and $O^\sim(n^5 + n^4\lambda)$ with fast arithmetic. Thus the estimate of Theorem 7.18 for fast arithmetic is better by a factor of about n. The estimate for classical arithmetic is about the same in the diagonal case where $n \approx \lambda$, but the exponent of n is smaller by 2. For comparison, Loos (1983) states in his Theorem 9 a running time estimate of $O(n^{10} + n^8\lambda^2)$ word operations for computing all rational zeroes of the resultant $\mathrm{res}_x(f(x), g(x + y))$. His analysis uses classical arithmetic, and the estimate of Theorem 7.18 is better by a factor of about n^4.*

(v) *It may be advantageous to choose p in step 1 of Algorithm 7.12 in such a way that $p - 1$ is divisible by a large power 2^t of two. Then there is a primitive 2^tth root of unity modulo p^k, and fast FFT-based polynomial arithmetic modulo p^k can be implemented very efficiently for degrees not exceeding 2^{t-1}.*

7.2 Computing All Integral Roots Via Factoring

In this section, we discuss an alternative to Algorithm 7.12, turning out to be faster. One disadvantage of Algorithm 7.12 is that the resultant $r = \operatorname{res}_x(f(x), g(x + y))$ has quadratic degree $(\deg f)(\deg g)$. Man & Wright (1994) analyze an algorithm which determines the roots of r by computing the irreducible factorizations of f and g. The idea is as follows. If p is an irreducible factor of f and q is an irreducible factor of g such that $q(x) = p(x + z)$ for some $z \in \mathbb{Z}$, then $\deg p = \deg q = m$, $\operatorname{lc}(p) = \operatorname{lc}(q)$, and $q_{m-1} = p_{m-1} + mz \operatorname{lc}(p)$, where p_{m-1} and q_{m-1} are the coefficients of x^{m-1} in p and q, respectively. Thus we check for all pairs (p, q) of irreducible factors with $\deg p = \deg q = m$ and $\operatorname{lc}(p) = \operatorname{lc}(q)$ whether $z = (q_{m-1} - p_{m-1})/m \operatorname{lc}(p)$ is an integer, and if so, whether $q(x) = p(x + z)$, and return all successful such z's.

The modern algorithms for factoring polynomials with integer coefficients use a modular approach, proceeding in three stages (see Chap. 15 and 16 in von zur Gathen & Gerhard 1999):

1. factorization modulo a small prime,
2. factorization modulo a prime power via Hensel lifting,
3. reconstruction of the factors in $\mathbb{Z}[x]$ from the modular factors.

The last stage is the dominant step in the worst case, and it is comparatively expensive. There are essentially two alternatives: *factor combination* and *short vectors*. It is well-known that there exist polynomials of degree n for which factor combination takes time at least $2^{n/2}$, for infinitely many n. Lenstra, Lenstra & Lovàsz' (1982) replace the factor combination stage by an algorithm for computing short vectors in lattices, taking $O^\sim(n^9 + n^7 \lambda^2)$ word operations with fast arithmetic if the coefficients of f are absolutely less than 2^λ. Schönhage (1984) improved this bound to $O^\sim(n^6 + n^4 \lambda^2)$. Recently, van Hoeij (2002) has found a very efficient algorithm by combining the ideas of factor combination and short vectors.

However, for our problem at hand, it is not necessary that we perform the third stage at all. We simply take the factorization modulo a prime power and execute Man & Wright's algorithm modulo that prime power. If the prime power is large enough, then we are guaranteed to find all integral roots of r, plus possibly some additional integers that are not roots of r. For our intended application in Chap. 8, these additional non-roots do not increase the worst case cost estimates. Note that Algorithm 7.12 from the previous section may also return such additional non-roots.

We assume that our polynomials have degree at most $2^{\omega-1}$, where ω is the word size of our processor, as usual.

Algorithm 7.20 (Prime power modular Man & Wright algorithm).
Input: Nonzero polynomials $f, g \in \mathbb{Z}[x]$ of degree at most $n \leq 2^{\omega-1}$ and max-norm less than 2^λ.
Output: A set of at most $(\deg f)(\deg g)$ integers containing all $z \in \mathbb{Z}$ such that $\gcd(f(x), g(x + z))$ is nonconstant.

1. compute the squarefree parts f_* and g_* of f and g, respectively
2. **repeat**

          choose a single precision prime p

    until p does not divide $\mathrm{lc}(f_* g_*)$ and f_* mod p and g_* mod p are squarefree in $\mathbb{F}_p[x]$

    $k \longleftarrow \lceil (\lambda + 3)/\log_2 p \rceil$
3. compute the irreducible factorizations

$$f_* \equiv \mathrm{lc}(f_*)f_1 \cdots f_s \bmod p, \quad g_* \equiv \mathrm{lc}(g_*)g_1 \cdots g_t \bmod p ,$$

    with monic polynomials $f_1, \ldots, f_s, g_1, \ldots, g_t \in \mathbb{Z}[x]$ of max-norm less than $p/2$, such that $f_1, \ldots, f_s$ are distinct and $g_1, \ldots, g_t$ are distinct and all are irreducible modulo p
4. use Hensel lifting to compute a factorization

$$f_* \equiv \mathrm{lc}(f_*)F_1 \cdots F_s \bmod p^k, \quad g_* \equiv \mathrm{lc}(g_*)G_1 \cdots G_t \bmod p^k ,$$

    with monic polynomials $F_1, \ldots, F_s, G_1, \ldots, G_t \in \mathbb{Z}[x]$ of max-norm less than $p^k/2$, such that $F_i \equiv f_i \bmod p$ for $1 \le i \le s$ and $G_j \equiv g_j \bmod p$ for $1 \le j \le t$
5. $S \longleftarrow \emptyset$

    for $1 \le i \le s$ and $1 \le j \le t$ **do**
6. $m \longleftarrow \deg F_i$

          if $\deg G_j = m$ **then**

                let $z_{ij} \in \mathbb{Z}$ with $|z_{ij}| < p^k/2$ be such that z_{ij} mod p^k is the coefficient of x^{m-1} in $((G_j - F_i)/m)$ mod p^k

                if $G_j(x) \equiv F_i(x + z_{ij})$ mod p^k **then** $S \longleftarrow S \cup \{z_{ij}\}$
7. **return** S

Theorem 7.21. *Algorithm 7.20 solves the integral root distances problem 7.10 correctly as specified. Let $C = (n+1)^{2n}2^{4n\lambda}$. The number of single precision primes that are unlucky with respect to the squarefree factorization of both f and g is at most $\lfloor (\log_2 C)/(\omega - 1) \rfloor \in \Theta(n(\lambda + \log n))$. If p is chosen uniformly at random from among at least $2\lfloor (\log_2 C)/(\omega - 1) \rfloor$ single precision primes, then the expected number of iterations of step 2 is at most two, and the expected cost for the algorithm is $O(n^3\lambda + n^2\lambda^2)$ word operations with classical arithmetic and*

$$O(n\,\mathsf{M}(n + \lambda)\log(n + \lambda) + \mathsf{M}(n^2)\log n + n\,\mathsf{M}(n)\,\mathsf{M}(\lambda) + n^2\,\mathsf{M}(\lambda)\log\lambda)$$

or $O^\sim(n^2\lambda)$ with fast arithmetic, if we ignore the cost for prime finding.

If we omit the check whether $G_j(x) \equiv F_i(x + z_{ij})$ mod p^k holds in step 6 and add z_{ij} to S anyway, then the algorithm is still correct, and the cost drops to $O(n^3 + n^2(\log n)\lambda + n^2\lambda^2)$ word operations with classical arithmetic and

$$O(n\,\mathsf{M}(n + \lambda)\log(n + \lambda) + \mathsf{M}(n^2)\log n + n^2(\mathsf{M}(\lambda) + \lambda\log n))$$

or $O^\sim(n^2\lambda)$ with fast arithmetic.

Proof. Let $z \in \mathbb{Z}$ be such that $\gcd(f(x), g(x + z))$ is nonconstant. Then also $h(x) = \gcd(f_*(x), g_*(x + z))$ is nonconstant. Let $i \in \{1, \ldots, s\}$ be such that f_i divides h modulo p. Then $f_i(x)$ divides $g_*(x + z)$ modulo p, and hence $f_i(x) \equiv g_j(x + z) \bmod p$ for a unique j. The uniqueness of Hensel lifting (Fact 3.20) implies that F_i and $G_j(x + z)$ both divide h modulo p^k. Now both $h \equiv F_i \cdot (h/F_i) \bmod p^k$ and $h \equiv G_j(x+z) \cdot (h/G_j(x+z)) \bmod p^k$ are liftings of the same factorization modulo p, and again the uniqueness of Hensel lifting yields $F_i(x) \equiv G_j(x + z) \bmod p^k$. Thus $\deg F_i = \deg G_j$ and $z \equiv z_{ij} \bmod p^k$, and since $|z_{ij}| < p^k/2$ and $|z| < 2^{\lambda+2} < p^k/2$, by Fact 7.11 (iii), we have $z = z_{ij} \in S$. This is true in any case, whether the check in step 6 is omitted or not.

Let $d = \deg \gcd(f, f')$ and $e = \deg \gcd(g, g')$. By Lemma 5.4, a prime p is lucky with respect to the squarefree factorization of both f and g if and only if the condition in step 2 is satisfied. Corollary 5.5 implies that p is lucky if and only if it divides neither the dth subresultant of f and f' nor the eth subresultant of g and g'. By Corollary 3.2 (i), the product of these two subresultants is absolutely at most C, and hence the number of unlucky single precision primes is at most $\lfloor \log_2 C/(\omega - 1) \rfloor$. By assumption, at most half of the single precision primes are unlucky. Thus the condition in step 2 is satisfied is with probability at least $1/2$.

Step 1 takes $O(n^3 + n\lambda^2)$ and $O(n \mathsf{M}(n + \lambda) \log(n + \lambda) + \lambda \mathsf{M}(n) \log n)$ word operations with classical and fast arithmetic, respectively, by Theorem 5.10. By Mignotte's bound 3.3, the coefficients of f_* and g_* are at most $(n + 1)^{1/2} 2^{n+\lambda}$ in absolute value, and reducing them modulo p in step 2 takes $O(n^2 + n\lambda)$ word operations. We check squarefreeness by computing $\gcd(f_* \bmod p, (f_*)' \bmod p)$ and $\gcd(g_* \bmod p, (g_*)' \bmod p)$, taking $O(n^2)$ and $O(\mathsf{M}(n) \log n)$ word operations with classical and fast arithmetic, respectively. Thus the expected cost of step 2 is $O(n^2 + n\lambda)$ word operations.

The modular factorization in step 3 takes $O(n^3)$ word operations with classical arithmetic and $O(\mathsf{M}(n^2) \log n)$ with fast arithmetic, by Fact 3.21 (ii). Hensel lifting in step 4 uses $O(n^2\lambda^2)$ and $O((\mathsf{M}(\lambda) + \log n)\mathsf{M}(n) \log n)$ word operations with classical and fast arithmetic, respectively, by Fact 3.20.

The cost for one execution of step 6 is dominated by the cost for the Taylor shift, taking $O(m^2)$ additions and $O(m)$ multiplications and divisions modulo p^k, together $O(m^2\lambda + m\lambda^2)$ word operations with classical arithmetic, by Theorem 4.3. With fast arithmetic, the cost is $O(\mathsf{M}(m) \mathsf{M}(\lambda) + m \mathsf{M}(\lambda) \log \lambda)$ word operations, by Theorem 4.5. Summing over all i and j, we find that the overall cost for steps 5 and 6 is $O(n^3\lambda + n^2\lambda^2)$ word operations with classical arithmetic and $O(n \mathsf{M}(n) \mathsf{M}(\lambda) + n^2 \mathsf{M}(\lambda) \log \lambda)$ with fast arithmetic.

If we omit the check in step 6, then the cost for one execution of step 6 is $O(\lambda^2 + \lambda \log n)$ word operations for the computation of z_{ij} with classical arithmetic, and $O(\mathsf{M}(\lambda) + \lambda \log n)$ with fast arithmetic. Then the overall cost for steps 5 and 6 is $O(n^2(\lambda^2 + \lambda \log n))$ and $O(n^2(\mathsf{M}(\lambda) + \lambda \log n))$ with classical and fast arithmetic, respectively. The $O^\sim$-estimates follow from taking $\mathsf{M}(n) = n \log n \log\log n$. □

In the diagonal case where $n \approx \lambda$, the cost estimates for classical arithmetic above are better by a factor of about n^2 than the estimate from Theorem 7.18.

For fast arithmetic, the estimates are better by a factor of about n, up to logarithmic factors. Another advantage of Algorithm 7.20 is that its output is always correct, while Algorithm 7.12 is of Monte-Carlo type: it may err with probability at most $1/2$.

Remark 7.22. *(i) Although the check in step 6 is not necessary for the algorithm to be correct, it sorts out some of the non-roots of $\mathrm{res}_x(f(x), g(x+y))$. A good compromise in practice is to replace this check by the cheaper check whether the constant coefficients $G_j(0)$ and $F_i(z_{ij})$ or the coefficients of x^{m-2} in G_j and $F_i(x + z_{ij})$ agree modulo p^k. Heuristically, this should rule out almost all non-roots as well, within the same time bound.*

(ii) We may decrease the cost for step 6 to $O(\lambda \log n)$ by slightly increasing the lifting bound k to $\lceil (\lambda + 3 + \log_2 n)/\log_2 p \rceil$. Then we would compute the coefficient of x^{m-1} in $G_j - F_i$, discard it if it is greater than $m2^{\lambda+2}$ or not divisible by m, and add its quotient by m to S otherwise. This increases the cost of step 4 to $n^2(\lambda^2 + \log^2 n)$ with classical arithmetic and $O(\mathsf{M}(n)\,\mathsf{M}(\lambda + \log n)\log n)$ with fast arithmetic.

(iii) In some applications, the polynomials f and g may already be given in factored form. Then the cost of the algorithm is essentially the cost of steps 5 and 6.

(iv) Man & Wright (1994) state an average case estimate of $O(n^4(n + \lambda)^2)$ word operations for their algorithm, based on classical arithmetic. In the diagonal case where $n \approx \lambda$, our estimate from Theorem 7.21 beats this by a factor of about n^2.

7.3 Application to Hyperexponential Integration

We now give an analog of Algorithm 7.12 for hyperexponential integration. Given $f, g \in \mathbb{Z}[x]$, the relevant resultant is $r = \mathrm{res}_x(g, f - yg') \in \mathbb{Z}[y]$.

Problem 7.23 (Integral residues). *Given nonzero coprime polynomials f, g in $\mathbb{Z}[x]$ of degree at most n, compute the at most n integers $z \in \mathbb{Z}$ such that $\gcd(g, f - zg')$ is nonconstant.*

The following factorization is a consequence of Fact 7.11.

Fact 7.24. *Let F be a field, f and $g = \mathrm{lc}(g)\prod_{1 \le i \le n}(x - \beta_i)$ be nonzero polynomials in $F[x]$, $m = \deg_x(f - yg')$, and $r = \mathrm{res}_x(g, f - yg') \in F[y]$.*

(i) $r = \mathrm{lc}(g)^m \displaystyle\prod_{1 \le i \le n} (f(\beta_i) - yg'(\beta_i))$.

(ii) $r(0) = \mathrm{lc}(g)^{m-\deg f}\mathrm{res}_x(g, f) = (-1)^{n \cdot \deg f}\mathrm{lc}(g)^{m-\deg f}\mathrm{res}_x(f, g)$.

(iii) Assume that $\beta_1, \ldots, \beta_d$ are precisely the simple roots of g, let $g_1 \in F[x]$ be a squarefree polynomial with roots $\beta_1, \ldots, \beta_d$, and let $h = g/g_1$. Then $r \ne 0$ if and only if f and h are coprime, and in that case we have $\deg r = d$ and

$$\mathrm{lc}(r) = (-1)^d \mathrm{lc}(g_1)^{m-\deg g'}\mathrm{lc}(h)^{m-\deg f}\mathrm{res}_x(g_1, g')\mathrm{res}_x(h, f) \,.$$

In particular, if g is squarefree, then $\deg r = n$ *and*

$$\mathrm{lc}(r) = (-1)^n \mathrm{lc}(g)^{m - \deg g'} \mathrm{res}_x(g, g') \, .$$

If f and g are coprime, then the factorization (i) implies that every root of r has the form $f(\beta)/g'(\beta)$ for a simple root β of g, i.e., the roots of r are precisely the residues of the rational function f/g at its simple poles.

There are several notable differences to the hypergeometric situation. If the degrees of f and g are less than n, then the degree of r, and hence also the number of roots of r, is at most n, in contrast to at most n^2 in the hypergeometric case. On the other hand, the word size of the integral roots of r is smaller than the size of the coefficients of r by a factor of about n in the hypergeometric case. In the hyperexponential case, however, we are not aware of a better bound for the word size of the roots of r than the bound on the word size of the coefficients of r. For that reason, we can afford to compute the coefficients of r exactly, by computing them modulo sufficiently many single precision primes and reconstructing them via the Chinese Remainder Theorem. As a consequence, we obtain an algorithm for finding all integral roots of r which is of Las Vegas type, with the exception of the prime finding stages. As usual, ω denotes the word size of our processor.

Algorithm 7.25 (Small primes modular integral residues). ▬▬▬
Input: Nonzero coprime polynomials $f, g \in \mathbb{Z}[x]$ with $\deg f, \deg g \leq n \leq 2^{\omega - 1}$ and $\|f\|_\infty, \|g\|_\infty < 2^\lambda$.
Output: The set of at most n integers $z \in \mathbb{Z}$ for which $\gcd(g, f - zg')$ is nonconstant.

1. $C \longleftarrow (2n)! (n+1)^{2n} 2^{2n\lambda}, \quad k \longleftarrow \lceil \log_2(2C)/(\omega - 1) \rceil$
   $s \longleftarrow \lceil \log_2(2\lambda)/(\omega - 1) \rceil$
   choose a set S_0 of $k + s$ single precision primes
   $S_1 \longleftarrow \{p \in S_0 : p \nmid \mathrm{lc}_x(fg)\}$
2. **for** $p \in S_1$ **do**
3.   **call** Algorithm 7.2 with input g, $f - yg'$, p, and 1, to compute $\sigma_p \in \mathbb{Z}[y]$
           with $\sigma_p \equiv \mathrm{res}_x(g, f - yg') \bmod p$, $\deg_y \sigma_p \leq n$, and $\|\sigma_p\|_\infty < p/2$
           if 0 does not occur in the degree sequence of the output
           then $\sigma_p \longleftarrow 0$
4. use the Chinese Remainder Algorithm to compute $\sigma \in \mathbb{Z}[y]$ with $\|\sigma\|_\infty < \frac{1}{2} \prod_{p \in S_1} p$ and $\sigma \equiv \sigma_p \bmod p$ for all $p \in S_1$
5. **call** Algorithm 5.6 to compute the squarefree part $\sigma_* \in \mathbb{Z}[y]$ of σ
6. choose an odd single precision prime $p \in \mathbb{N}$ not dividing $\mathrm{lc}(\sigma_*)$ and such that $\sigma_* \bmod p$ is squarefree
7. use root finding over finite fields to compute all roots $a_1, \ldots, a_t \in \mathbb{Z}$ of σ_* modulo p, with $0 \leq t \leq \deg \sigma_*$ and $|a_i| < p/2$ for all i, such that

   $$\sigma_* \equiv (y - a_1) \cdots (y - a_t) v \bmod p \, ,$$

   where $v \in \mathbb{Z}[y]$ has no roots modulo p
8. use Hensel lifting to lift the factorization from 6 to a factorization $\sigma_* \equiv (y - z_1) \cdots (y - z_t) w \bmod p^k$, with $z_i \in \mathbb{Z}$ absolutely less than $p^k/2$ for all i
9. **return** $\{z_i : 1 \leq i \leq t \text{ and } \sigma(z_i) = 0\}$ ▬▬▬

In fact, one might compute the complete squarefree decomposition of σ in step 6 and then perform steps 7 through 9 for each of the factors in the squarefree decomposition. This is more efficient in practice, but it does not improve the order of magnitude of our running time estimate, and therefore we only discuss the simpler algorithm above.

Theorem 7.26. *If there are at least $k + s \in \Theta(n(\lambda + \log n))$ single precision primes, then Algorithm 7.25 solves the integral residues problem 7.23 correctly as specified. If we choose the prime p in step 6 uniformly at random from among at least $\lfloor n \log_2(nC^2)/(\omega - 1) \rfloor \in \Theta(n^2(\lambda + \log n))$ single precision primes, then the expected number of iterations of step 6 is at most two, and the cost for steps 2 through 9 is $O(n^4(\lambda^2 + \log^2 n))$ word operations with classical arithmetic and*

$$O(n^2(\mathsf{M}(n(\lambda + \log n)) + (\lambda + \log n)\mathsf{M}(n) \log n)$$
$$+ \mathsf{M}(n(\lambda + \log n))(\mathsf{M}(n) \log n + n \log \lambda))$$

or $O^\sim(n^3\lambda)$ word operations with fast arithmetic.

Proof. We first note that with the modifications mentioned in Remark 7.6, the call to Algorithm 7.2 in step 3 of Algorithm 7.25 does not return "FAIL". We have $|\mathrm{lc}_x(fg)| < 2^{2\lambda} \leq 2^{(\omega-1)s}$, and hence S_1 contains at least k primes. Let $r = \mathrm{res}(g, f - yg') \in \mathbb{Z}[y]$ and $p \in S_1$. By Lemma 3.11 (i), 0 occurs in the degree sequence of the Euclidean Algorithm for $g \bmod p$ and $f - yg' \bmod p$ if and only if $p \nmid r$. Thus we have $r \equiv \sigma_p \bmod p$ for all $p \in S_1$. Let $m = \prod_{p \in S_1} p$. Then $r \equiv \sigma \bmod m$. The maximal one-norm of a coefficient of g or $f - yg'$, regarded as polynomials in x, is less than $(n + 1)2^\lambda$. We have $\|r\|_\infty \leq \|r\|_1 < C \leq 2^{(\omega-1)k-1} < m/2$, by Corollary 3.2 (iii), and $\|\sigma\|_\infty < m/2$, by construction, and hence $r = \sigma$.

Now let $z \in \mathbb{Z}$ be a root of r. Then $z \equiv a_i \bmod p$ for a unique $i \in \{1, \ldots, t\}$. By the choice of p, the factorization in step 7 is a factorization into coprime polynomials modulo p, and the uniqueness of Hensel lifting (Fact 3.20) implies that $z \equiv z_i \bmod p^k$. We have $|z_i| < p^k/2$, by construction. Since z is an integral root of r, it divides the constant coefficient of r. Fact 7.24 (ii) implies that this constant coefficient is $\mathrm{res}(g, f)$ times a power of $\mathrm{lc}(g)$, and since f and g are coprime, Corollary 3.10 implies that it is nonzero. Thus $|z| \leq \|r\|_\infty \leq 2^{(\omega-1)k-1} < p^k/2$, and hence $z = z_i$. Moreover, Fact 7.24 (i) implies that $t \leq \deg r \leq n$, and the correctness follows from Corollary 3.12.

In step 2, we first reduce all coefficients of f and g modulo all primes in S_1. With classical arithmetic, this takes $O(\lambda)$ word operations for each coefficient and each prime, in total $O(nk\lambda)$. With fast arithmetic, the cost is $O(\mathsf{M}(k) \log k)$ per coefficient, or $O(n \mathsf{M}(k) \log k)$ in total.

One execution of step 3 takes $O(n^3)$ word operations with classical arithmetic and $O(n \mathsf{M}(n) \log n)$ with fast arithmetic, by Corollary 7.9, and there are $O(k)$ iterations. Step 4 takes $O(k^2)$ operations per coefficient of σ with classical arithmetic and $O(\mathsf{M}(k) \log k)$ with fast arithmetic, and there are at most $n+1$ coefficients. The coefficients of σ are of word size $O(k)$. Since $n \in O(k)$, computing the squarefree part in step 5 takes $O(k^2 n)$ with classical and $O(\mathsf{M}(k) \log k \cdot n)$ with fast arithmetic, by Theorem 5.10.

By Mignotte's bound 3.3, the coefficients of σ_* are of word size $O(k)$. Reducing all coefficients modulo p thus takes $O(kn)$ word operations. The cost for computing the gcd of σ_* mod p with its derivative in step 6 is $O(n^2)$ and $O(\mathsf{M}(n) \log n)$ word operations with classical and fast arithmetic, respectively.

Lemma 5.4 implies that a single precision prime p is lucky with respect to the squarefree factorization of $r = \sigma$ if and only if $p \nmid \mathrm{lc}(r)$ and σ_* mod p is squarefree. If $\delta = \deg \gcd(r, r')$, then Corollary 5.5 implies that p is lucky if and only if it does not divide the δth subresultant of r and r'. Since $\deg r \leq n$, $\|r\|_2 \leq \|r\|_1 < C$, and $\|r'\|_2 \leq n \cdot \|r\|_2 < nC$, this subresultant is nonzero of absolute value at most $n^n C^{2n}$, by Corollary 3.2, and the number of unlucky single precision primes is at most $\lfloor n \log_2(nC^2)/(\omega - 1) \rfloor$. By assumption, there are at least twice as many single precision primes, and hence the expected number of iterations of step 6 is at most two.

The root finding in step 7 takes $O(n^2 \log n)$ word operations with classical arithmetic and $O(\mathsf{M}(n) \log^2 n)$ with fast arithmetic, by Fact 3.21 (i). The cost for the Hensel lifting in step 8 is $O(k^2 n^2)$ and $O((\mathsf{M}(k) + \log n)\mathsf{M}(n) \log n)$ word operations with classical and fast arithmetic, respectively, by Fact 3.20.

In step 9, for each i we divide σ by $x - z_i$ with remainder to check whether $\sigma(z_i) = 0$. If this is the case, then the coefficients of the quotient $\sigma/(x - z_i)$ are at most nC in absolute value, by Exercise 14.21 in von zur Gathen & Gerhard (1999). Thus we may abort the computation as soon as we encounter a coefficient that exceeds this bound. Then the cost for one check is $O(nk^2)$ word operations with classical arithmetic and $O(n\,\mathsf{M}(k))$ with fast arithmetic, and there are $O(n)$ checks. The claims now follow from $k \in O(n(\lambda + \log n))$ by summing up costs. $\square$

We might employ fast multipoint evaluation techniques in step 9. This would reduce the cost for step 9 to $O^{\sim}(n^2 \lambda)$ word operations with fast arithmetic, but the overall estimate would still be dominated by the $O^{\sim}(n^3 \lambda)$ word operations for step 3.

We note that the cost estimates for Algorithm 7.25 agree with those for Algorithm 7.12, up to logarithmic factors. Since the product of all integral roots of r divides its constant coefficient, the output size of Algorithm 7.25 is $O(n(\lambda + \log n))$. By a similar argument, the output size of Algorithm 7.12 is $O(n^2 + n\lambda)$.

Remark 7.27. *(i) In practice, one would test whether z_i divides the constant coefficient of σ_* before checking whether z_i is a root of σ. Heuristically, this should already rule out all non-roots.*

(ii) By Fact 7.24 (ii), the constant coefficient of r is $\mathrm{lc}(g)^{\min\{0, \deg g - \deg f - 1\}} \cdot \mathrm{res}_x(g, f)$. Since each integral root z of r divides this constant coefficient, Corollary 3.2 (i) implies that $|z| \leq (n+1)^n 2^{2n\lambda}$. This bound is much smaller than the bound C on the absolute value of the coefficients of r from step 1 of Algorithm 7.25. Thus it makes sense to compute the resultant not exactly, but only modulo a prime power $p^k > 2(n+1)^n 2^{2n\lambda}$, similarly to Algorithm 7.12.

Since the word size of the smaller bound differs from the word size of C only by a constant factor, we do not analyze this variant.

(iii) *We give an example showing that the word size of at least one of the integral roots of r may be $\Omega(n)$ times the word size of the input polynomials. Let β, n be integers with $\beta \geq 2$ and $n \geq 4$, $f = \beta^2 x^{n-1}$, and $g = x^{n-2}(x - \beta)^2 + \beta(x - \beta)$. Then $\|f\|_\infty = \|g\|_\infty = \beta^2$. Now $g(\beta) = 0$, $g'(\beta) = \beta$, and $f(\beta) = \beta^{n+1}$. Thus $\gamma = \beta^n$ is a root of r, by Fact 7.24 (i), and $\|\gamma\|_\infty = \|g\|_\infty^{n/2}$.*

7.4 Modular LRT Algorithm

The final algorithm in this chapter is a modular variant of the algorithm by Lazard, Rioboo and Trager to compute the logarithmic part of the symbolic integral of a rational function. We briefly recapitulate the idea of the algorithm.

Let $f, g \in \mathbb{Z}[x]$ be nonzero with $\deg f < \deg g = n$ and g squarefree, and let $r = \operatorname{res}_x(g, f - yg')$ (this is the same resultant as in the previous section). Fact 7.24 (iii) implies that $\deg r = n$. Let $g_\gamma = \gcd(g, f - \gamma g')$, for all $\gamma \in \mathbb{C}$. Rothstein (1976, 1977) and Trager (1976) proved that

$$\frac{f}{g} = \sum_{r(\gamma)=0} \gamma \frac{g_\gamma'}{g_\gamma} .$$

Moreover, Lazard & Rioboo (1990) have shown that the degree of g_γ is equal to the multiplicity of γ as a root of r, for all γ. Let γ be a root of r, let $i = \deg g_\gamma$, and let $\tau_i \in \mathbb{Z}[y]$ be the ith subresultant of g and $f - yg'$. Lemma 3.11 (i) implies that $\tau_i(\gamma) \neq 0$, i occurs in the degree sequence of the monic EEA of g and $f - yg'$ in $\mathbb{Q}(x)[y]$, $\tau_i \pi_i \in \mathbb{Z}[x, y]$ holds for the monic remainder $\pi_i \in \mathbb{Q}(y)[x]$ of degree i in the monic EEA of g and $f - yg'$, and g_γ and $(\tau_i \pi_i)(x, \gamma)$ agree up to leading coefficients. If $r = c \prod_{1 \leq i \leq n} r_i^i$ is a squarefree decomposition of r, with squarefree and pairwise coprime polynomials $r_1, \ldots, r_n \in \mathbb{Z}[y]$, then the preceding discussion implies that

$$\frac{f}{g} = \sum_{1 \leq i \leq n} \sum_{r_i(\gamma)=0} \gamma \frac{(\tau_i \pi_i)(x, \gamma)'}{(\tau_i \pi_i)(x, \gamma)} .$$

This formula was found independently by Lazard & Rioboo (1990) and Trager (unpublished). See also Mulders 1997.

Problem 7.28 (Rational integration with squarefree denominator). *Given nonzero coprime polynomials $f, g \in \mathbb{Z}[x]$ with g squarefree and $\deg f < \deg g = n$, compute squarefree and pairwise coprime polynomials $r_1, \ldots, r_n \in \mathbb{Z}[y]$ and polynomials $v_1, \ldots, v_n \in \mathbb{Z}[x, y]$ satisfying*

$$\frac{f}{g} = \sum_{1 \leq i \leq n} \sum_{r_i(\gamma)=0} \gamma \frac{v(x, \gamma)'}{v(x, \gamma)} ,$$

where $'$ denotes the formal derivative with respect to x.

Our modular algorithm proceeds as follows. As in the previous section, we first compute the resultant exactly. However, we do not need the integral roots of r, but instead compute its squarefree decomposition. Finally, we use Algorithm 7.2 to compute all required τ_i and $\tau_i \pi_i$ modulo several primes p and reconstruct them via Chinese remaindering.

Algorithm 7.29 (Small primes modular LRT algorithm).

Input: Nonzero polynomials $f, g \in \mathbb{Z}[x]$ such that g is squarefree, with $\deg f <$
$\deg g = n$ and $\|f\|_\infty, \|g\|_\infty < 2^\lambda$.

Output: Polynomials $r_1, \dots, r_n \in \mathbb{Z}[x]$ and $v_1, \dots, v_n \in \mathbb{Z}[x, y]$ such that the r_i are
squarefree and pairwise coprime, $\operatorname{res}(g, f - yg') = c \prod_{1 \le i \le n} r_i^i$ for some nonzero
$c \in \mathbb{Z}$, and

$$\frac{f}{g} = \sum_{1 \le i \le n} \sum_{r_i(\gamma)=0} \gamma \frac{v_i(x, \gamma)'}{v_i(x, \gamma)} .$$

1. $C \longleftarrow (2n)!(n+1)^{2n} 2^{2n\lambda}, \quad k \longleftarrow \lceil \log_2(2C)/(\omega - 1) \rceil,$
   $s \longleftarrow \lceil \log_2(2\lambda)/(\omega - 1) \rceil$
   choose a set S_0 of $2k + s$ single precision primes
   $S_1 \longleftarrow \{p \in S_0 : p \nmid \operatorname{lc}_x(fg)\}$
2. **for** $p \in S_1$ **do**
3.   **call** Algorithm 7.2 with input g, $f - yg'$, p, and 1, to compute $\sigma_p \in \mathbb{Z}[y]$
         with $\deg \sigma_p \le n$, $\|\sigma_p\|_\infty < p/2$, and $\sigma_p \equiv \operatorname{res}_x(g, f - yg') \bmod p$
         if 0 does not occur in the degree sequence of the output
         then $\sigma_p \longleftarrow 0$
4. use the Chinese Remainder Algorithm to compute $\sigma \in \mathbb{Z}[y]$ with $\|\sigma\|_\infty <$
   $\frac{1}{2} \prod_{p \in S_1} p$ and $\sigma \equiv \sigma_p \bmod p$ for all $p \in S_1$
5. **call** Algorithm 5.6 to compute the normalized squarefree decomposition
   $\operatorname{normal}(\sigma) = \prod_{1 \le i \le n} r_i^i$ of σ
6. **for** $p \in S_1$ **do**
7.   **for** $1 \le i \le n$ such that $\deg r_i > 0$ **do**
8.   **call** Algorithm 7.2 with input g, $f - yg'$, p, and 1, to compute
               $v_{i,p} \in \mathbb{Z}[x, y]$ with $\deg_y v_{i,p} \le n - i$, $\deg_x v_{i,p} \le i$, $\|v_{i,p}\|_\infty <$
               $p/2$, and such that $\operatorname{lc}_x(v_{i,p}) \bmod p$ is the ith subresultant and
               $v_{i,p}/\operatorname{lc}_x(v_{i,p}) \bmod p$ is the monic remainder of degree i in the
               EEA of $g \bmod p$ and $f - yg' \bmod p$ with respect to x
               if i does not occur in the degree sequence **then** $v_{i,p} \longleftarrow 0$
9. **for** $1 \le i \le n$ **do**
10.   **if** $\deg r_i = 0$ **then** $v_i \longleftarrow 1$
         else
               $T_i \longleftarrow \{p \in S_1 : v_{i,p} \ne 0\}$
               use the Chinese Remainder Algorithm to compute $v_i \in \mathbb{Z}[x, y]$
               with $\|v_{i,p}\|_\infty < \frac{1}{2} \prod_{p \in T_i} p$ and $v_i \equiv v_{i,p} \bmod p$ for all $p \in T_i$
11. **return** $r_1, \dots, r_n$ and $v_1, \dots, v_n$

Theorem 7.30 (Correctness of Algorithm 7.29). *If there are at least $2k + s \in \Theta(n(\lambda + \log n))$ single precision primes, then Algorithm 7.29 solves the problem 7.28 of rational integration with squarefree denominator correctly as specified.*

Proof. Let $r = \text{res}_x(g, f - yg') \in \mathbb{Z}[y]$. With the modifications mentioned in Remark 7.6, the calls to Algorithm 7.2 in steps 3 and 8 of Algorithm 7.29 do not return "FAIL". As in the proof of Theorem 7.26, we find that S_1 contains at least $2k$ single precision primes, and that $r = \sigma$ in step 4.

Now let $1 \leq i \leq n$ be such that $\deg r_i > 0$, let $\tau_i \in \mathbb{Z}[y] \setminus \{0\}$ be the ith subresultant of g and $f - yg'$, and let $\pi_i \in \mathbb{Q}(y)[x]$ be the monic remainder of degree i in the EEA of g and $f - yg'$ with respect to x. By the discussion preceding the algorithm, all we have to show for the correctness is that $v_i = \tau_i \pi_i$.

Let $p \in S_1$. Lemma 3.11 (i) implies that i occurs in the degree sequence of the EEA of $g \bmod p$ and $f - yg' \bmod p$ if and only if $p \nmid \tau_i$, and in that case we have $v_{i,p} \equiv \tau_i \pi_i \bmod p$. Otherwise, if $p \mid \tau_i$, we have $v_{i,p} = 0$ after step 8. Thus $T_i = \{p \in S_1 : p \nmid \tau_i\}$, $v_{i,p} \equiv \tau_i \pi_i \bmod p$ for all $p \in T_i$, and hence $v_i \equiv \tau_i \pi_i \bmod m$, where $m = \prod_{p \in T_i} p$. Now all primes in $S_1 \setminus T_i$ divide the leading coefficient of τ_i, which has absolute value at most C, by Corollary 3.2, and hence T_i contains at least k primes. The polynomial $\tau_i \pi_i$ is equal to the ith "subresultant" of g and $f - yg'$ in the sense of Collins (1966, 1967) and Brown & Traub (1971) (see, e.g., Proposition 3.11 in von zur Gathen & Lücking 2002). Therefore each coefficient of $\tau_i \pi_i$ can be written as the determinant of a certain square $(n + \deg_x(f - yg') - 2i)$-dimensional submatrix of the Sylvester matrix of g and $f - yg'$, and Lemma 3.1 (iii) implies that $\|\tau_i \pi_i\|_\infty \leq C < m/2$. Thus both sides of the congruence $v_i \equiv \tau_i \pi_i \bmod m$ are absolutely less than $m/2$, and hence they are equal. $\square$

Theorem 7.31 (Cost of Algorithm 7.29). *The cost for steps 2 through 11 of Algorithm 7.29 is $O(n^4(\lambda^2 + \log^2 n))$ word operations with classical arithmetic and $O(n^2 \mathsf{M}(n(\lambda + \log n)) \log(n\lambda))$ or $O^\sim(n^3 \lambda)$ with fast arithmetic.*

Proof. As in the proof of Theorem 7.26, the cost for steps 2–5 is $O(n^3 k + nk^2)$ word operations with classical arithmetic and $O(kn \mathsf{M}(n) \log n + n \mathsf{M}(k) \log k)$ with fast arithmetic.

In steps 7 and 8, we modify Algorithm 7.2 so as to compute only those intermediate results that are needed. Since $\sum_{\deg r_i > 0} \deg r_i \leq \deg \sigma = n$, the total number of coefficients in $\mathbb{Z}[y]$ that have to be reconstructed is only $O(n)$, and not $O(n^2)$, as for the coefficients of *all* intermediate results. With classical arithmetic, the cost estimate for this is the same as for computing one row in the EEA, namely $O(n^3)$ word operations, by the proof of Theorem 7.7.

With fast arithmetic, we proceed as follows. Let $n \geq d_1 > d_2 > \cdots > d_t \geq 1$ be such that $\{d_1, \ldots, d_t\} = \{i : 1 \leq i \leq n \text{ and } \deg r_i > 0\}$. We may assume that we have already computed all subresultants in step 3 of Algorithm 7.29; the additional cost for this is negligible. In step 4 of Algorithm 7.2, we use the fast Euclidean Algorithm to compute first π_{d_1} and the subsequent remainder of degree less than d_1. Then we call the fast Euclidean Algorithm with these two polynomials as input to compute π_{d_2} and the subsequent remainder of degree less than d_2, and so on. In this way, step 4 of Algorithm 7.2 takes

$$O(\mathsf{M}(n)\log n + \mathsf{M}(d_1)\log d_1 + \mathsf{M}(d_2)\log d_2 + \cdots)$$

word operations. Since $d_1 + d_2 + \cdots \le n$, this yields a total of $O(\mathsf{M}(n)\log n)$. This is the same estimate as for computing only one specific row in the EEA, and hence the cost estimate for steps 7 and 8 of Algorithm 7.29 with fast arithmetic is the same as the cost estimate for computing only one row in the EEA of g and $f - yg'$, namely $O(n\,\mathsf{M}(n)\log n)$ word operations, by Corollary 7.9.

Thus the overall cost for steps 6 through 8 is $O(kn^3)$ word operations with classical arithmetic and $O(kn\,\mathsf{M}(n)\log n)$ with fast arithmetic. Finally, the cost for reconstructing one coefficient in $\mathbb{Z}$ of some v_i is $O(k^2)$ word operations with classical arithmetic and $O(\mathsf{M}(k)\log k)$ with fast arithmetic. Since $\deg_x v_i = i$ and $\deg_y v_i \le n - i$, by Lemma 3.2 (ii), we have at most $(i+1)(n-i+1)$ coefficients for each i with $\deg r_i > 0$, and the total number of coefficients is

$$\sum_{\deg r_i > 0} (i+1)(n-i+1) \le n \sum_{\deg r_i > 0} (i+1) \le 2n^2 \ .$$

Thus steps 9 and 10 take $O(n^2 k^2)$ word operations with classical arithmetic and $O(n^2\,\mathsf{M}(k)\log k)$ with fast arithmetic. The claims follow from $k \in O(n(\lambda + \log n))$ by adding up costs. $\square$

All r_i together have $O(n)$ coefficients in $\mathbb{Z}$, of word size $O(n(\lambda + \log n))$, by Mignotte's bound 3.3. The proof above shows that all v_i together have $O(n^2)$ coefficients in $\mathbb{Z}$, each of word size $O(n(\lambda + \log n))$. Thus the output size of Algorithm 7.29 is $O(n^3(\lambda + \log n))$ words, and the algorithm with fast arithmetic is – up to logarithmic factors – asymptotically optimal for those inputs where the output size is close to the upper bound.

The final result of this chapter is a cost estimate for a modular variant of the complete algorithm for symbolic integration of rational functions.

Theorem 7.32. *Let $f, g \in \mathbb{Z}[x]$ be nonzero polynomials with $\deg f, \deg g \le n$ and $\|f\|_\infty, \|g\|_\infty < 2^\lambda$. We can compute a symbolic integral of the rational function f/g using $O(n^8 + n^6\lambda^2)$ word operations with classical arithmetic and $O(n^2\,\mathsf{M}(n^3 + n^2\lambda)\log(n\lambda))$ or $O^\sim(n^5 + n^4\lambda)$ with fast arithmetic.*

Proof. The algorithm works as follows. We first compute the squarefree decomposition $g = \prod_{1 \le i \le n} g_i^i$ of g, taking $O(n^3 + n\lambda^2)$ word operations with classical arithmetic and $O(n\,\mathsf{M}(n + \lambda)\log(n + \lambda) + \lambda\,\mathsf{M}(n)\log n)$ with fast arithmetic, by Theorem 5.10.

Then we call Algorithm 6.4 to perform the Hermite reduction, leaving the task to compute symbolic integrals of a_i/g_i for $1 \le i \le n$, where $\deg a_i < \deg g_i$ and a_i is of the form p_i/q_i for some $p_i \in \mathbb{Z}[x]$ and $q_i \in \mathbb{Z}$ with coefficients of word size $O(n^2 + n\lambda)$, by Lemma 6.3. The p_i and q_i can be computed from the a_i via $O(n)$ divisions and gcd computations in $\mathbb{Z}$, on integers of word size $O(n^2 + n\lambda)$, and the cost for this is dominated by the cost for the other steps. The cost for the Hermite reduction is $O(n^5 + n^3\lambda^2)$ word operations with classical arithmetic and $O(n \, \mathsf{M}(n^2 + n\lambda) \log(n\lambda))$ with fast arithmetic, by Theorem 6.6.

Finally, for $1 \leq i \leq n$ we call Algorithm 7.29 with input p_i and $q_i g_i$. If we let $d_i = \deg g_i$, then Theorem 7.31, with n and λ replaced by d_i and $O(n^2 + n\lambda)$, respectively, implies that this takes $O(d_i^4(n^4 + n^2\lambda^2))$ word operations with classical arithmetic and $O(d_i^2 \, \mathsf{M}(d_i(n^2 + n\lambda)) \log(n\lambda))$ with fast arithmetic, for each i. Since $\sum_{1 \leq i \leq n} d_i \leq n$, the overall cost for this is $O(n^8 + n^6\lambda^2)$ with classical arithmetic and $O(n^2 \, \mathsf{M}(n^3 + n^2\lambda) \log(n\lambda))$ with fast arithmetic. This dominates the cost for the other steps. $\square$

The algorithm described above is probabilistic, and the cost estimate ignores the cost for prime finding and assumes that there are sufficiently many single precision primes. There may be better bounds for the coefficient sizes of the output than the ones implied by the proof above.

8. Modular Algorithms for the Gosper-Petkovšek Form

Let F be a field. The *shift operator* E acts on polynomials in $F[x]$ via $(Ef)(x) = f(x + 1)$. For $i \in \mathbb{Z}$, the ith power of E acts via $(E^i f)(x) = f(x + i)$.

Definition 8.1. *Let F be a field and $f, g \in F[x] \setminus \{0\}$ be polynomials. A triple $(a, b, c) \in F[x]^3$ is called a* Gosper form *of the rational function f/g if*

$$\frac{f}{g} = \frac{a}{b}\frac{Ec}{c} \text{ and } \gcd(a, E^i b) = 1 \text{ for } i \in \mathbb{N} .$$

If in addition

$$\gcd(a, c) = \gcd(b, Ec) = 1 ,$$

then (a, b, c) is a Gosper-Petkovšek form *of f/g.*

Gosper (1978) has first given the algorithm below to compute a Gosper form. Petkovšek (1992) has shown that it even computes a Gosper-Petkovšek form, and that the latter is unique up to multiplication by constants: if (A, B, C) is another Gosper-Petkovšek form for f/g, then $\mathrm{lc}(A)a = \mathrm{lc}(a)A$, $\mathrm{lc}(B)b = \mathrm{lc}(b)B$, and $\mathrm{lc}(C)c = \mathrm{lc}(c)C$. See also Paule (1995) and Petkovšek, Wilf & Zeilberger (1996), Chap. 5.

Abramov & Petkovšek (2001, 2002a) extend the definition of the Gosper-Petkovšek form and define a *strict rational normal form* of a rational function $f/g \in \mathbb{Q}(x)$ as a quintuple $(z, r, s, u, v) \in F \times F[x]^4$, such that $\gcd(r, E^i s) = 1$ for all $i \in \mathbb{Z}$, $\gcd(r, u \cdot Ev) = \gcd(s, (Eu)v) = 1$, the polynomials r, s, u, v are monic, and

$$\frac{f}{g} = z\frac{r}{s}\frac{E(u/v)}{u/v} .$$

In contrast to the Gosper-Petkovšek form, however, this normal form is not unique; see Abramov & Petkovšek (2001) for examples.

Algorithm 8.2 (Gosper-Petkovšek form).
Input: Nonconstant polynomials $f, g \in \mathbb{Z}[x]$ of degree at most n and max-norm less than 2^λ.
Output: A Gosper-Petkovšek form of f/g.

1. compute a list $z_1 < \ldots < z_t < 2^{\lambda+2}$ of nonnegative integers, with $t \le n^2$, containing all integers $z \in \mathbb{N}$ such that $\gcd(f, E^z g)$ is nonconstant

J. Gerhard: Modular Algorithms, LNCS 3218, pp. 121–148, 2004.
© Springer-Verlag Berlin Heidelberg 2004

2. $f_0 \longleftarrow \mathrm{normal}(f), \quad g_0 \longleftarrow \mathrm{normal}(g)$
3. **for** $1 \leq i \leq t$ **do**
4. $\qquad h_i \longleftarrow \gcd(f_{i-1}, E^{z_i} g_{i-1}), \quad f_i \longleftarrow \dfrac{f_{i-1}}{h_i}, \quad g_i \longleftarrow \dfrac{g_{i-1}}{E^{-z_i} h_i}$
5. **return** $a = \mathrm{lu}(f) f_t$, $b = \mathrm{lu}(g) g_t$, and $c = E^{-1}(h_1^{\frac{z_1}{2}} \cdots h_t^{\frac{z_t}{2}})$

Correctness of the algorithm has been established in the literature. It is a crucial step in Gosper's hypergeometric summation algorithm. Clearly, the algorithm may be stopped as soon as $f_i = 1$ or $g_i = 1$ in step 4.

We now discuss two new modular versions of Algorithm 8.2. Both use the modular Man & Wright algorithm 7.20 in step 1. The first variant uses reduction modulo several single precision primes and the Chinese Remainder Theorem for each gcd in step 4 separately. The second variant is an all-modular approach, where f_0 and g_0 are reduced modulo several primes only once before step 3 and steps 3 and 4 are executed independently modulo each prime. This variant is advantageous since we need not compute the intermediate results $f_1, g_1, \ldots, f_{t-1}, g_{t-1}$.

Problem 8.3 (Shift gcd). *Given two nonzero polynomials $f, g \in \mathbb{Z}[x]$ of degree at most n and $z \in \mathbb{Z}$, compute*

$$ h = \gcd(f, E^z g), \quad w = E^{-z} h, \quad u = \frac{f}{h}, \quad \text{and} \quad v = \frac{g}{w}. $$

Problem 8.4 (Gosper-Petkovšek form). *Given two nonzero normalized polynomials $f = f_0$ and $g = g_0$ in $\mathbb{Z}[x]$ of degree at most n and a list $z_1 < z_2 < \cdots < z_t$ of $t \leq n^2$ nonnegative integers containing all $z \in \mathbb{N}$ such that $\gcd(f, E^z g)$ is nonconstant, compute*

$$ h_i = \gcd(f_{i-1}, E^{z_i} g_{i-1}), \quad f_i = \frac{f_{i-1}}{h_i}, \quad g_i = \frac{g_{i-1}}{E^{-z_i} h_i} $$

for $1 \leq i \leq t$.

We start by analyzing a modular algorithm that computes $\gcd(f, E^z g)$ for given polynomials $f, g \in \mathbb{Z}[x]$ and $z \in \mathbb{Z}$. The algorithm is an extension of the well-known modular gcd algorithm appearing in Brown (1971) (see also Algorithm 6.38 in von zur Gathen & Gerhard 1999), which essentially corresponds to the case $z = 0$. In addition to the gcd h, our algorithm also computes $w = E^{-z} h$ and the cofactors f/h and g/w. In order to obtain a Las Vegas algorithm, it is necessary to check that $w = E^{-z} h$. In general, the word size of the coefficients of $E^{-z} h$ is larger than the word size of the coefficients of h by a factor of $O(\deg h)$, by Lemma 4.2. The following lemma says that if both h and $E^{-z} h$ have "small" coefficients, then z is "small" as well.

Lemma 8.5. *Let $h \in \mathbb{Z}[x]$ be nonzero of degree d, $z \in \mathbb{Z}$, $w = E^{-z} h$, and $\|h\|_2, \|w\|_2 \leq B$. Then $|z|^d \leq 2^d B^2$.*

Proof. We may assume that $d \geq 1$ and $|z| \geq 2$, and write $h = c \prod_{1 \leq i \leq d}(x - a_i)$, with $c = \mathrm{lc}(h) \in \mathbb{Z}$ and all $a_i \in \mathbb{C}$. We partition the roots of h into those of "big" and those of "small" absolute value, by letting $I_1 = \{1 \leq i \leq d : |a_i| \geq |z|/2\}$ and $I_2 = \{1 \leq i \leq d : |a_i| < |z|/2\}$. Moreover, we employ the *Mahler measure* (named after Mahler 1960) of h defined by $M(h) = |c| \prod_{1 \leq i \leq d} \max\{1, |a_i|\}$. Using Landau's (1905) inequality (Theorem 6.31 in von zur Gathen & Gerhard 1999), we find

$$B \geq \|h\|_2 \geq M(h) \geq \prod_{i \in I_1} \max\{1, |a_i|\} \geq \left(\frac{|z|}{2}\right)^{\#I_1}.$$

The roots of w are $a_i + z$ for $1 \leq i \leq d$, and hence

$$
\begin{aligned}
B &\geq \|w\|_2 \geq M(w) = |c| \prod_{1 \leq i \leq d} \max\{1, |a_i + z|\} \geq \prod_{i \in I_2} \max\{1, |a_i + z|\} \\
&> \left(\frac{|z|}{2}\right)^{\#I_2}.
\end{aligned}
$$

Now one of I_1 and I_2 has at least $d/2$ elements, so that $B \geq (|z|/2)^{d/2}$, and the claim follows. $\square$

In the following algorithm, we assume that ω is the word size of our processor, as usual.

Algorithm 8.6 (Small primes modular shift gcd). ▬▬▬▬▬▬▬▬▬▬▬
Input: Nonconstant normalized polynomials $f, g \in \mathbb{Z}[x]$ of degrees at most n and
  max norm less than 2^λ, and $z \subset \mathbb{Z}$ with $|z| < 2^{\lambda+2}$.
Output: Normalized polynomials $h, u, v, w \in \mathbb{Z}[x]$ such that $h = \gcd(f, E^z g)$,
  $w = E^{-z}h$, $f = uh$, and $g = vw$.

1. $c \longleftarrow \gcd(\mathrm{lc}(f), \mathrm{lc}(g))$, $B \longleftarrow \lfloor (n+1)^{1/2} 2^{n+\lambda} \rfloor$
   $k \longleftarrow \lceil \log_2(2^{2n+1} cB^3)/(\omega - 1) \rceil$
2. choose a set S_0 of $2k$ odd single precision primes
   $S_1 \longleftarrow \{p \in S_0 : p \nmid c\}$
3. **for** all $p \in S_1$ compute the polynomials $h_p, u_p, v_p, w_p \in \mathbb{Z}[x]$ of max-norms
   less than $p/2$ and with $\mathrm{lc}(u_p) \equiv \mathrm{lc}(f) \bmod p$, $\mathrm{lc}(v_p) \equiv \mathrm{lc}(g) \bmod p$, and
   $\mathrm{lc}(h_p) \equiv \mathrm{lc}(w_p) \equiv c \bmod p$, such that $c^{-1}h_p \bmod p$ is the monic gcd of
   $f \bmod p$ and $E^z g \bmod p$, $w_p \equiv E^{-z}h_p \bmod p$, $cf \equiv u_p h_p \bmod p$, and
   $cg \equiv v_p w_p \bmod p$
4. $d \longleftarrow \min\{\deg h_p : p \in S_1\}$, $S_2 \longleftarrow \{p \in S_1 : \deg h_p = d\}$
   if $\#S_2 \geq k$ and $|z|^d \leq 2^d B^2$
   then remove $\#S_2 - k$ primes from S_2 **else goto** 2
5. use the Chinese Remainder Algorithm to compute $h^*, u^*, v^*, w^* \in \mathbb{Z}[x]$ of
   max-norm less than $(\prod_{p \in S_2} p)/2$ such that

$$u^* \equiv u_p \bmod p, \quad v^* \equiv v_p \bmod p, \quad h^* \equiv h_p \bmod p, \quad w^* \equiv w_p \bmod p$$

holds for all $p \in S_2$

6. **if** $\|u^*\|_1\|h^*\|_1 \le cB$ and $\|v^*\|_1\|w^*\|_1 \le cB$
   then return normal(h^*), normal(u^*), normal(v^*), normal(w^*)
   else goto 2

Definition 8.7. *Let $f, g \in \mathbb{Z}[x]$ be nonzero and $z \in \mathbb{Z}$. A prime $p \in \mathbb{N}$ is* lucky *with respect to the shift gcd problem 8.3, if $p \nmid \gcd(\mathrm{lc}(f), \mathrm{lc}(g))$ and $\deg \gcd(f, E^z g) = \deg \gcd(f \bmod p, E^z g \bmod p)$.*

Theorem 8.8 (Correctness of Algorithm 8.6). *The algorithm terminates if and only if at most $k \in \Theta(n + \lambda)$ primes in S_0 are unlucky with respect to the shift gcd problem, and in that case, it returns the correct results.*

Proof. We first show that the condition in step 6 is false if and only if $u^* h^* = cf$ and $v^* w^* = cg$. The "if" part follows from Mignotte's bound (Fact 3.3). For the converse, let $m = \prod_{p \in S_2} p$. Then $u^* h^* \equiv cf \bmod m$, and

$$\|u^* h^*\|_\infty \le \|u^* h^*\|_1 \le \|u^*\|_1\|h^*\|_1 \le cB \le 2^{(\omega-1)k-1} < \frac{m}{2}$$

implies that the coefficients of both sides of the congruence are less than $m/2$ in absolute value, so that they are equal. Similarly, we find that $v^* w^* = cg$.

Let h be the normalized gcd of f and $E^z g$ in $\mathbb{Z}[x]$ and $w = E^{-z} h$. We note that $\mathrm{lc}(h) = \mathrm{lc}(w)$ divides both $\mathrm{lc}(f)$ and $\mathrm{lc}(g)$, so that it divides c, and let $u = f/h$ and $v = g/w$. Since f, g, h are normalized, so are u, v, w.

Assume now first that at least k primes in S_0 are lucky. Then these primes are contained in S_1 as well. It is clear that $\deg h_p \ge \deg h$ for all $p \in S_1$, and hence also $d \ge \deg h$. By definition, we have $\deg h_p = \deg h$ for all lucky primes in S_1, and therefore $d = \deg h$ and $\#S_2 \ge k$. Since $h \mid f$ and $E^{-z}h \mid g$, Mignotte's bound (Fact 3.3) implies that $\|h\|_2, \|E^{-z}h\|_2 \le B$, and Lemma 8.5 yields $|z|^d \le 2^d B^2$. By Corollary 3.12, we have $h_p \equiv (c/\mathrm{lc}(h))h \bmod p$ for all $p \in S_2$, and hence $h^* \equiv (c/\mathrm{lc}(h))h \bmod m$, $w^* \equiv (c/\mathrm{lc}(h))E^{-z}h \bmod m$, $u^* \equiv (\mathrm{lc}(f)/\mathrm{lc}(u))u \bmod m$, and $v^* \equiv (\mathrm{lc}(g)/\mathrm{lc}(v))v \bmod m$. The left hand sides of these congruences have max-norms less than $m/2$ by construction, the right hand sides enjoy the same property by Mignotte's bound, and hence all four congruences are in fact equalities. However, then $u^* h^* = cf$ and $v^* w^* = cg$, and the condition in step 6 is false.

Conversely, if the condition in step 6 is false, the initial discussion shows that $u^* h^* = cf$ and $v^* w^* = cg$. Mignotte's bound (Fact 3.3) yields $\|h^*\|_1, \|w^*\|_1 \le cB$, and we have

$$\begin{aligned}
\|E^{-z}h^*\|_\infty &= \|h^*(x - z)\|_\infty \le (|z| + 1)^d\|h^*\|_1 \le |2z|^d\|h^*\|_1 \\
&\le 2^{2d}cB^3 \le 2^{(\omega-1)k-1} < m/2 \,,
\end{aligned}$$

if $z \ne 0$, by Lemma 4.2 and the condition in step 4, and trivially $\|E^{-z}h^*\|_\infty \le cB < m/2$ if $z = 0$. Now $w^* \equiv E^{-z}h^* \bmod m$, both sides of the congruence have max-norms less than $m/2$, and hence they are equal. Thus h^* is a common divisor of cf and $cE^z g$, so that $d = \deg h^* \leq \deg h$, and in fact $d = \deg h$ and $h^* = (c/\mathrm{lc}(h))h$. This also implies that $w^* = (c/\mathrm{lc}(h))E^{-z}h$, $u^* = (\mathrm{lc}(f)/\mathrm{lc}(u))u$, and $v^* = (\mathrm{lc}(g)/\mathrm{lc}(v))v$. Moreover, we have $\deg h = \deg h^* = \deg h_p$ for all $p \in S_2$, so that at least k of the initial primes in S_0 are lucky. Finally, we find that $h = \mathrm{normal}(h) = \mathrm{normal}((c/\mathrm{lc}(h))h) = \mathrm{normal}(h^*)$, and similarly also $u = \mathrm{normal}(u^*)$, $v = \mathrm{normal}(v^*)$, and $w = \mathrm{normal}(w^*)$. $\square$

Theorem 8.9 (Success probability of Algorithm 8.6). *Let*

$$C = (n+1)^{3n/2} 2^{n(n\lambda + 2n + 2\lambda)} \text{ and } s = \lfloor (\log_2 C)/(\omega - 1) \rfloor \in \Theta(n^2 \lambda) .$$

The number of single precision primes that are unlucky with respect to the shift gcd problem 8.3 is at most s. Suppose that $k \leq s$ and that the number of single precision primes, between $2^{\omega-1}$ and 2^ω, is at least $2s$. If we choose the set S_0 in step 2 uniformly at random from among at least $2s$ single precision primes, then the expected number of iterations of the algorithm is at most two.

Proof. Let h be the normalized gcd of f and $E^z g$ in $\mathbb{Z}[x]$ and $\sigma \in \mathbb{Z}$ the $(\deg h)$th subresultant of f and $E^z g$. Exercise 6.41 in von zur Gathen & Gerhard (1999) and Corollary 3.12 imply that a prime $p \in \mathbb{N}$ is unlucky if and only if it divides σ.

By Theorem 8.8, the choice of S_0 in step 2 is successful if and only if at least k primes in S_0 do not divide σ. By Lemma 3.9 (i), σ is a nonzero integer. We have $\|f\|_2 < (n+1)^{1/2} 2^\lambda$ and

$$\|E^z g\|_2 = \|g(x+z)\|_2 \leq (|z|+1)^n \|g\|_1 < (n+1)2^{n\lambda + 2n + \lambda} ,$$

by Lemma 4.2, and Corollary 3.2 implies that $0 < |\sigma| \leq C$. Thus the number of single precision primes dividing σ, i.e., the number of unlucky single precision primes, is at most s. Since there are at least twice as many of them, the probability that least half of the primes in S_0 are lucky is at least $1/2$, by Lemma 3.24. $\square$

Theorem 8.10 (Cost of Algorithm 8.6). *The cost for one execution of steps 2–6 of Algorithm 8.6 is $O(n^3 + n\lambda^2)$ word operations with classical arithmetic and $O(\lambda \,\mathsf{M}(n) \log n + n \,\mathsf{M}(n + \lambda) \log(n + \lambda))$ or $O^\sim(n^2 + n\lambda)$ with fast arithmetic.*

Proof. The cost for reducing $\mathrm{lc}(f)$ and $\mathrm{lc}(g)$ modulo all primes in S_0 in step 2 is dominated by the cost for step 3. There, we first reduce f, g, and z modulo all primes in S_1. This takes $O(nk\lambda)$ word operations with classical arithmetic and $O(n \,\mathsf{M}(k) \log k)$ with fast arithmetic. Then, for each p, we compute $g(x+z) \bmod p$, taking $O(n^2)$ word operations with classical arithmetic and $O(\mathsf{M}(n))$ with fast arithmetic, by Theorems 4.3 and 4.5. The cost for the EEA to calculate h_p and for computing u_p, v_p, w_p is $O(n^2)$ word operations with classical arithmetic and $O(\mathsf{M}(n) \log n)$ with fast arithmetic. Thus the overall cost of step 3 is $O(kn(n+\lambda))$ word operations with classical arithmetic and $O(n \,\mathsf{M}(k) \log k + k \,\mathsf{M}(n) \log n)$ with fast arithmetic. The Chinese remaindering in step 5 takes $O(k^2)$ word operations per coefficient with classical arithmetic and $O(\mathsf{M}(k) \log k)$ with fast arithmetic. There are $O(n)$ coefficients, and hence the cost for step 5 is $O(nk^2)$ with classical arithmetic and $O(n \,\mathsf{M}(k) \log k)$ with fast arithmetic. Finally, in step 6 we compute $O(n)$

gcd's of k-word integers, each taking $O(k^2)$ or $O(\mathsf{M}(k) \log k)$ with classical or fast arithmetic, respectively, and this cost is dominated by the cost for step 5. Thus the overall cost for one pass through steps 2 through 6 is $O(nk(n+k))$ word operations with classical arithmetic and $O(k \mathsf{M}(n) \log n + n \mathsf{M}(k) \log k)$ with fast arithmetic. Now the claims follow from $k \in O(n + \lambda)$. □

We note that Algorithm 8.6 is more efficient than the direct way of computing $E^z g$ and using a standard modular algorithm to compute the gcd, since the coefficients of $E^z g$ are in general much larger than the coefficients of the output of Algorithm 8.6, by Lemma 4.2 and the discussion following it.

Corollary 8.11. *Let f, g, z be as in Algorithm 8.6, and assume that $\gcd(f, E^z g)$ is constant. Under the assumptions of Theorem 8.9, Algorithm 8.6 can detect this with an expected number of $O(n^2 + n\lambda)$ word operations with classical arithmetic and $O(n \mathsf{M}(n + \lambda) \log(n + \lambda))$ or $O^\sim(n^2 + n\lambda)$ with fast arithmetic.*

Proof. Let σ be as in Theorem 8.8. We know that the gcd is constant and we may stop the algorithm as soon as h_p is constant for some $p \in S_1$, which happens if and only if $p \nmid \sigma$. If at most half of the single precision primes to choose from divide σ, as in Theorem 8.9, then we expect this to happen after at most two primes. Thus with classical arithmetic, the expected cost is $O(n^2 + n\lambda)$ word operations. With fast arithmetic, we still have the cost of $O(n \mathsf{M}(k) \log k)$ for reducing z and all coefficients of f and g simultaneously modulo all primes in S_1, but the expected cost until we find $\deg h_p = 0$ is only $O(\mathsf{M}(n) \log n)$. □

It is surprising that the cost estimate in Corollary 8.11 for classical arithmetic is *smaller* than for fast arithmetic. In practice, one would first reduce f, g, and z modulo some small number of the primes in S_0, using classical arithmetic, and then compute the gcd of f and $E^z g$ modulo these small primes. The additional cost involved is then $O(n^2 + n\lambda)$, and with high probability, at least one of the modular gcd's will be constant.

Before analyzing Algorithm 8.2 when step 4 is performed by Algorithm 8.6, we present a variant for steps 3 and 4 where modular reduction and Chinese remaindering takes place only once.

Algorithm 8.12 (Small primes modular Gosper-Petkovšek form).
Input: Nonconstant normalized polynomials $f, g \in \mathbb{Z}[x]$ with $\deg g \leq \deg f \leq n$ and $\|f\|_\infty, \|g\|_\infty < 2^\lambda$, and distinct integers $z_1, \ldots, z_t \in \mathbb{Z}$ of absolute value less than $2^{\lambda+2}$, where $t \leq n^2$.
Output: Normalized polynomials $f_t, g_t, h_1, \ldots, h_t \in \mathbb{Z}[x]$ with $f = h_1 \cdots h_t \cdot f_t$, $g = (E^{-z_1} h_1) \cdots (E^{-z_t} h_t) \cdot g_t$, and

$$h_i = \gcd\left(\frac{f}{h_1 \cdots h_{i-1}}, E^{z_i}\left(\frac{g}{(E^{-z_1} h_1) \cdots (E^{-z_{i-1}} h_{i-1})}\right)\right) \quad (8.1)$$

for $1 \leq i \leq t$, or "FAIL".

1. $c \longleftarrow \gcd(\mathrm{lc}(f), \mathrm{lc}(g)), \quad B \longleftarrow \lfloor (n+1)^{1/2} 2^{n+\lambda} \rfloor$
   $k \longleftarrow \lceil \log_2(2^{2n+\lambda+1} B^3)/(\omega - 1) \rceil$
2. choose a set S of $2k$ odd single precision primes
   $S_0 \longleftarrow \{p \in S : p \nmid \mathrm{lc}(fg)\}$
3. **for all** $p \in S_0$ **do**
   $$f_{0,p} \longleftarrow f \text{ rem } p, \quad g_{0,p} \longleftarrow g \text{ rem } p$$
4. **for** $1 \leq i \leq t$ **do**
5.   **for all** $p \in S_{i-1}$ **do**
6.   compute polynomials $f_{i,p}, g_{i,p}, h_{i,p}$ in $\mathbb{Z}[x]$ of max-norm less than $p/2$ such that $h_{i,p} \bmod p$ is the monic gcd of $f_{i-1,p} \bmod p$ and $E^{z_i} g_{i-1,p} \bmod p$ in $\mathbb{F}_p[x]$, $f_{i-1,p} \equiv h_{i,p} f_{i,p} \bmod p$, and $g_{i-1,p} \equiv (E^{-z_i} h_{i,p}) g_{i,p} \bmod p$
7. $d_i \longleftarrow \min\{\deg h_{i,p} : p \in S_{i-1}\}$
           if $|z_i|^{d_i} > 2^{d_i} B^2$ **then return** "FAIL"
8. $S_i \longleftarrow \{p \in S_{i-1} : \deg h_{i,p} = d_i\}$
           if $\#S_{i-1} < k$ **then return** "FAIL"
9.   use the Chinese Remainder Algorithm to compute $h_i^*, w_i^* \in \mathbb{Z}[x]$ of max-norm less than $(\prod_{p \in S_2} p)/2$ such that $h_i^* \equiv c h_{i,p} \bmod p$ and $w_i^* \equiv c \cdot E^{-z_i} h_{i,p} \bmod p$ for all $p \in S_i$
   $$H_i \longleftarrow \mathrm{normal}(h_i^*), \quad W_i \longleftarrow \mathrm{normal}(w_i^*)$$
10. use the Chinese Remainder Algorithm to compute $f_t^*, g_t^* \in \mathbb{Z}[x]$ of max-norm less than $(\prod_{p \in S_2} p)/2$ such that $f_t^* \equiv f_{t,p} \bmod p$ and $g_t^* \equiv g_{t,p} \bmod p$ for all $p \in S_t$
    $$F_t \longleftarrow \mathrm{normal}(f_t^*), \quad G_t \longleftarrow \mathrm{normal}(g_t^*)$$
11. **if** $\mathrm{lc}(H_1 \cdots H_t \cdot F_t) \neq \mathrm{lc}(f)$ or $\mathrm{lc}(W_1 \cdots W_t \cdot G_t) \neq \mathrm{lc}(g)$
    then return "FAIL"
12. **if** $\|H_1\|_1 \cdots \|H_t\|_1 \cdot \|F_t\|_1 \leq B$ and $\|W_1\|_1 \cdots \|W_t\|_1 \cdot \|G_t\|_1 \leq B$
    then return $F_t, G_t, H_t, \ldots, H_t$ **else return** "FAIL"

Definition 8.13. Let $f, g \in \mathbb{Z}[x] \setminus \{0\}$ be normalized, $r = \mathrm{res}_x(f(x), g(x+y)) \in \mathbb{Z}[y]$, and $r_1 = r \cdot \prod_{1 \leq i \leq t}(y - z_i)$. We say that a prime $p \in \mathbb{N}$ is lucky with respect to the Gosper-Petkovšek form problem 8.4 if $p \nmid \mathrm{lc}(fg)$ and $\deg \gcd(r_1, r_1') = \deg \gcd(r_1 \bmod p, r_1' \bmod p)$.

The leading coefficient of r and r_1 is $\mathrm{lc}(f)^{\deg g} \mathrm{lc}(g)^{\deg f}$, by Fact 7.11 (ii). Together with Lemma 5.4, the following hold for a lucky prime:

- $r \bmod p = \mathrm{res}_x(f(x) \bmod p, g(x+y) \bmod p)$,
- any two distinct roots of r_1 remain distinct modulo p,
- the multiplicities of the roots of r_1 do not change modulo p, and
- $r(z_i) = 0$ if and only if $r(z_i) \equiv 0 \bmod p$, for all i.

Theorem 8.14 (Correctness of Algorithm 8.12). *If the set S_0 in step 2 contains at least $k \in \Theta(n + \lambda)$ primes that are lucky with respect to the Gosper-Petkovšek form problem 8.4, then Algorithm 8.12 does not return "FAIL". If the algorithm does not return "FAIL", then it returns the correct result.*

Proof. Let h_1 be the normalized gcd of f and $E^{z_1}g$, and let $p \in S_0$ be a lucky prime. We first show that $h_1 \equiv \mathrm{lc}(h_1)h_{1,p} \bmod p$. If $r(z_1) \neq 0$, then $h_1 = 1$, by Corollary 3.12, and the discussion preceding the theorem implies that $r(z_1) \not\equiv 0 \bmod p$ and $h_{1,p} = 1$ as well. Thus we may assume that $r(z_1) = 0$ and h_1 is nonconstant. The following proof uses some valuation theory; see Fröhlich (1967) for proofs of some facts that we use.

Let $K \subseteq \mathbb{C}$ be the splitting field of fg over $\mathbb{Q}$, and let $\alpha_1, \ldots, \alpha_l \in K$ and $\beta_1, \ldots, \beta_m \in K$ be the roots of f and g, respectively, repeated with multiplicities. Let $w \colon K \to \mathbb{Q}$ be any valuation extending the p-adic valuation on $\mathbb{Q}$. Since $p \nmid \mathrm{lc}(fg)$, we have $w(\alpha_u), w(\beta_u) \geq 0$ for all u. Thus all α_u and all β_u belong to the valuation ring $K_w \subseteq K$ of all elements of K with nonnegative value of w. Similarly, let $\mathbb{Q}_w$ be the ring of all rational numbers with denominator not divisible by p. We can extend the canonical epimorphism $\bar{} \colon \mathbb{Q}_w \longrightarrow \mathbb{F}_p$ to an epimorphism $K_w \longrightarrow E$ onto an extension field E of $\mathbb{F}_p$ over which fg splits into linear factors. We denote this extended morphism by $\bar{}$ as well. Thus $\overline{\alpha}_1, \ldots, \overline{\alpha}_l$ and $\overline{\beta}_1, \ldots, \overline{\beta}_m$ are the roots of $\overline{f}$ and $\overline{g}$, respectively, repeated with multiplicities.

For a polynomial $h \in \mathbb{Z}[y]$ and a number $\gamma \in K$, we denote by $\mu_\gamma(h)$ the multiplicity of γ as a root of h. In particular, $\mu_\gamma(h) = 0$ if $h(\gamma) \neq 0$. We define $\overline{\mu}_\gamma(h)$ for $h \in \mathbb{F}_p[y]$ and $\gamma \in E$ accordingly. By Fact 7.11 (ii), r splits in $K_w[y]$ as well. Let $\gamma \in K_w$ be a root of r_1,

$$S_\gamma = \{(u,v) \colon \alpha_u - \beta_v = \gamma\}, \text{ and } \overline{S}_{\overline{\gamma}} = \{(u,v) \colon \overline{\alpha}_u - \overline{\beta}_v = \overline{\gamma}\}.$$

Clearly $S_\gamma \subseteq \overline{S}_{\overline{\gamma}}$, and by Fact 7.11 (ii) and the discussion preceding the theorem, we have

$$\#S_\gamma = \mu_\gamma(r) = \overline{\mu}_{\overline{\gamma}}(\overline{r}) = \#\overline{S}_{\overline{\gamma}},$$

and hence $S_\gamma = \overline{S}_{\overline{\gamma}}$. Thus, for any fixed root $\gamma \in K_w$ of r_1,

$$\alpha_u - \beta_v = \gamma \iff \overline{\alpha}_u - \overline{\beta}_v = \overline{\gamma} \tag{8.2}$$

holds for all $1 \leq u \leq l$ and $1 \leq v \leq m$.

We claim that the images in E of any two distinct α_u are distinct, and similarly for the β_u. Assume to the contrary that $\overline{\alpha}_u = \overline{\alpha}_v$ for some u, v with $\alpha_u \neq \alpha_v$, and let $\gamma = \alpha_u - \beta_1$. Then $\alpha_v - \beta_1 \neq \gamma$ and $\overline{\alpha}_v - \overline{\beta}_1 = \overline{\gamma}$, contradicting (8.2). By symmetry, the same also holds for the β_u, and the claim is proved. Similarly, we find that

$$g(\alpha_u - \gamma) = 0 \iff \overline{g}(\overline{\alpha}_u - \overline{\gamma}) = 0$$

holds for all $1 \leq u \leq l$ and all roots $\gamma \in K_w$ of r_1. Thus

$$\mu_{\alpha_u}(f) = \overline{\mu}_{\overline{\alpha}_u}(\overline{f}) \text{ and } \mu_{\alpha_u - \gamma}(g) = \overline{\mu}_{\overline{\alpha}_u - \overline{\gamma}}(\overline{g}) \tag{8.3}$$

for all $1 \leq u \leq l$ and all roots $\gamma \in K_w$ of r_1.

By reordering the α_u if necessary, we may assume that $\alpha_1, \ldots, \alpha_q$ are precisely the distinct roots of f, for some $q \leq l$. Then (8.3) implies that $\overline{\alpha}_1, \ldots, \overline{\alpha}_q$ are the distinct roots of $\overline{f}$. Finally, we have

$$h_1 = \gcd(f(x), g(x + z_1)) = \mathrm{lc}(h_1) \prod_{1 \le u \le q} (x - \alpha_u)^{\min\{\mu_{\alpha_u}(f), \mu_{\alpha_u - z_1}(g)\}} ,$$

$$h_{1,p} = \gcd(\overline{f}(x), \overline{g}(x + \overline{z}_1)) = \prod_{1 \le u \le q} (x - \overline{\alpha}_u)^{\min\{\overline{\mu}_{\overline{\alpha}_u}(\overline{f}), \overline{\mu}_{\overline{\alpha}_u - \overline{z}_1}(\overline{g})\}} ,$$

and (8.3) for $\gamma = z_1$ implies that $h_1 \equiv \mathrm{lc}(h_1)h_{1,p} \bmod p$.

Let $h_1, \ldots, h_t, f_t, g_t$ be as in (8.1). Proceeding inductively with f/h_1 and $g/E^{-z_1}h_1$ instead of f and g, we find that

$$h_j \equiv \mathrm{lc}(h_j)h_{j,p} \bmod p \text{ for } 1 \le j \le t ,$$
$$\mathrm{lc}(f)f_t \equiv \mathrm{lc}(f_t)f_{t,p} \bmod p, \text{ and } \mathrm{lc}(g)g_t \equiv \mathrm{lc}(g_t)g_{t,p} \bmod p$$

for each lucky prime p. In particular, if S contains at least one lucky prime, then

$$d_j = \deg h_j = \deg h_{j,p} \text{ for } 1 \le j \le i \tag{8.4}$$

holds for all lucky primes $p \in S_i$ and $1 \le i \le t$. Conversely, if (8.4) holds for a prime $p \in S_i$, then this implies

$$h_j \equiv \mathrm{lc}(h_j)h_{j,p} \bmod p \text{ for } 1 \le j \le i . \tag{8.5}$$

Now assume that S contains at least k lucky primes. Then S_i contains at least k lucky primes as well, for all i, and (8.4) and (8.5) hold for all $p \in S_i$. In particular, the algorithm does not return "FAIL" in step 8. Moreover, since $h_i \mid f$ and $E^{-z_i}h_i \mid g$, Mignotte's bound implies that $\|h_i\|_1, \|E^{-z_i}h_i\|_2 \le B$, and the algorithm does not return "FAIL" in step 7 either, by Lemma 8.5.

Let $1 \le i \le t$ and $m = \prod_{p \in S_i} p$. Then

$$c/\mathrm{lc}(h_i) \cdot h_i \equiv h_i^* \bmod m \text{ and } c/\mathrm{lc}(h_i) \cdot E^{-z_i}h_i \equiv w_i^* \bmod m .$$

The right hand sides in these congruences have max-norms less than $m/2$, by construction. The left hand sides are divisors of cf, and hence they have max-norms at most $2^\lambda B \le 2^{(\omega-1)k-1} < m/2$ as well, by Mignotte's bound 3.3. Thus both congruences are in fact equalities. It follows that $h_i = \mathrm{normal}(h_i^*) = H_i$ and $E^{-z_i}h_i = \mathrm{normal}(w_i^*) = W_i$ for all i. By similar arguments, we find $f_t = \mathrm{normal}(f_t^*) = F_t$ and $g_t = \mathrm{normal}(g_t^*) = G_t$. Thus the algorithm does not return "FAIL" in step 11. By Mignotte's bound, it neither returns "FAIL" in step 12.

On the other hand, if the algorithm does not return "FAIL" in steps 7 and 8, then the polynomials $H_1 \cdots H_t \cdot F_t$ and f agree modulo $m = \prod_{p \in S_t} p$ up to a multiplicative constant. If the algorithm does not return "FAIL" in step 11, then this constant is 1. If, in addition, it does not return "FAIL" in step 12, then

$$\|H_1 \cdots H_t \cdot F_t\|_1 \le \|H_1\|_1 \cdots \|H_t\|_1 \cdot \|F_t\|_1 \le B \le 2^{(\omega-1)k-1} < m/2 ,$$

and hence the congruence $H_1 \cdots H_t \cdot F_t \equiv f \bmod m$ is in fact an equality. Similarly, we find $W_1 \cdots W_t \cdot G_t = g$. Let $1 \le i \le t$. By construction, we have

$W_i \equiv E^{-z_i} H_i \bmod m$. Since the algorithm does not return "FAIL" in step 7, Lemma 4.2 implies that

$$\|E^{-z_i} H_i\|_\infty \leq (|z_i| + 1)^{d_i} \|H_i\|_1 \leq |2z_i|^{d_i} \|H_i\|_1 \leq 2^{2d_i} B^3 \leq 2^{(\omega-1)k-1} < m/2$$

if $z_i \neq 0$, and trivially $\|E^{-z_i} H_i\|_\infty \leq B < m/2$ if $z_i = 0$. Thus the congruence $W_i \equiv E^{-z_i} H_i \bmod m$ is in fact an equality. Finally, the gcd conditions (8.1) hold for all i, since they are valid modulo at least one prime in S_t. □

Theorem 8.15 (Success probability of Algorithm 8.12). *Let*

$$C = n^{8n^3 + 4n^2} 2^{4n^4 \lambda + 16n^4 + 8n^3 \lambda + 12n^3 + 2n^2}$$

and $s = \lfloor (\log_2 C)/(\omega - 1) \rfloor \in \Theta(n^4 \lambda)$. The number of single precision primes that are unlucky with respect to the Gosper-Petkovšek problem 8.4 is at most s. Suppose that the number of single precision primes, between $2^{\omega-1}$ and 2^ω, is at least $2s$. If we choose the set S in step 2 uniformly at random from among at least $2s$ single precision primes, then the success probability of the algorithm is at least $1/2$.

Proof. The resultant r is a nonzero polynomial of degree at most n^2, by Fact 7.11, and r_1 is a nonzero polynomial of degree at most $2n^2$. Lemma 7.15 yields $\|r\|_1 < (n2^{n+\lambda+3/2})^{2n}$. Thus

$$\|r_1\|_1 \leq \|r\|_1 \cdot \prod_{1 \leq i \leq t} \|y - z_i\|_1 \leq (n2^{n+\lambda+3/2})^{2n} \cdot 2^{n^2(\lambda+2)}$$
$$= n^{2n} 2^{n^2 \lambda + 4n^2 + 2n\lambda + 3n} .$$

Let $d = \deg \gcd(r, r')$, and let $\sigma \in \mathbb{Z} \setminus \{0\}$ be the dth subresultant of r and r'. By Corollary 3.2, we have

$$|\sigma| \leq \|r_1\|_2^{2n^2-d} \|r_1'\|_2^{2n^2-d} \leq (2n^2 \|r_1\|_2^2)^{2n^2} \leq C .$$

The leading coefficient of r is $\text{lc}(f)^{\deg g} \text{lc}(g)^{\deg f}$, by Fact 7.11 (ii). Thus, by Corollary 5.5, a prime is unlucky if and only if it divides σ_d, and the number of unlucky single precision primes is at most s. Hence the probability that a uniformly randomly chosen single precision prime is unlucky is as most $1/2$, and the claim follows from Lemma 3.24 (i). □

Theorem 8.16 (Cost of Algorithm 8.12). *If we ignore the cost for prime finding, then Algorithm 8.12 takes $O(n^4 + n^3 \lambda + n\lambda^2)$ word operations with classical arithmetic and $O(n^2 \, \mathsf{M}(n + \lambda) \log(n + \lambda) + n\lambda \, \mathsf{M}(n) \log n)$ or $O^\sim(n^3 + n^2 \lambda)$ with fast arithmetic.*

Proof. The cost for reducing all coefficients of f and g modulo all primes p in S and S_0 in steps 2 and 3, respectively, is $O(kn\lambda)$ word operations with classical arithmetic and $O(n \, \mathsf{M}(k) \log k)$ with fast arithmetic.

In step 6, we perform a gcd computation, two divisions, and two Taylor shifts. Using Fact 3.13 and Theorems 4.3 and 4.5, this takes $O(n^2)$ operations in $\mathbb{F}_p$ with classical arithmetic and $O(\mathsf{M}(n) \log n)$ with fast arithmetic. Reducing one z_i modulo p takes $O(\lambda)$ with classical arithmetic, and reducing it modulo all primes in S_{i-1} with fast arithmetic takes $O(\mathsf{M}(k) \log k)$.

At most n of the t gcd's are nontrivial, and we can expect to detect a trivial gcd with a constant number of primes, by modifying the algorithm in a similar way as described in the proof of Corollary 8.11. Thus the expected cost for steps 5 and 6 with classical arithmetic is $O(n^2 + \lambda)$ if the gcd is constant, and $O(k(n^2 + \lambda))$ otherwise. With fast arithmetic, the expected cost for steps 5 and 6 is $O(\mathsf{M}(k) \log k + \mathsf{M}(n) \log n)$ when the gcd is trivial, and $O(\mathsf{M}(k) \log k + k \mathsf{M}(n) \log n)$ otherwise. Thus the expected total cost of steps 5 and 6 for all gcd's is $O((n^2 + nk)(n^2 + \lambda))$ word operations with classical arithmetic and $O((n^2 + nk) \mathsf{M}(n) \log n + n^2 \mathsf{M}(k) \log k)$ with fast arithmetic.

The cost for the Chinese remaindering and the subsequent normalizations in steps 9 and 10 is $O(k^2)$ and $O(\mathsf{M}(k) \log k)$ word operations per coefficient with classical and fast arithmetic, respectively, and there are $O(n)$ coefficients in total. The claims now follow from $k \in O(n + \lambda)$ by adding up costs. □

We are now ready to analyze the two modular variants of Algorithm 8.2. The following parameter, which has been introduced by Abramov (1971), enters our cost analysis.

Definition 8.17. *Let $f, g \in \mathbb{Z}[x]$ be nonzero. The* dispersion *of f and g is*

$$\mathrm{dis}(f, g) = \max\{i \in \mathbb{N}: i = 0 \text{ or } \mathrm{res}(f, E^i g) = 0\} .$$

By Corollary 3.12, the dispersion is 0 or the maximal nonnegative integer root of $\mathrm{res}_x(f(x), g(x + y)) \in \mathbb{Z}[y]$, or equivalently, the maximal nonnegative integer distance between a root of f and a root of g. The example $f = x$ and $g = x - e$, for a positive integer e, shows that the dispersion, which is e in this case, can be exponential in the size of the coefficients of f and g in the worst case.

Theorem 8.18 (Cost of Algorithm 8.2). *(i) If we use Algorithm 7.20 in step 1 and Algorithm 8.6 in step 4, then the expected cost of steps 1 through 4 is $O(n^4 + n^2\lambda^2)$ word operations with classical arithmetic and*

$$O(n^3 \mathsf{M}(n + \lambda) \log(n + \lambda) + n\lambda \mathsf{M}(n) \log n)$$

or $O^\sim(n^3(n + \lambda))$ with fast arithmetic.
(ii) If we use Algorithm 7.20 in step 1 and Algorithm 8.12 in steps 3 and 4, then the cost of steps 1 through 4 is $O(n^4 + n^2\lambda^2)$ word operations with classical arithmetic and

$$O(n^2 \mathsf{M}(n + \lambda) \log(n + \lambda) + n\lambda \mathsf{M}(n) \log n + \mathsf{M}(n^2) \log n)$$

or $O^\sim(n^2(n + \lambda))$ with fast arithmetic.

(iii) If $e = \operatorname{dis}(f, g)$ is the dispersion of f and g, then $e < 2^{\lambda+2}$, and the cost for step 5 is $O(e^3(n^3 + n\lambda^2))$ word operations with classical arithmetic and

$$O((en\,\mathsf{M}(e(n+\lambda)) + e\lambda\,\mathsf{M}(en))\log(e(n+\lambda)))$$

or $O^\sim(e^2 n(n+\lambda))$ with fast arithmetic. The polynomial c has degree at most en and its coefficients are of word size $O(e(n+\lambda))$.

Proof. Step 1 takes $O(n^3 + n^2(\log n)\lambda + n^2\lambda^2)$ word operations with classical arithmetic and $O(n\,\mathsf{M}(n+\lambda)\log(n+\lambda) + \mathsf{M}(n^2)\log n + n^2(\mathsf{M}(\lambda)+\lambda\log n))$ with fast arithmetic, by Theorem 7.21. The cost for step 2 is $O(n\lambda^2)$ or $O(n\,\mathsf{M}(\lambda)\log\lambda)$ with classical or fast arithmetic, respectively. Inductively, we find that

$$f_i h_i \cdots h_1 = f_0 \quad \text{and} \quad g_i(E^{-z_i}h_i)\cdots(E^{-z_1}h_1) = g_0 \tag{8.6}$$

for all i. Thus $\deg f_{i-1}, \deg g_{i-1} \le n$ and $\|f_{i-1}\|_\infty, \|g_{i-1}\|_\infty < (n+1)^{1/2}2^{n+\lambda}$ in step 4, by Mignotte's bound (Fact 3.3), and Theorem 8.10 implies that the expected cost for one execution of step 4 with Algorithm 8.6 is $O(n^3 + n\lambda^2)$ word operations with classical arithmetic and $O(\lambda\,\mathsf{M}(n)\log n + n\,\mathsf{M}(n+\lambda)\log(n+\lambda))$ with fast arithmetic. In fact, at most n of the h_i are nonconstant, and for all other i we detect that h_i is constant with an expected number of only $O(n^2 + n\lambda)$ word operations with classical arithmetic and $O(n\,\mathsf{M}(n+\lambda)\log(n+\lambda))$ with fast arithmetic, by Corollary 8.11. This proves (i). The estimate (ii) follows from Theorem 8.16.

Fact 7.11 (iii) yields the upper bound on e. By definition of e, we have $z_i \le e$ for all i such that h_i is nonconstant. Let $n_i = \deg h_i$ for $1 \le i \le t$. Then

$$\deg c = \sum_{1 \le i \le t} z_i n_i \le e \sum_{1 \le i \le t} n_i \le en\,.$$

Let $1 \le i \le t$ such that h_i is nonconstant, $1 \le j \le z_i$, and $B_i = \|h_i\|_1 \cdot \|E^{-z_i}h_i\|_1$. Then Lemma 8.5 implies that $j^{n_i} \le 2^{n_i}B_i^2$, and (8.6) and Mignotte's bound show that $\prod_{1 \le i \le t} B_i \le (n+1)4^{n+\lambda}$. Lemma 4.2 yields

$$\|E^{-j}h_i\|_1 = \|h_i(x-j)\|_1 \le (j+1)^{n_i}\|h_i\|_1 \le (2j)^{n_i}B_i \le 2^{2n_i}B_i^3\,.$$

Thus

$$\begin{aligned}
\log_2\|c\|_\infty &\le \log_2\|h\|_1 \le \sum_{\substack{1 \le i \le t \\ 1 \le j \le z_i}} \log_2\|E^{-j}h_i\|_1 \le \sum_{1 \le i \le t} z_i \cdot (2n_i + 3\log_2 B_i) \\
&\le e\sum_{1 \le i \le t}(2n_i + 3\log_2 B_i) \in O(e(n+\lambda))\,.
\end{aligned}$$

For each i, the cost for computing all $E^{-j}h_i = h_i(x-j)$ for $1 \le j \le z_i$ by iterated application of E^{-1} is $O(en_i^2(n_i + \log B_i))$ word operations with classical arithmetic and $O(e(n_i\,\mathsf{M}(n_i + \log B_i)\log(n_i + \log B_i) + \log B_i \cdot \mathsf{M}(n_i)))$ with fast arithmetic, by Theorems 4.3 and 4.8, in total $O(en^2(n+\lambda))$ with classical or $O(e(n\,\mathsf{M}(n+\lambda)\log(n+\lambda) + \lambda\,\mathsf{M}(n)))$ with fast arithmetic. Now Lemma 3.15 implies the cost estimate for step 5. □

The estimate (ii) with fast arithmetic is better by a factor of about n than the corresponding estimate in (i). This is mainly due to the fact that modular reduction and Chinese remaindering happens only once in the variant (ii), namely at the beginning and at the end of Algorithm 8.12, while it is performed each time when Algorithm 8.6 is called in the variant (i).

Using Algorithm 8.2, we can compute rational solutions of homogeneous linear first order difference equations with polynomial coefficients.

Corollary 8.19. *Given two nonconstant polynomials $f, g \in \mathbb{Z}[x]$ of degree at most n and max-norm less than 2^λ, we can decide whether the homogeneous linear first order difference equation*

$$g \cdot E\rho - f \cdot \rho = 0$$

has a nonzero solution $\rho \in \mathbb{Q}(x)$, and if so, compute one, using $O(n^4 + n^2\lambda^2)$ word operations with classical arithmetic and

$$O(n^2 \, \mathsf{M}(n + \lambda) \log(n + \lambda) + n\lambda \, \mathsf{M}(n) \log n + \mathsf{M}(n^2) \log n)$$

or $O^\sim(n^2(n + \lambda))$ with fast arithmetic.

Proof. We call Algorithm 8.2 to compute a Gosper-Petkovšek form $(a, b, u) \in \mathbb{Z}[x]^3$ of the rational function f/g. Then we call the algorithm again to compute a Gosper-Petkovšek form $(b^*, a^*, v) \in \mathbb{Z}[x]^3$ of the rational function b/a. (In fact, if we compute not only the nonnegative integers $z \in \mathbb{Z}$ such that $\gcd(f, E^z g) \neq 1$ in step 1, but also the negative ones and adapt the remaining steps suitably, then we can compute both u and v with only one call to the modified algorithm.)

If $a^*/b^* = 1$, then

$$g \cdot E\frac{u}{v} = \frac{gu}{v} \cdot \frac{v}{Ev} \cdot \frac{Eu}{u} = \frac{gu}{v} \cdot \frac{a}{b} \cdot \frac{Eu}{u} = \frac{gu}{v} \cdot \frac{f}{g} = f \cdot \frac{u}{v} \, ,$$

and $\rho = u/v$ solves the difference equation.

Conversely, assume that $\rho \in \mathbb{Q}(x) \setminus \{0\}$ solves the difference equation. Then a similar calculation as above shows that $\rho^* = \rho \cdot v/u$ satisfies

$$\frac{E\rho^*}{\rho^*} = \frac{a^*}{b^*} \, .$$

If we write $\rho^* = u^*/v^*$, with coprime polynomials $u^*, v^* \in \mathbb{Z}[x]$, then

$$\frac{a^*}{b^*} = \frac{Eu^*}{u^*} \cdot \frac{v^*}{Ev^*} \, .$$

The properties of the Gosper-Petkovšek form imply that $\gcd(a^*, E^i b^*) = 1$ for all $i \in \mathbb{Z}$. If u^* is not constant, then it has a nonconstant irreducible factor $p \in \mathbb{Z}[x]$ such that $E^{-1}p \nmid u^*$. Then, since u^* and v^* are coprime, p does not divide $(Eu^*)v^*$, and hence p divides b^*. Now let $i \geq 1$ be maximal such that $E^{i-1}p \mid u^*$. Then $E^i p \mid Eu^*$, but $E^i p \nmid u^* \cdot Ev^*$, and hence $E^i p \mid a^*$. We arrive at the contradiction that the nonconstant polynomial $E^i p$ divides $\gcd(a^*, E^i b^*) = 1$, and conclude that u^* is constant. A similar argument shows that v^* is constant as well, whence

$$\frac{a^*}{b^*} = \frac{E\rho^*}{\rho^*} = 1 .$$

Thus the difference equation is unsolvable if $a^*/b^* \neq 1$.

If we do not expand the product representation of u step 5 of Algorithm 8.2, then the cost for computing (a, b, u) is given by Theorem 8.18 (ii). Since $a \mid f$ and $b \mid g$, Mignotte's bound (Fact 3.3) implies that their coefficients are of word size $O(n+\lambda)$, and another application of the same theorem, with λ replaced by $O(n + \lambda)$, shows that we can compute (a^*, b^*, v) within the same time bound. $\square$

The proof also shows that if the homogenous difference equation has a rational solution $\rho \in \mathbb{Q}(x)$, then any other solution is a constant multiple of ρ, i.e., the solution space is $\mathbb{Q}\rho$.

8.1 Modular GP′-Form Computation

In this section, we discuss the differential analog of Algorithm 8.2. The prime ′ denotes the usual differential operator.

Definition 8.20. *Let F be a field and $f, g \in F[x]$ be nonzero polynomials. In the style of Definition 8.1, we say that a triple $(a, b, c) \in F[x]^3$ is a differential Gosper form or G'-form of the rational function f/g if*

$$\frac{f}{g} = \frac{a}{b} + \frac{c'}{c} \text{ and } \gcd(b, a - ib') = 1 \text{ for all } i \in \mathbb{N} . \tag{8.7}$$

If in addition $\gcd(b, c) = 1$, then (a, b, c) is a differential Gosper-Petkovšek form or GP′-form of f/g.

In particular, the condition (8.7) for $i = 0$ implies that a and b are coprime.

Bronstein (1990, 1997) calls a rational function a/b with $\gcd(b, a - ib') = 1$ for all $i \in \mathbb{N}$ *weakly normalized*. Before we state an algorithm for computing a GP′-form, we show that the GP′-form is unique up to multiplication by constants. The following concepts are useful for the proof.

Definition 8.21. *Let F be a field, $f, g \in F[x]$ nonzero, and $p \in F[x]$ irreducible.*

(i) The p-adic valuation of f is

$$v_p(f) = \max\{e \in \mathbb{N} : p^e \mid f\} ,$$

and $v_p(0) = \infty$. The p-adic valuation of the rational function $f/g \in F(x)$ is $v_p(f/g) = v_p(f) - v_p(g)$.

(ii) We can expand the rational function f/g as a unique generalized p-adic Laurent series

$$\frac{f}{g} = \sum_{i \geq m} a_i p^i \in F(x)_{(p)} ,$$

with $m = v_p(f/g) \in \mathbb{Z}$ and all $a_i \in F[x]$ of degree less than $\deg p$. Here $F(x)_{(p)}$ is the p-adic completion of $F(x)$. The p-adic residue of f/g is the coefficient of p^{-1} in this expansion: $\operatorname{Res}_p(f/g) = a_{-1}$. In particular, $\operatorname{Res}_p(0) = 0$.

If $F = \mathbb{C}$ and $p = x - \lambda$ is a linear polynomial, then $\operatorname{Res}_{x-\lambda}(f/g)$ is just the residue of f/g at the point $\lambda \in \mathbb{C}$ known from complex analysis. To avoid confusion with the resultant, we denote the residue by Res and the resultant by res. The valuation and the residue have some well-known properties:

Lemma 8.22. *Let F be a field of characteristic c, $f, g \in F(x)$, and $p \in F[x]$ irreducible.*

  (i) $v_p(1) = 0$, $v_p(f) = \infty \iff f = 0$.
 (ii) $v_p(f + g) \geq \min\{v_p(f), v_p(g)\}$, with equality if $v_p(f) \neq v_p(g)$.
(iii) $v_p(fg) = v_p(f) + v_p(g)$, and $v_p(f/g) = v_p(f) - v_p(g)$ if $g \neq 0$.
 (iv) $v_p(f') = v_p(f) - 1$ if $c \nmid v_p(f)$, and $v_p(f') \geq v_p(f)$ if $c \mid v_p(f)$.
  (v) $\operatorname{Res}_p(f) = 0$ if $v_p(f) \geq 0$.
 (vi) $\operatorname{Res}_p(f + g) = \operatorname{Res}_p(f) + \operatorname{Res}_p(g)$.
(vii) $\operatorname{Res}_p(fg) \equiv f \cdot \operatorname{Res}_p(g) \bmod p$ if $v_p(g) = -1$ and $v_p(f) \geq 0$.
(viii) $\operatorname{Res}_p(f'/f) = (v_p(f) \bmod c)$ if $f \neq 0$.
 (ix) If f, g are nonzero polynomials such that $v_p(f) \geq 0$ and $v_p(g) = 1$, then

$$r = \operatorname{Res}_p(f/g) \iff p \mid \gcd(g, f - rg') ,$$

for all $r \in F[x]$ of degree less than $\deg p$.

Proof. We only show the last three claims and refer to Chap. 4 in Bronstein (1997) for a proof of the other statements. For the proof of (vii), let $r = \operatorname{Res}_p(g)$ and $q = g - rp^{-1}$. Then $v_p(q) \geq 0$, $v_p(fr) \geq 0$, and $v_p(fq) \geq 0$. There exist unique $s \in F(x)$ and $t \in F[x]$ such that $fr = sp + t$, $v_p(s) \geq 0$, and $\deg t < \deg p$, and hence

$$
\begin{aligned}
\operatorname{Res}_p(fg) &= \operatorname{Res}_p(fq + frp^{-1}) = \operatorname{Res}_p(fq + s + tp^{-1}) \\
&= \operatorname{Res}_p(fq) + \operatorname{Res}_p(s) + \operatorname{Res}_p(tp^{-1}) = t \equiv fr \bmod p ,
\end{aligned}
$$

by (v). This proves (vii).

For (viii) let $e = v_p(f) \in \mathbb{Z}$ and $h = fp^{-e}$. Then $v_p(h) = v_p(f) + v_p(p^{-e}) = 0$. Thus

$$
\begin{aligned}
\operatorname{Res}_p\left(\frac{f'}{f}\right) &= \operatorname{Res}_p\left(\frac{(hp^e)'}{hp^e}\right) = \operatorname{Res}_p\left(\frac{h'}{h}\right) + \operatorname{Res}_p\left(\frac{(p^e)'}{p^e}\right) \\
&= 0 + \operatorname{Res}_p(ep^{-1}) = e \text{ in } F[x] ,
\end{aligned}
$$

where we have used the linearity of the logarithmic derivative and the fact that $v_p(h'/h) = v_p(h') - v_p(h) = v_p(h') \geq 0$. This proves (viii).

Finally, for (ix) we have $v_p(g'/g) = -1$ and

$$\operatorname{Res}_p\left(\frac{f - rg'}{g}\right) \equiv \operatorname{Res}_p\left(\frac{f}{g}\right) - r \cdot \operatorname{Res}_p\left(\frac{g'}{g}\right) = \operatorname{Res}_p\left(\frac{f}{g}\right) - r \bmod p , \quad (8.8)$$

by (vi), (vii), and (viii). Since the degree of both sides is less than $\deg p$, the congruence is in fact an equality. If $p \mid (f - rg')$, then $v_p((f - rg')/g) \geq 0$, and (v) implies that both sides in (8.8) are zero. Conversely, if $r = \operatorname{Res}_p(f/g)$, then (8.8) and

$$v_p\left(\frac{f - rg'}{g}\right) \geq \max\{v_p(f) - v_p(g), v_p(r) + v_p(g') - v_p(g)\} \geq -1$$

imply that $v_p((f - rg')/g) \geq 0$. Hence p is a common divisor of g and $f - rg'$. □

Lemma 8.23. *Let F be a field of characteristic zero and (a, b, c), (A, B, C) in $F[x]^3$ be two GP'-forms of the same rational function, and assume that b, B, c, C are monic. Then $(a, b, c) = (A, B, C)$.*

Proof. We have

$$\frac{a}{b} + \frac{c'}{c} = \frac{A}{B} + \frac{C'}{C} . \quad (8.9)$$

Condition (8.7) for $i = 0$ implies that both a, b and A, B are coprime. Let $p \in \mathbb{Z}[x]$ be an irreducible factor of c of multiplicity $e = v_p(c) \geq 1$. Since b and c are coprime, we have $v_p(b) = 0$, $v_p(a/b) \geq 0$, and

$$e = \operatorname{Res}_p\left(\frac{a}{b}\right) + \operatorname{Res}_p\left(\frac{c'}{c}\right) = \operatorname{Res}_p\left(\frac{A}{B}\right) + \operatorname{Res}_p\left(\frac{C'}{C}\right) ,$$

by Lemma 8.22 (v) and (viii). Assume that $p \mid B$. Then $v_p(A) = 0$. Since B and C are coprime, we have $v_p(C) = 0$, $v_p(C'/C) \geq 0$, and $\operatorname{Res}_p(C'/C) = 0$. Thus $\operatorname{Res}_p(A/B) = e$. If $v_p(B) > 1$, then $v_p(c'/c) = -1$ and $v_p(C'/C) \geq 0$ imply that

$$\begin{aligned}
v_p\left(\frac{a}{b}\right) &= v_p\left(\frac{A}{B} + \frac{C'}{C} - \frac{c'}{c}\right) \\
&= \min\left\{v_p\left(\frac{A}{B}\right), v_p\left(\frac{C'}{C}\right), v_p\left(\frac{c'}{c}\right)\right\} = v_p\left(\frac{A}{B}\right) = -v_p(B) ,
\end{aligned}$$

contradicting $v_p(a/b) \geq 0$. Thus $v_p(B) = 1$. Then Lemma 8.22 (xi) shows that $p \mid \gcd(B, A - eB')$, which contradicts (8.7). Thus our assumption was wrong and $p \nmid B$, $v_p(A/B) \geq 0$, $\operatorname{Res}_p(A/B) = 0$, and $v_p(C) = \operatorname{Res}_p(C'/C) = e$. Since this holds for all irreducible factors of c, we conclude that $c \mid C$. By a symmetric reasoning also $C \mid c$, and since both polynomials are monic, they are equal. Finally, (8.9) implies that $a/b = A/B$. Since both a, b and A, B are coprime and b, B are monic, we conclude that $b = B$ and $a = A$. □

The uniqueness statement is wrong in positive characteristic: for example, both $(0, 1, 1)$ and $(0, 1, x^p)$ are GP′-forms of the rational function 1.

Let $f, g \in F[x] \setminus \{0\}$ be coprime. In analogy to Abramov & Petkovšek (2001) and Geddes, Le & Li (2004), we might call a quadruple $(a, b, c, d) \in \mathbb{Z}[x]^4$ satisfying

$$\frac{f}{g} = \frac{a}{b} + \frac{c'}{c} - \frac{d'}{d}$$

a strict differential rational normal form of the rational function f/g, if b, c, d are pairwise coprime and $\gcd(b, a - ib') = 1$ for all $i \in \mathbb{Z}$. In contrast to the difference case, however, similar arguments as in the proof of Lemma 8.23 show that this normal form is unique up to multiplication by constants.

The following decomposition of a rational function is a generalization of the GP′-form.

Definition 8.24. *Let F be a field and $f, g \in F[x]$ be nonzero coprime polynomials. A GP′-refinement of the rational function f/g is a decomposition*

$$\frac{f}{g} = \frac{a}{b} + \sum_{1 \leq i \leq t} z_i \frac{h_i'}{h_i} \,, \tag{8.10}$$

where

(i) $z_1, \ldots, z_t \in F$ are nonzero and distinct,
(ii) $h_1, \ldots, h_t \in F[x]$ are nonconstant, squarefree, and pairwise coprime,
(iii) $a, b \in F[x]$ and b is coprime to $h_1, \ldots, h_t$,
(iv) $g = bh_1 \cdots h_t$.

Lemma 8.25. *With the assumptions as in Definition 8.24, the polynomials a and b are coprime.*

Proof. Let $p \in F[x]$ be a common divisor of a and b. Then p divides g and $f - \sum_{1 \leq i \leq t} z_i h_i' g / h_i$. By (iii), p is coprime to h_i for all i, so that it divides all g/h_i. Hence $p \mid f$, and the coprimality of f and g implies that p is constant. $\square$

Similarly to Lemma 8.23, one can show that the GP′-refinement for given $z_1, \ldots, z_t$ – if one exists – is unique, up to leading coefficients, but we will not need this in what follows. Lemma 8.22 implies that z_i is the p-adic residue of f/g for every irreducible factor p of h_i, and that h_i divides $\gcd(g, f - z_i g')$. Thus a necessary condition for the existence of a GP′-refinement is that all of these gcd's are nonconstant. These observations lead to the following algorithm for computing a GP′-refinement, due to Almkvist & Zeilberger (1990) (see also Algorithm Weak-Normalizer in §6.1 of Bronstein 1997 and Algorithm 11.1 in Koepf 1998).

Algorithm 8.26 (GP′-refinement computation). ▬▬▬▬▬▬
Input: Nonzero coprime polynomials $f, g \in F[x]$ of degrees at most n, where F is a field of characteristic greater than n, and a list of distinct nonzero elements $z_1, \ldots, z_t \in F$ such that $\gcd(g, f - z_i g')$ is nonconstant for all i.
Output: A GP′-refinement of f/g as in Definition 8.24.

1. **for** $1 \leq i \leq t$ compute $h_i \longleftarrow \gcd(g, f - z_i g')$
2. $b \longleftarrow \dfrac{g}{h_1 \cdots h_t}, \quad a \longleftarrow \dfrac{f - \sum_{1 \leq i \leq t} z_i h_i' g / h_i}{h_1 \cdots h_t}$
3. **return** $a, b, h_1, \ldots, h_t$

In practice, it is more efficient to replace g and f by g/h_i and $(f - z_i g')/h_i$ after computing h_i. Bronstein's Algorithm WeakNormalizer first performs a partial fraction decomposition

$$\frac{f}{g} = \frac{f_1}{g_1} + \frac{h}{g/g_1} \, ,$$

where g_1 is the product of all irreducible factors of g of multiplicity one. Since these modifications do not improve the order of magnitude of the worst case running time estimate, we only discuss the simpler algorithm above.

Theorem 8.27. *Algorithm 8.26 works correctly as specified. It takes $O(tn^2)$ arithmetic operations in F with classical arithmetic and $O(t\, \mathsf{M}(n) \log n)$ with fast arithmetic.*

Proof. By assumption, all h_i are nonconstant. Equation (8.10) and property (iv) in Definition 8.24 are satisfied by construction; but we still need to show that a and b are in fact polynomials. Let $1 \leq i \leq t$ and $p \in F[x]$ be an irreducible factor of h_i. If $p^2 \mid g$, then $p \mid z_i g'$, and hence $p \mid f$. Since f and g are coprime, this is a contradiction, and we conclude that $v_p(g) = 1$. This in turn implies that h_i is squarefree and coprime to g/h_i. Now assume that p divides h_j for some $j \neq i$. Then p divides $f - z_j g' - (f - z_i g') = (z_i - z_j)g'$. Since $v_p(g) = 1$ and the characteristic of F is greater than $\deg g$, we have $p \nmid g'$, and hence p divides the nonzero field element $z_i - z_j$. This contradiction proves that h_i and h_j are coprime, which concludes the proof of (ii). Moreover, we conclude that $h_1 \cdots h_t \mid g$, and b is a polynomial coprime to all h_i. By construction, h_i divides $f - z_i g' = f - z_i h_i' g/h_i - z_i h_i (g/h_i)'$, and hence it divides $f - z_i h_i' g/h_i$. Since h_i and h_j are coprime for $i \neq j$, h_i divides g/h_j for all such j, and we see that $h_i \mid (f - \sum_{1 \leq i \leq t} z_i h_i' g/h_i)$. Again the coprimality of the h_i implies that a is a polynomial.

The cost for step 1 is $O(tn^2)$ with classical and $O(t\, \mathsf{M}(n) \log n)$ with fast arithmetic. In step 2, we first compute $h_1 \cdots h_t$. Since this product divides f, it has degree at most n, and the cost is $O(n^2)$ and $O(\mathsf{M}(n) \log t)$ with classical and fast arithmetic, respectively. Using classical arithmetic, we can compute g/h_i and multiply the result by h_i' with $O(n \cdot \deg h_i)$ field operations for each i, together $O(n^2)$. The linear combination of these products with the z_i can be computed within the same time bound. With fast arithmetic, Algorithm 10.20 in von zur Gathen & Gerhard (1999) takes $O(\mathsf{M}(n) \log t)$ operations to compute the linear combination. Finally, the cost for dividing by $h_1 \cdots h_t$ to compute b and a is another $O(n^2)$ operations with classical and $O(\mathsf{M}(n))$ with fast arithmetic. Thus the cost for step 2 is $O(n^2)$ and $O(\mathsf{M}(n) \log t)$, respectively. This is dominated by the cost for step 1, and the claim follows. $\square$

Using Algorithm 8.26, we obtain the following algorithm for computing a G'-form in $\mathbb{Z}[x]$. In fact, it even computes a GP'-form.

Algorithm 8.28 (GP'-form computation).
Input: Nonzero coprime polynomials $f, g \in \mathbb{Z}[x]$ of degree at most n and max-norm less than 2^λ.
Output: A GP'-form of f/g.

1. compute a list $0 < z_1 < \cdots < z_t \in \mathbb{N}$, with $t \leq n$, of all positive integers $k \in \mathbb{N}$ such that $\gcd(g, f - kg')$ is nonconstant
2. compute a GP'-refinement of f/g with respect to $z_1, \ldots, z_t$, as in Definition 8.24
3. **return** a, b, and $c = h_1^{z_1} \cdots h_t^{z_t}$

Theorem 8.29 (Correctness of Algorithm 8.28). *Algorithm 8.28 correctly computes a GP'-form of f/g.*

Proof. We first note that the z_i are roots of the resultant $\mathrm{res}_x(g, f - yg') \in \mathbb{Z}[y]$, by Corollary 3.12, and Fact 7.24 implies that $t \leq n$. The decomposition

$$\frac{f}{g} = \frac{a}{b} + \frac{c'}{c}$$

follows from the linearity of the logarithmic derivative. Property (iii) in Definition 8.24 implies that b and c are coprime. It remains to show that a/b is weakly normalized. Let $z \in \mathbb{N}$, and assume that $p \in \mathbb{Z}[x]$ is an irreducible common divisor of b and $a - zb'$. Then Lemma 8.25 implies that $z > 0$, and Lemma 8.22 yields

$$z = \mathrm{Res}_p\left(\frac{a}{b}\right) = \mathrm{Res}_p\left(\frac{f}{g}\right) - \sum_{1 \leq i \leq t} z_i \mathrm{Res}_p\left(\frac{h_i'}{h_i}\right) = \mathrm{Res}_p\left(\frac{f}{g}\right),$$

since $p \nmid h_i$ for all i. By the same lemma, we have $p \mid \gcd(g, f - zg')$, and hence $z = z_i$ and $p \mid h_i$ for some $1 \leq i \leq t$. This contradiction proves that b and $a - zb'$ are coprime. $\square$

Corollary 8.30. *Let F be a field of characteristic zero and $f, g \in F[x]$ coprime polynomials such that $g \neq 0$ is monic. There exists a nonzero polynomial $h \in F[x]$ with $f/g = h'/h$ if and only if $\deg f < \deg g$ and $g = \prod_{i \geq 1} \gcd(g, f - ig')$. In fact, we may choose $h = \prod_{i \geq 1} \gcd(g, f - ig')^i$.*

Proof. The GP'-form of f/g is $(0, 1, h)$, and the claims follow from Theorems 8.27 and 8.29, since the latter theorem holds for an arbitrary field of characteristic zero. $\square$

The following fact is a consequence of Fact 7.24 (i).

Fact 8.31. *Let $f, g \in F[x] \setminus \{0\}$ be coprime polynomials, where F is a perfect field of characteristic zero or greater than $\deg f$. Let $g_1 \in F[x]$ be the product of all monic irreducible factors of multiplicity 1 in g, $r = \mathrm{res}_x(g, f - yg) \in F[y]$, and suppose that r splits into linear factors in F and $z_1, \ldots, z_t \in F$ are the distinct roots of r. Then*

$$g_1 = \prod_{1 \leq i \leq t} \gcd(g, f - z_ig') .$$

We now analyze two modular variants of Algorithm 8.26. The first one uses Chinese remainder techniques, and the second one uses Hensel lifting. We assume that our polynomials have degree at most $2^{\omega-1}$, where ω is the word size of our processor.

Algorithm 8.32 (Small primes modular GP$'$-refinement).
Input: Nonzero coprime polynomials $f, g \in \mathbb{Z}[x]$ with $\deg f, \deg g \leq n \leq 2^{\omega-1}$ and $\|f\|_\infty, \|g\|_\infty < 2^\lambda$, and a list of distinct positive integers $z_1, \ldots, z_t$, where $t \leq n$ and $\gcd(g, f - z_i g')$ is nonconstant for all i.
Output: A GP$'$-refinement of f/g as in Definition 8.24, or otherwise "FAIL".

1. $B \longleftarrow \lfloor (n+1)^{1/2} 2^{n+\lambda} \rfloor, \quad C \longleftarrow (n+1)^n 2^{2n\lambda}, \quad D \longleftarrow (n^2+1) 2^n BC$
   $k \longleftarrow \lceil \log_2(2^{\lambda+1}B)/(\omega-1) \rceil, \quad s \longleftarrow \lceil \log_2(2D)/(\omega-1) \rceil$
2. choose a set S_0 of $2k$ odd single precision primes
   $S_1 \longleftarrow \{p \in S_0 : p \nmid \text{lc}(fg) \text{ and } \gcd(f \bmod p, g \bmod p) = 1 \text{ and } z_i \not\equiv z_j \bmod p \text{ for all } i \neq j\}$
3. **for** $1 \leq i \leq t$ and all $p \in S_1$ compute $h_{i,p} \in \mathbb{Z}[x]$ of max-norm less than $p/2$ such that $h_{i,p} \bmod p$ is the monic gcd of $g \bmod p$ and $(f - z_i g') \bmod p$ in $\mathbb{F}_p[x]$
4. $d \longleftarrow \min\{\deg(h_{1,p} \cdots h_{t,p}) : p \in S_1\}$
   $S_2 \longleftarrow \{p \in S_1 : \deg(h_{1,p} \cdots h_{t,p}) = d\}$
   if $\#S_2 < k$ **then return** "FAIL"
5. **for** $1 \leq i \leq t$ **do**
6.   use the Chinese Remainder Algorithm to compute $h_i^* \in \mathbb{Z}[x]$ of max-norm less than $(\prod_{p \in S_2} p)/2$ such that $h_i^* \equiv \text{lc}(g) h_{i,p} \bmod p$ for all $p \in S_2$
   $h_i \longleftarrow \text{normal}(h_i^*)$
7. use the Chinese Remainder Algorithm to compute $b \in \mathbb{Z}[x]$ of max-norm less than $(\prod_{p \in S_2} p)/2$ such that

$$g \equiv b \cdot h_1 \cdots h_t \bmod p \text{ for all } p \in S_2$$

8. **if** $\|b\|_1 \cdot \|h_1\|_1 \cdots \|h_t\|_1 > B$ **then return** "FAIL"
9. choose a set S_3 of s single precision primes not dividing $\text{lc}(g)$
10. reduce $f, g, z_1, \ldots, z_t, h_1, \ldots, h_t$ modulo all primes in S_3
11. **for** all $p \in S_3$ compute $a_p \in \mathbb{Z}[x]$ of max-norm less than $p/2$ such that

$$a_p h_1 \cdots h_t \equiv f - \sum_{1 \leq i \leq t} z_i h_i' \frac{g}{h_i} \bmod p$$

12. use the Chinese Remainder Algorithm to compute $a \in \mathbb{Z}[x]$ of max-norm less than $(\prod_{p \in S_3} p)/2$ such that $a \equiv a_p \bmod p$ for all $p \in S_3$
13. **if** $\|a\|_1 \cdot \|h_1\|_1 \cdots \|h_t\|_1 > D$ **then return** "FAIL"
14. **return** $a, b, h_1, \ldots, h_t$

Definition 8.33. *Let $f, g \in \mathbb{Z}[x]$ be nonzero coprime, $z_1, \ldots, z_t \in \mathbb{Z}$ nonzero and distinct, and let $r = \text{res}_x(g, f - yg') \in \mathbb{Z}[y]$. We say that a single precision prime $p \in \mathbb{N}$ is lucky with respect to GP$'$-refinement if it satisfies the following conditions:*

a. $p \nmid \mathrm{lc}(fg)$,
b. $f \bmod p$ and $g \bmod p$ are coprime in $\mathbb{F}_p[x]$,
c. $\deg \gcd(g, g') = \deg \gcd(g \bmod p, g' \bmod p)$, i.e., p is lucky with respect to the squarefree factorization of g,
d. $\deg \gcd(r, r') = \deg \gcd(r \bmod p, r' \bmod p)$, i.e., p is lucky with respect to the squarefree factorization of r.

The third condition implies that any two distinct roots of g remain distinct modulo p, so that the squarefree decomposition of $g \bmod p$ is the image modulo p of the squarefree decomposition of g in $\mathbb{Z}[x]$ (Lemma 5.4). Similarly, the last condition says that any two distinct roots of r remain distinct modulo p. In particular, since $r(z_i) = 0$ for all i, by Corollary 3.12, the z_i remain distinct modulo p.

Theorem 8.34 (Correctness of Algorithm 8.32). *If the set S_0 in step 2 contains at least $k \in \Theta(n + \lambda)$ lucky primes, then Algorithm 8.32 does not return "FAIL". If the algorithm does not return "FAIL", then it returns the correct result.*

Proof. For $1 \le i \le t$, let $u_i \in \mathbb{Q}[x]$ be the monic gcd of g and $f - z_i g'$. We claim that $\mathrm{lc}(h_i)u_i = h_i$ for all i. Let $p \in S_0$ be a lucky prime, and denote reduction modulo p by a bar. Then $p \in S_1$, by the discussion preceding the theorem. Since $p > 2^{\omega - 1} \ge n$, p divides neither $\mathrm{lc}_x(g)$ nor $\mathrm{lc}_x(f - yg')$, and hence $\overline{r}$ is the resultant of $\overline{g}$ and $\overline{f} - y\overline{g}'$ in $\mathbb{F}_p[x]$. We claim that the decomposition from Fact 8.31 corresponding to $\overline{f}$ and $\overline{g}$ is the modular image of the analogous decomposition corresponding to f and g. The precise formulation of this statement and its proof, which now follows, are somewhat technical, since they take place in the splitting fields of $\overline{r}$ and r, respectively. The proof is quite similar to the proof of Theorem 8.14.

Let $K \subseteq \mathbb{C}$ be the splitting field of r, and let $z_{t+1}, \dots, z_\tau \in K$ be such that $z_1, \dots, z_\tau$ are the distinct complex roots of r. Moreover, let $u_i \in K[x]$ be the monic gcd of g and $f - z_i g'$ for $t < i \le \tau$. Let $v: K \to \mathbb{Q}$ be any valuation on K extending the p-adic valuation on $\mathbb{Q}$. Conditions b) and c) together with Fact 7.24 (iii) imply that the leading coefficient of r is nonzero modulo p, and hence $v(z_i) \ge 0$ for all i. Condition d) implies that p does not divide the discriminant of K, so that it is unramified in K and $v(K) = v(\mathbb{Q}) = \mathbb{Z}$. Let K_v be the ring of all elements of K with nonnegative v-valuation, and similarly let $\mathbb{Q}_v$ be the ring of all rational numbers whose denominator is not divisible by p. We can extend the epimorphism $\overline{}: \mathbb{Q}_v \to \mathbb{F}_p$ to an epimorphism from K_v onto a finite extension E of $\mathbb{F}_p$, which we also denote by $\overline{}$. Since r splits into linear factors in $K_v[x]$, E contains the splitting field of $\overline{r}$. The ring K_v is local, so in particular a UFD. Gauß' Theorem (Theorem 6.8 in von zur Gathen & Gerhard 1999) implies that $K_v[x]$ is a UFD as well, and that a nonconstant irreducible polynomial in $K_v[x]$ remains irreducible in $K[x]$. The leading coefficient of g is not divisible by p, and hence it is a unit in K_v. Thus the irreducible factorizations of $g/\mathrm{lc}(g)$ in $K_v[x]$ and in $K[x]$ coincide. In particular, all u_i lie in $K_v[x]$.

Now let $h_{i,p} \in E[x]$ be the monic gcd of $\overline{g}$ and $\overline{f} - \overline{z}_i \overline{g}'$ for $t < i \le \tau$. Moreover, let $g_1 \in \mathbb{Q}_v[x]$ be the product of all monic irreducible factors of multiplicity 1 in g. Condition c) implies that the product of all monic irreducible factors of multiplicity 1 in $\bar{g}$ is equal to $\bar{g}_1$. Fact 8.31 says that $g_1 = \prod_{1 \leq i \leq \tau} u_i$ in $K_v[x]$ and $\bar{g}_1 = \prod_{1 \leq i \leq \tau} \bar{h}_{i,p}$ in $E[x]$. Thus

$$\sum_{1 \leq i \leq \tau} \deg u_i = \deg g_1 = \sum_{1 \leq i \leq \tau} \deg h_{i,p} . \tag{8.11}$$

On the other hand, $\bar{u}_i$ divides $\bar{h}_{i,p}$, for all i. This together with (8.11) implies that $\bar{u}_i = \bar{h}_{i,p}$ for all i, and hence in particular $u_i \equiv h_{i,p} \bmod p$ for $1 \leq i \leq t$.

What we just have shown is true for all lucky primes $p \in S_0$. Since S_0 contains at least one lucky prime, we have $d = \deg(u_1 \cdots u_t)$ and $u_i \equiv h_{i,p} \bmod p$ for all $p \in S_2$ in step 4. Moreover, all lucky primes in S_0 are also in S_2, and the algorithm does not return "FAIL" in step 4. Let $m = \prod_{p \in S_2} p$. Then $h_i^* \equiv \mathrm{lc}(g)u_i \bmod m$ in step 6. For all i, the polynomial $\mathrm{lc}(g)u_i$ is in $\mathbb{Z}[x]$ and divides $\mathrm{lc}(g)g$, and hence $\|\mathrm{lc}(g)u_i\|_\infty \leq |\mathrm{lc}(g)|B \leq 2^{(\omega-1)k-1} < m/2$, by Mignotte's bound (Fact 3.3). The polynomial h_i^* has max-norm less than $m/2$ as well, by construction, and hence $h_i^* = \mathrm{lc}(g)u_i$ and $h_i = \mathrm{lc}(h_i)u_i \in \mathbb{Z}[x]$. The latter polynomial is primitive and divides g. By a similar reasoning, we have $b = g/h_1 \cdots h_t$, and again Mignotte's bound implies that the algorithm does not return "FAIL" in step 8.

Now let $f^* = f - \sum_{1 \leq i \leq t} z_i h_i' g/h_i \in \mathbb{Z}[x]$ and $m^* = \prod_{p \in S_3} p$. Definition 8.24 implies that f^* is divisible by $h_1 \cdots h_t$. Each z_i is a root of r and divides $r(0)$, and hence $|z_i| \leq |r(0)| < C$ for all i, by Corollary 3.2 (i). Thus

$$\|z_i h_i' g/h_i\|_2 \leq |z_i| \cdot \|h_i'\|_1 \|g/h_i\|_1 \leq nBC ,$$

by Mignotte's bound. Thus $\|f^*\|_2 < (n+1)^{1/2}2^\lambda + n^2 BC \leq (n^2+1)BC$, and Mignotte's bound implies that

$$\|f^*/h_1 \cdots h_t\|_\infty \leq 2^n \|f^*\|_2 \leq D \leq 2^{(\omega-1)s} < m^*/2 .$$

We have $a \equiv f^*/h_1 \cdots h_t \bmod m^*$ after step 12, and since both sides of the congruence have max-norms less than $m^*/2$, they are equal. Finally, Mignotte's bound implies that

$$\|a\|_1 \cdot \|h_1\|_1 \cdots \|h_t\|_1 \leq 2^n \|f^*\|_2 \leq D ,$$

and the algorithm does not return "FAIL" in step 13.

It remains to prove the second part of the theorem. Since the algorithm does not return "FAIL" in step 4, we have $B \leq 2^{(\omega-1)k-1} < m/2$. Since it does not return "FAIL" in step 8, we have

$$\|bh_1 \cdots h_t\|_\infty \leq \|bh_1 \cdots h_t\|_1 \leq \|b\|_1 \cdot \|h_1\|_1 \cdots \|h_t\|_1 \leq B < m/2 .$$

Thus the congruence $bh_1 \cdots h_t \equiv g \bmod m$ is in fact an equality. Similarly, the fact that the algorithm does not return "FAIL" in step 13 implies that

$$\begin{aligned}\|ah_1 \cdots h_t\|_\infty &\leq \|ah_1 \cdots h_t\|_1 \leq \|a\|_1 \cdot \|h_1\|_1 \cdots \|h_t\|_1 \leq D \leq 2^{(\omega-1)s-1} \\ &< m^*/2 .\end{aligned}$$

Thus the congruence $ah_1 \cdots h_r \equiv f^* \bmod m^*$ is in fact an equality as well. The h_i are nonconstant, squarefree, pairwise coprime, and coprime to b, since they are so modulo at least one in S_2, by Theorem 8.27. $\square$

Theorem 8.35 (Success probability of Algorithm 8.32). *Let*

$$A = (2n)!^{2n}(n+1)^{4n^2+4n}2^{(4n^2+4n+2)\lambda}$$

and $w = \lfloor (\log_2 A)/(\omega - 1) \rfloor \in \Theta(n^2(\lambda + \log n))$. *There are at most w single precision primes that are unlucky with respect to GP'-refinement. Suppose that the number of single precision primes, between $2^{\omega-1}$ and 2^ω, is at least $2w$. If we choose the set S_0 in step 2 uniformly at random from among at least $2w$ single precision primes, then the failure probability of the algorithm is at most $1/2$.*

Proof. Let $r = \mathrm{res}_x(g, f - yg') \in \mathbb{Z}[y] \setminus \{0\}$. Then Fact 7.24 (i) implies that $\deg r \le n$. Let $\varphi \in \mathbb{Z} \setminus \{0\}$ be the resultant of f and g, let $\gamma \in \mathbb{Z} \setminus \{0\}$ be the $\deg \gcd(g, g')$-th subresultant of g and g', and let $\rho \in \mathbb{Z} \setminus \{0\}$ be the $\deg \gcd(r, r')$-th subresultant of r and r'. Finally, let $R = \mathrm{lc}(fg)\varphi\gamma\rho \in \mathbb{Z} \setminus \{0\}$. We now show that a prime $p \in \mathbb{N}$ not dividing R is a lucky prime, $|R| \le A$, and the number of unlucky single precision primes is at most w.

Let p be a prime not dividing R. Then $\deg f = \deg(f \bmod p)$ and $\deg g = \deg(g \bmod p)$. Thus $p \nmid \varphi$ and Corollary 3.12 imply that $f \bmod p$ and $g \bmod p$ are coprime. Similarly, since $p > 2^{\omega-1} \ge n$, we have $\deg(g' \bmod p) = \deg g'$, and $p \nmid \gamma$ and Corollary 3.12 imply that $\deg \gcd(g \bmod p, g' \bmod p) = \deg \gcd(g, g')$. Finally, $\mathrm{lc}_y(r)$ is a divisor of $\mathrm{lc}(g)^{|\deg g - \deg f - 1|}\varphi\gamma$, by Fact 7.24 (iii), so that $\deg(r \bmod p) = \deg r$ and $\deg(r' \bmod p) = \deg r'$, and again $p \nmid \rho$ and Corollary 3.12 imply that $\deg \gcd(r \bmod p, r' \bmod p) = \deg \gcd(r, r')$. Thus p is a lucky prime.

The one-norm of any coefficient of g or $f - yg'$ in $\mathbb{Z}[y]$ is less than $(n+1)2^\lambda$. Corollary 3.2 shows that $\|r\|_2 \le \|r\|_1 \le (2n)!(n+1)^{2n}2^{2n\lambda}$ and

$$|\rho| \le \|r\|_2^n \cdot \|r'\|_2^n \le n^n\|r\|_2^{2n} \le (2n)!^{2n}(n+1)^{4n^2+n}2^{4n^2\lambda} .$$

The same corollary also shows that $|\varphi| \le (n+1)^n2^{2n\lambda}$ and $|\gamma| \le (n^2+n)^n2^{2n\lambda}$. Thus $|\mathrm{lc}(fg)\varphi\gamma| \le (n+1)^{3n}2^{(4n+2)\lambda}$, and we conclude that $|R| \le A$. We clearly have $k \le w$, so that we can choose $2k$ single precision primes in step 2. By the choice of w, the number of single precision primes dividing R, and hence also the number of unlucky single precision primes, is at most w. Thus the probability that a uniformly randomly chosen single precision prime is unlucky is at most $1/2$, and the last claim follows from Lemma 3.24 (i) and Theorem 8.34. $\square$

Theorem 8.36 (Cost of Algorithm 8.32). *If we ignore the cost for prime finding, then Algorithm 8.32 takes $O(n^4 + n^3\lambda^2)$ word operations with classical arithmetic and $O(n^2 \mathsf{M}(n) \log n + n \mathsf{M}(n(\lambda + \log n)) \log(n\lambda))$ or $O^{\sim}(n^3 + n^2\lambda)$ with fast arithmetic.*

Proof. Let $r = \text{res}_x(g, f - yg') \in \mathbb{Z}[y]$. Then the product of all z_i divides the constant coefficient of r, which is of absolute value at most C. Thus the sum of the word sizes of all z_i is $O(\log C)$ or $O(s)$.

We reduce f, g, and the z_i modulo all primes in S_0, at a cost of $O(k(n\lambda + s))$ word operations with classical arithmetic and $O(n\,\mathsf{M}(k)\log k + \mathsf{M}(s)\log s)$ with fast arithmetic. Computing the set S_1 in step 2 takes $O(kn^2)$ and $O(k\,\mathsf{M}(n) \cdot \log n)$ word operations with classical and fast arithmetic, respectively; checking whether $z_i \not\equiv z_j \bmod p$ is done by sorting the residues of the z_i. For each i and each prime, step 3 takes $O(n^2)$ and $O(\mathsf{M}(n)\log n)$ word operations with classical and fast arithmetic, respectively, and the number of iterations is $O(nk)$. The Chinese remaindering in steps 6 and 7 uses $O(k^2)$ word operations with classical arithmetic for each coefficient of some h_i^* or of b except the leading coefficients, and there are at most n coefficients in total. Thus the overall cost for steps 5 through 7 with classical arithmetic is $O(nk^2)$ word operations, including the cost for the normalization. The cost with fast arithmetic is $O(n\,\mathsf{M}(k)\log k)$.

The cost for step 10 is $O(snk + s^2)$ word operations with classical arithmetic and $O(n\,\mathsf{M}(s)\log s)$ with fast arithmetic. The modular computation in step 11 with classical arithmetic takes $O(n^2)$ for each of the s primes, as in the proof of Theorem 8.27. The cost with fast arithmetic is $O(s\,\mathsf{M}(n)\log n)$. Finally, step 12 takes $O(s^2)$ and $O(\mathsf{M}(s)\log s)$ per coefficient with classical and fast arithmetic, respectively, and there are $O(n)$ coefficients. The claims now follow from $k \in O(n + \lambda)$ and $s \in O(n(\lambda + \log n))$ by adding up costs. $\square$

We note that the coefficients of a are of word size $O(s)$, and hence the output size of the algorithm is $O(n^2(\lambda + \log n))$. In the diagonal case where $n \approx \lambda$, Algorithm 8.32 is asymptotically optimal – up to logarithmic factors – for those inputs where the upper bound on the output size is achieved.

Algorithm 8.37 (Prime power modular GP$'$-refinement). ▬▬▬▬▬
Input: Nonzero coprime polynomials $f, g \in \mathbb{Z}[x]$ with $\deg f, \deg g \leq n \leq 2^{\omega-1}$ and $\|f\|_\infty, \|g\|_\infty < 2^\lambda$, and a list of distinct positive integers $z_1, \ldots, z_t$, where $t \leq n$ and $\gcd(g, f - z_i g')$ is nonconstant for all i.
Output: A GP$'$-refinement of f/g as in Definition 8.24, or otherwise "FAIL".

1. $B \longleftarrow \lfloor (n+1)^{1/2} 2^{n+\lambda} \rfloor$, $\quad k \longleftarrow \lceil \log_2(2^{\lambda+1}B)/(\omega - 1) \rceil$
2. choose an odd single precision prime $p \in \mathbb{N}$
   if $p \mid \text{lc}(fg)$ or $\gcd(f \bmod p, g \bmod p) \neq 1$ or $z_i \equiv z_j \bmod p$ for some $i \neq j$
   then return "FAIL"
3. **for** $1 \leq i \leq t$ compute $\overline{h}_i \in \mathbb{Z}[x]$ of max-norm less than $p/2$ such that $\text{lc}(\overline{h}_i) = \text{lc}(g)$ and $\text{lc}(g)^{-1}\overline{h}_i \bmod p$ is the monic gcd in $\mathbb{F}_p[x]$ of $g \bmod p$ and $(f - z_i g') \bmod p$
4. compute $\overline{b} \in \mathbb{Z}[x]$ of max-norm less than $p/2$ with $\overline{b}\,\overline{h}_1 \cdots \overline{h}_t \equiv g \bmod p$
5. use Hensel lifting to lift the factorization of g from step 4 to a factorization $b^* h_1^* \cdots h_t^* \equiv g \bmod p^k$, with $b^*, h_1^*, \ldots, h_t^* \in \mathbb{Z}[x]$ of max-norm less than $p^k/2$ that are congruent modulo p to $\overline{b}, \overline{h}_1, \ldots, \overline{h}_t$, respectively
6. **for** $1 \leq i \leq t$ **do** $h_i \longleftarrow \text{normal}(h_i^*)$

7. compute $b \in \mathbb{Z}[x]$ of max-norm less than $p^k/2$ with $bh_1 \cdots h_r \equiv g \bmod p^k$
   if $\|b\|_1 \cdot \|h_1\|_1 \cdots \|h_t\|_1 > B$ **then return** "FAIL"
8. compute a such that $ah_1 \cdots h_t = f - \displaystyle\sum_{1 \le i \le t} z_i h_i' \frac{g}{h_i}$
9. **if** $a \notin \mathbb{Z}[x]$ **then return** "FAIL"
10. **return** $a, b, h_1, \ldots, h_t$

Theorem 8.38 (Correctness of Algorithm 8.37). *If p is lucky with respect to GP'-refinement, then Algorithm 8.37 does not return "FAIL". If the algorithm does not return "FAIL", then it returns the correct result.*

Proof. As in the proof of Theorem 8.34, we find that the algorithm does not return "FAIL" in step 2, and that $\mathrm{lc}(g)u_i \equiv \overline{h}_i \bmod p$ for $1 \le i \le t$, where u_i is the monic gcd of g and $f - z_i g'$ in $\mathbb{Q}[x]$. Since p is a lucky prime, the $\overline{h}_i$ are pairwise coprime and coprime to $\overline{b}$, and Hensel lifting can be applied in step 5. The uniqueness of Hensel lifting (Fact 3.20) implies that $\mathrm{lc}(g)u_i \equiv h_i^* \bmod p^k$ for $1 \le i \le t$. As in the proof of Theorem 8.34, Mignotte's bound implies that both sides of the congruence above are equal. Thus $h_i^* = \mathrm{lc}(g)u_i$, and $h_i = \mathrm{lc}(h_i)u_i \in \mathbb{Z}[x]$ is a normalized divisor of g, for $1 \le i \le t$. Similarly, we find that $b = g/h_1 \cdots h_t$, and Mignotte's bound implies that the algorithm does not return "FAIL" in step 7. By Theorem 8.29, the algorithm neither returns "FAIL" in step 9. The proof of the second claim is analogous to the proof of Theorem 8.34. $\square$

The proof of the following theorem is completely analogous to the proof of Theorem 8.35.

Theorem 8.39 (Success probability of Algorithm 8.37). *Let*

$$A = (2n)!^{2n}(n+1)^{4n^2+4n}2^{(4n^2+4n+2)\lambda} ,$$

$w = \lfloor (\log_2 A)/(\omega - 1) \rfloor \in O(n^2(\lambda + \log n))$, *and suppose that the number of single precision primes, between $2^{\omega-1}$ and 2^{ω}, is at least $2w$. If we choose p in step 2 uniformly at random from among at least $2w$ single precision primes, then the failure probability is at most $1/2$.*

Theorem 8.40 (Cost of Algorithm 8.37). *If we ignore the cost for prime finding, then Algorithm 8.37 takes $O(n^3(n \log n + n\lambda + \lambda^2))$ word operations with classical arithmetic and*

$$O(\mathsf{M}(n^2(\lambda + \log n)) \log n + n\,\mathsf{M}(n(\lambda + \log n)) \log \lambda + \mathsf{M}(n + \lambda)\,\mathsf{M}(n) \log n)$$

or $O^\sim(n^2\lambda)$ with fast arithmetic.

Proof. Let $D = (n^2 + 1)(n+1)^n 2^{2n\lambda+n}B$ and $s = \lceil \log_2(2D)/(\omega - 1) \rceil$, as in Algorithm 8.32. As in the proof of Theorem 8.36, the sum of the word sizes of all z_i is $O(s)$. Thus the cost for reducing all coefficients of f and g and all z_i modulo p is $O(n\lambda + s)$ word operations. As in the proof of Theorem 8.36, the expected cost for steps 2 through 4 is $O(n^3)$ word operations with classical arithmetic and $O(n \, \mathsf{M}(n) \log n)$ with fast arithmetic.

Hensel lifting in step 5 takes $O(k^2 n^2)$ word operations with classical arithmetic and $O((\mathsf{M}(k) + \log n) \, \mathsf{M}(n) \log n)$ with fast arithmetic, by Fact 3.20. The normalization in step 6 takes $O(nk^2)$ operations with classical arithmetic and $O(n \, \mathsf{M}(k) \log k)$ with fast arithmetic. In step 7, we just multiply b by a constant modulo p^k; this is dominated by the cost for step 6.

With classical arithmetic, we first compute g/h_i and $h_i' g/h_i$ for all i in step 8. By Mignotte's bound 3.3, the coefficients of h_i and g/h_i have word size $O(k)$, and hence this takes $O(n(\deg h_i)k^2)$ word operations, together $O(n^2 k^2)$ for all i. The same bound suffices to compute $h_1 \cdots h_t$, by Lemma 3.15. Let $f^* = f - \sum_{1 \le i \le t} z_i h_i' g/h_i$. As in the proof of Theorem 8.34, we have $|z_i|, \|f_*\|_\infty \le D$. Thus multiplying $h_i' g/h_i$ by z_i for all i, adding all t factors, and subtracting the result from f to get f^* takes $O(n^2 ks)$ word operations. Finally, dividing the polynomial f^*, with coefficients of word size $O(s)$, by the polynomial $h_1 \cdots h_t$, with coefficients of word size $O(k)$, takes $O(n^2 ks)$ word operations as well.

With fast arithmetic, we first compute $z_i h_i'$ for all i, taking $O(n \, \mathsf{M}(s))$ word operations. Using Algorithm 10.20 in von zur Gathen & Gerhard (1999) and Remark 3.14, we can compute f^* at a cost of $O(\mathsf{M}(n \log(nD)) \log n)$ word operations. By Fact 3.15, we can compute $h_1 \cdots h_k$ with $O(\mathsf{M}(n(k+\log n)) \log n)$ word operations. Using a modular algorithm based on Chinese remaindering similar to steps 10 through 12 of Algorithm 8.32 (see also Exercise 10.21 in von zur Gathen & Gerhard 1999), we can check whether $h_1 \cdots h_t \mid f^*$, and if so, compute the quotient a using $O(n \, \mathsf{M}(s) \log s + s \, \mathsf{M}(n))$ word operations.

The claims now follow from $k \in O(n + \lambda)$ and $s \in O(n(\lambda + \log n))$. $\square$

The classical estimate above for the variant with Hensel lifting is slightly worse than the corresponding estimate from Theorem 8.36 for the variant with Chinese remaindering. Up to logarithmic factors, the exponent of n in the fast arithmetic estimates is lower by one for Hensel lifting, although both estimates are about cubic in the diagonal case where $n \approx \lambda$.

We now analyze Algorithm 8.28 when we use Algorithm 7.25 in step 1 and Algorithm 8.32 or 8.37 for step 2.

Definition 8.41. *Let $f, g \in \mathbb{Z}[x]$ be nonzero and coprime. In analogy to the definition of the dispersion, we let*

$$\varepsilon(f, g) = \max\{i \in \mathbb{N} : i = 0 \text{ or } \mathrm{res}(g, f - ig') = 0\} \ .$$

By Corollary 3.12, $\varepsilon(f, g)$ is 0 or the maximal nonnegative integer root of $\mathrm{res}_x(g, f - yg') \in \mathbb{Z}[y]$, or equivalently, the maximal positive integer residue of the rational function f/g at a simple pole. The example $f = e$ and $g = x$, for $e \in \mathbb{N}$, shows that $\varepsilon(f, g)$ may be exponential in the size of the coefficients of f and g in the worst case.

Theorem 8.42 (Cost of Algorithm 8.28). *If we use Algorithm 7.25 in step 1 and Algorithm 8.32 (or 8.37) in step 2, then the cost for steps 1 and 2 of*

Algorithm 8.28 is $O(n^4(\lambda^2 + \log^2 n))$ word operations with classical arithmetic and

$$O(n^2(\mathsf{M}(n(\lambda + \log n)) + (\lambda + \log n)\mathsf{M}(n)\log n)$$
$$+\mathsf{M}(n(\lambda + \log n))(\mathsf{M}(n)\log n + n\log\lambda))$$

or $O^{\sim}(n^3\lambda)$ with fast arithmetic. If $e = \varepsilon(f, g)$, then $e \leq (n + 1)^n 2^{2n\lambda}$, and the cost for step 3 is $O(e^3(n^3 + n\lambda^2))$ word operations with classical arithmetic and

$$O((en\,\mathsf{M}(e(n + \lambda)) + e\lambda\,\mathsf{M}(en))\log(e(n + \lambda)))$$

or $O^{\sim}(e^2 n(n + \lambda))$ with fast arithmetic. The polynomial c has degree at most en and its coefficients are of word size $O(e(n + \lambda))$.

Proof. The first statement follows from Theorems 7.26, 8.36, and 8.40. Let $r = \mathrm{res}_x(g, f - yg') \in \mathbb{Z}[y]$. If $e \neq 0$, then it is a root of r, and hence it divides the constant coefficient of r. The upper bound on e now follows from Fact 7.24 (ii) and Corollary 3.2 (i). Let $B = (n + 1)^{1/2} 2^{n+\lambda}$. By Mignotte's bound 3.3, we have $\|h_i\|_1 \leq B$ for all i. Thus $\|h_i^z\|_\infty \leq \|h_i^z\|_1 \leq \|h_i\|_1^z \leq B^z$ for all $z \in \mathbb{N}$. By using a similar modular approach as in the proof of Lemma 3.15, the computation of $h_i^{z_i}$ takes $O(z_i^3 \log^2 B \cdot \deg h_i)$ word operations with classical arithmetic and

$$O(z_i(\deg h_i)\mathsf{M}(z_i\log B)\log(z_i\log B) + z_i(\log B)\mathsf{M}(z_i\deg h_i))$$

with fast arithmetic, together $O(e^3 n(n^2 + \lambda^2))$ and

$$O(en\,\mathsf{M}(e(n + \lambda))\log(e(n + \lambda)) + e\lambda\,\mathsf{M}(en))\,,$$

respectively. Finally, we have $\deg c = \sum_{1 \leq i \leq t} z_i \deg h_i \leq en$ and

$$\|c\|_\infty \leq \|c\|_1 \leq \prod_{1 \leq i \leq t} \|h_i^{z_i}\|_1 \leq \left(\prod_{1 \leq i \leq t} \|h_i\|\right)^e \leq B^e\,,$$

again by Mignotte's bound, and Lemma 3.15 yields the cost estimates. □

One might think about combining steps 1 and 2 of Algorithm 8.2 in a similar way as described in Sect. 7.4, by computing not only the resultant of g and $f - yg'$, but also some appropriate subresultants, and plugging in $y = z_i$ to obtain h_i. However, a comparison of the estimates above with those from Theorem 7.31 indicates that there would be no benefit when using classical arithmetic, and at most a gain of logarithmic factors when using fast arithmetic.

As in the difference case, we can use Algorithm 8.28 to compute rational solutions of homogeneous linear first order differential equations with polynomial coefficients (see Corollary 8.30).

Corollary 8.43. *Given two nonconstant coprime polynomials $f, g \in \mathbb{Z}[x]$ of degree at most n and max-norm less than 2^λ, we can decide whether the homogeneous linear first order differential equation*

$$g \cdot \rho' - f \cdot \rho = 0$$

has a solution $\rho \in \mathbb{Q}(x) \setminus \{0\}$, and if so, compute one, using $O(n^4(\lambda^2 + \log^2 n))$ word operations with classical arithmetic and

$$O(n^2(\mathsf{M}(n(\lambda + \log n)) + (\lambda + \log n)\mathsf{M}(n) \log n)$$
$$+\mathsf{M}(n(\lambda + \log n))(\mathsf{M}(n) \log n + n \log \lambda))$$

or $O^\sim(n^3\lambda)$ with fast arithmetic.

Proof. We first call Algorithm 8.28 to compute a GP$'$-form $(a, b, u) \in \mathbb{Z}[x]^3$ of the rational function f/g. Then we call the same algorithm to compute a GP$'$-form $(A, B, v) \in \mathbb{Z}[x]^3$ of the rational function $-a/b$. If $A = 0$, then

$$g \left(\frac{u}{v}\right)' = \frac{gu}{v} \cdot \left(\frac{u'}{u} - \frac{v'}{v}\right) = \frac{gu}{v} \left(\frac{u'}{u} + \frac{a}{b}\right) = \frac{gu}{v} \frac{f}{g} = f \cdot \frac{u}{v},$$

and $\rho = u/v$ solves the differential equation.

Conversely, if $\rho \in \mathbb{Q}(x) \setminus \{0\}$ solves the differential equation and we let $\sigma = \rho \cdot v/u$, then a similar calculation as above shows that

$$\frac{\sigma'}{\sigma} = \frac{A}{B} .$$

The properties of the GP$'$-form imply that $\gcd(B, A - iB') = 1$ for all $i \in \mathbb{Z}$, i.e., the rational function A/B has no integer residues at simple poles. Let $\sigma = U/V$ with coprime polynomials $u, v \in \mathbb{Z}[x]$. Then

$$\frac{A}{B} = \frac{U'}{U} - \frac{V'}{V} .$$

Assume first that U is not constant, and let $p \in \mathbb{Z}[x]$ be a nonconstant irreducible factor, of multiplicity $i \in \mathbb{N}$. Then $p \nmid V$, since U and V are coprime, and Lemma 8.22 (viii) shows that $\mathrm{Res}_p(V'/V) = 0$. However, then $\mathrm{Res}_p(A/B) = \mathrm{Res}_p(U'/U) = i$, and part (ix) of the same Lemma leads to the contradiction that $p \mid \gcd(B, A - iB')$. A similar argument shows that V is constant as well, and hence $A = B\sigma'/\sigma = 0$. Thus the differential equation is unsolvable if $A \neq 0$.

To prove the cost estimate, we do not proceed as described above, but instead compute *all* integers $z \in \mathbb{Z}$ such that $\gcd(g, f - zg')$ is nonconstant in step 1 of Algorithm 8.28. The modified algorithm directly computes $A, B, u, v \in \mathbb{Z}[x]$ such that

$$\frac{f}{g} = \frac{A}{B} + \frac{u'}{u} - \frac{v'}{v},$$

within the same time bound as given by the first part of Theorem 8.42, if we do not expand the product representations of u and v in step 5 of the modified algorithm. $\square$

9. Polynomial Solutions of Linear First Order Equations

In this chapter, we discuss several algorithms for computing polynomial solutions of linear first order differential and difference equations with polynomial coefficients. We start with several non-modular algorithms, in historical order. Finally, in section 9.8, we present new modular algorithms.

Let F be a field. A *linear first order differential equation with polynomial coefficients* is an equation of the form $a\,Du + bu = c$, where D is the differential operator $Du = u'$ and $a, b, c \in F[x]$ are given polynomials. We are looking for a solution u which may – in principle – be an arbitrary function, but we will confine ourselves to polynomials in this chapter. A *linear first order difference equation with polynomial coefficients* is an equation of the form $a\,\Delta u + bu = c$, where $a, b, c \in F[x]$ are given polynomials and $u \in F[x]$ is sought. Here, Δ denotes the difference operator $(\Delta u)(x) = u(x+1) - u(x)$. Often difference equations are given in the equivalent form of *recurrences*, in terms of the shift operator E with $(Eu)(x) = u(x+1)$. For example, the difference equation above is equivalent to the recurrence $a\,Eu + (b - a)u = c$. Algorithms for finding polynomial solutions of linear first order difference or differential equations play an important role in symbolic summation and symbolic integration algorithms; see Chap. 10.

There are strong analogies between the theory of linear differential equations and the theory of linear difference equations, which become most clearly visible if we think of the difference operator Δ as being the discrete analog of the differential operator. These analogies lead to the concepts of *Ore rings* and *pseudo-linear algebra* (Ore 1932a, 1932b, 1933; Jacobson 1937; Bronstein & Petkovšek1994, 1996; Abramov, Bronstein & Petkovšek 1995; Bronstein 2000). In order to cover the similarities, we rephrase both differential and difference equations in terms of linear operators on polynomials. In the differential case, we let $L = aD + b$ be the linear operator with $Lu = au' + bu$, and in the difference case, we let $L = a\Delta + b$, with $Lu = a\,\Delta u + bu$, or equivalently, $L = aE + b - a$. Then both the differential equation and the difference equation can be conveniently written as $Lu = c$. We denote the solution space in $F[x]$ of the latter equation by $L^{-1}c$. In particular, $\ker L = L^{-1}0$ is the kernel of L. The following result is well-known for characteristic zero (see, e.g., Lisoněk, Paule & Strehl 1993).

Lemma 9.1. *Let F be a field of characteristic p, $a, b \in F[x]$ not both zero, and $L = aD + b$ or $L = a\Delta + b$. Moreover, let $\ker L = \{u \in F[x]: Lu = 0\}$ denote the kernel of L.*

J. Gerhard: Modular Algorithms, LNCS 3218, pp. 149-193, 2004.
© Springer-Verlag Berlin Heidelberg 2004

(i) *If $p = 0$, then $\ker L$ is either $\{0\}$ or a one-dimensional F-subspace of $F[x]$.*
(ii) *If $p > 0$, we let $V = \{u \in F[x] : \deg u < p\}$. If $V \cap \ker L \neq \{0\}$, then it is a one-dimensional F-subspace of V.*

Proof. If $a = 0$, then $\ker L = \{0\}$, and we may assume that $a \neq 0$. We first assume that we can choose a point $s \in F$ such that $a(x + s)$ has no integer roots (this is always possible if $p = 0$), and consider the linear map $\varphi \colon \ker L \longrightarrow F$ which evaluates a polynomial at $x = s$. We show that φ is injective on $\ker L$ if $p = 0$, and on $V \cap \ker L$ if $p > 0$. Let $u \in \ker L$ be such that $\varphi(u) = u(s) = 0$. We claim that $u^{(i)}(s) = 0$ for all $i \in \mathbb{N}$ in the differential case, and $u(s+i) = 0$ for all $i \in \mathbb{N}$ in the difference case. The proof proceeds by induction on i. The case $i = 0$ is clear, and we assume that $i \geq 1$. In the differential case, we differentiate the equation $Lu = 0$ $(i - 1)$ times and find polynomials $g_0, \ldots, g_{i-1} \in F[x]$ such that

$$au^{(i)} + \sum_{0 \leq j < i} g_j u^{(j)} = 0 \, .$$

Substituting $x = s$ yields $a(s)u^{(i)}(s) = 0$, by the induction hypothesis, and the assumption $a(s) \neq 0$ finishes the induction step. In the difference case, we substitute $x = s + i - 1$ in the equation $Lu = 0$ and obtain

$$a(s + i - 1)u(s + i) + (b(s + i - 1) - a(s + i - 1))u(s + i - 1) = 0 \, .$$

By induction, we have $u(s + i - 1) = 0$, and the assumption that $a(s + i - 1) \neq 0$ yields $u(s - i) = 0$. This proves the claim.

If $p = 0$, then a nonzero polynomial $u \in F[x]$ has $u^{(\deg u)}(s) \neq 0$ and at most $\deg u$ roots, and hence the only element $u \in \ker L$ with $\varphi(u) = u(s) = 0$ is the zero polynomial. Thus $\dim \ker L = \dim \ker \varphi + \dim \operatorname{im} \varphi \leq 1$.

If $p > 0$, then a nonzero polynomial $u \in F[x]$ of degree less than p has $u^{(\deg u)}(s) \neq 0$, and since $s, s + 1, \ldots, s + \deg u$ are distinct elements of F, at least one of them is not a root of u. If we denote by φ_V the restriction of φ to the subspace V, then we see that $\dim \ker \varphi_V = \dim(V \cap \ker \varphi) = 0$, and hence $\dim(V \cap \ker L) = \dim \ker \varphi_V + \dim \operatorname{im} \varphi_V \leq 1$.

It remains to prove the case where $p > 0$ and there is no $s \in F$ such that $a(x+s)$ has no integer roots. However, we can always find such an s in an extension field E of F (for example, any proper algebraic extension of the splitting field of a over F will do). If $V_E = \{u \in E[x] : \deg u < p\}$ and $\ker L_E$ denotes the kernel of the operator L over $E[x]$, then $\dim_E(V_E \cap \ker L_E) \leq 1$, by what we just have shown. However, any F-basis of $V \cap \ker L$ is also linearly independent over E, and hence $\dim_F(V \cap \ker L) \leq 1$ as well. $\square$

We note that $\ker L$ may be infinite-dimensional in positive characteristic. For example, if F is a field of characteristic $p > 0$, then the solution space of the differential equation $u' = 0$ in $F[x]$ is $F[x^p]$.

Definition 9.2. *Let $a, b \in F[x]$ be polynomials, not both zero. The degree $\deg L$ of the linear differential (or difference) operator $L = aD + b$ (or $L = a\Delta + b$) is $\max\{\deg a - 1, \deg b\}$.*

The motivation for this definition of degree is given by the following well-known lemma. We use the convention that the zero polynomial has degree $-\infty$ and leading coefficient 1.

Lemma 9.3. *Let F be a field of arbitrary characteristic, $a, b \in F[x]$ not both zero, $u \in F[x]$ nonzero, $L = aD + b$ or $L = a\Delta + b$, $n = \deg L$, and let $a_{n+1}, b_n \in F$ be the coefficients of x^{n+1} in a and of x^n in b, respectively. Then $\deg(Lu) \leq n + \deg u$ and the coefficient of $x^{n+\deg u}$ in Lu is $\mathrm{lc}(u)(a_{n+1} \deg u + b_n)$.*

Proof. We only give the proof in the difference case; the differential case is completely analogous. We have

$$
\begin{aligned}
\deg(Lu) &= \deg(a\, \Delta u + bu) \\
&\leq \max\{\deg a + \deg u - 1, \deg b + \deg u\} = n + \deg u\ .
\end{aligned}
$$

Write $a = \sum_{0 \leq i \leq n+1} a_i x^i$, $b = \sum_{0 \leq i \leq n} b_i x^i$, and $u = \sum_{0 \leq i \leq d} u_i x^i$, with $d = \deg u$ and $u_d = \mathrm{lc}(u) \neq 0$. We note that by definition of n, at least one of a_{n+1} and b_n is nonzero. Multiplying out and using the fact that Δx^d has degree at most $d - 1$ and the coefficient of x^{d-1} is d, we find that

$$
\begin{aligned}
Lu &= a\, \Delta u + bu = a_{n+1} x^{n+1} \cdot d u_d x^{d-1} + \cdots + b_n x^n \cdot u_d x^d + \cdots \\
&= u_d(a_{n+1} d + b_n) x^{n+d} + \cdots\ ,
\end{aligned}
$$

where the dots hide terms of degree less than $n + d$. $\square$

Definition 9.4. *Let $a, b \in F[x]$ be polynomials, not both zero, and $L = aD + b$ or $L = a\Delta + b$ with $n = \deg L$. We associate to L an element $\delta_L \in F \cup \{\infty\}$, as follows:*

$$
\delta_L = \begin{cases}
0 & \text{if } \deg a - 1 = n > \deg b\ , \\[2mm]
-\dfrac{\mathrm{lc}(b)}{\mathrm{lc}(a)} & \text{if } \deg a - 1 = n = \deg b\ , \\[2mm]
\infty & \text{if } \deg a - 1 < n = \deg b\ .
\end{cases}
$$

The equation $a_{n+1}d + b_n = 0$, with a_{n+1}, b_n as in Lemma 9.3, is called the *indicial equation* at infinity of the linear operator L in the theory of linear differential or difference equations (see, e.g, Part A, Sect. 18.1 in Kamke 1977 or Chap. VII in Ince 1926 for the differential case). Thus δ_L is the unique root of this indicial equation, if $\delta_L \neq \infty$. If δ_L is different from 0 and ∞, then it can also be considered as the residue at infinity of the rational function $-b/a$, i.e., the coefficient of x in its x^{-1}-adic expansion.

Corollary 9.5. *Let F be a field of characteristic p, $L = aD + b$ or $L = a\Delta + b$ a linear differential or difference operator, respectively, with $a, b \in F[x]$ not both zero, and $u \in F[x]$ nonzero. Then*

$$
\begin{aligned}
\deg(Lu) &\leq \deg L + \deg u\ , \\
\deg(Lu) &< \deg L + \deg u \iff (\deg u) \bmod p = \delta_L \text{ in } F\ .
\end{aligned}
$$

The following well-known corollary (see, e.g., Gosper 1978 for the difference case) gives a degree bound for a polynomial solution $u \in F[x]$ of $Lu = c$ in characteristic zero.

Corollary 9.6. *Let F be a field of characteristic zero and $L = aD+b$ or $L = a\Delta+b$ be a linear differential or difference operator, respectively, with $a, b \in F[x]$ not both zero, and $u, c \in F[x]$ nonzero polynomials satisfying $Lu = c$. Then $\delta_L \neq \deg c - \deg L$. Moreover,*

(i) $\deg u = \deg c - \deg L$ if $\delta_L \notin \mathbb{N}$ or $\delta_L < \deg c - \deg L$,
(ii) $\deg u \in \{\deg c - \deg L, \delta_L\}$ if $\delta_L \in \mathbb{N}$ and $\delta_L > \deg c - \deg L$.

In particular, we have the degree bound $\deg u \leq \max(\{\deg c - \deg L, \delta_L\} \cap \mathbb{Z})$. If $c = 0$, then $\deg u = \delta_L \in \mathbb{N}$.

We note that the degree bound of Corollary 9.6 may be exponentially large in the bit size of a, b, and c. For example, if $a = x$ and $b = -t$ for some $t \in \mathbb{N}_{\geq 1}$ and c has "small" degree and "small" coefficients as well, then $\deg L = 0$, $\delta_L = t$, and the bound is $d = \max\{\deg c, \delta\} = \delta = t$. If $F = \mathbb{Q}$, then this is exponential in the bit size of a and b, which is about $\log t$.

Here is the analog of Corollary 9.6 in positive characteristic.

Corollary 9.7. *Let F be a field of characteristic $p > 0$ and $L = aD + b$ or $L = a\Delta + b$ be a linear differential or difference operator, respectively, with $a, b \in F[x]$ not both zero, and $u, c \in F[x]$ nonzero polynomials satisfying $Lu = c$. Then $\delta_L \neq (\deg c - \deg L) \bmod p$. Moreover,*

(i) $\deg u = \deg c - \deg L$ if $\delta_L \notin \mathbb{F}_p$,
(ii) $\deg u = \deg c - \deg L$ or $(\deg u) \bmod p = \delta_L$ if $\delta_L \in \mathbb{F}_p$.

If $c = 0$, then $(\deg u) \bmod p = \delta_L \in \mathbb{F}_p$.

The following lemma says how the degree and the indicial equation change when we multiply a linear operator L by a polynomial.

Lemma 9.8. *Let F be a field and $L = aD + b$ or $L = a\Delta + b$ a linear differential or difference operator, respectively, with $a, b \in F[x]$ not both zero. Let $z \in F[x]$ be nonzero and define new operators M and K by $Mu = L(zu)$ and $Ku = zLu$ for all $u \in F[x]$. Then $\deg M = \deg K = \deg L + \deg z$ and $\delta_M + \deg z = \delta_K = \delta_L$, with the convention that $\delta_M = \infty$ if $\delta_L = \infty$.*

Proof. We carry out the proof only for the differential case; the difference case is shown analogously. First, we note that

$$Mu = aD(zu) + bzu = azDu + (az' + bz)u = azDu + Lz \cdot u \,,$$

holds for all polynomials u, and hence $M = (az)D + Lz$. Thus

$$\begin{aligned} \deg M &= \max\{(\deg az) - 1, \deg Lz\} \\ &\leq \max\{\deg a - 1 + \deg z, \deg L + \deg z\} = \deg L + \deg z \,, \end{aligned} \tag{9.1}$$

with equality if $\deg z \neq \delta_L$ in F, by Corollary 9.5. If $\deg z = \delta_L$, then in particular $\delta_L \neq \infty$, and hence $\deg L = \deg a - 1$ and we have equality in (9.1) as well. Similarly, $K = (az)D + bz$, and hence

$$
\begin{aligned}
\deg K &= \max\{(\deg az) - 1, \deg bz\} = \max\{\deg a - 1, \deg b\} + \deg z \\
&= \deg L + \deg z \ .
\end{aligned}
$$

If $\delta_L = \infty$, then $\deg a - 1 < \deg L$, $\deg Lz = \deg L + \deg z$, and hence $(\deg az) - 1 < \deg L + \deg z = \deg M = \deg K$ and $\delta_M = \delta_K = \infty$. Otherwise, if $\delta_L \neq \infty$, then $\deg a - 1 = \deg L$. The coefficient of $x^{\deg L + \deg z}$ in Lz is $\mathrm{lc}(az)(\deg z - \delta_L)$, by Lemma 9.3. Dividing this equality by $-\mathrm{lc}(az)$, we find that $\delta_M = \delta_L - \deg z$. Similarly, the coefficient of $x^{\deg L + \deg z}$ in bz is $\mathrm{lc}(z)$ times the coefficient of $x^{\deg L}$ in b, i.e., $-\mathrm{lc}(az)\delta_L$, and hence $\delta_K = \delta_L$. $\square$

The next two lemmas give cost estimates for applying a linear difference or differential operator, respectively.

Lemma 9.9. *Let $L = aE + b$ be a linear difference operator, with $a, b \in F[x]$ not both zero, $u \in F[x]$, $n = \deg L$, $d = \deg u$, and $k \leq n + d$. If F has characteristic zero or greater than k, then we can compute the lowest k coefficients of Lu using $O((d + \min\{k, n\}) \cdot \min\{k, d\})$ arithmetic operations in F with classical arithmetic and $O(\mathsf{M}(k) + \mathsf{M}(d))$ with fast arithmetic.*

Proof. Computing Eu takes $O(d^2)$ and $O(\mathsf{M}(d))$ with classical and fast arithmetic, respectively, by Theorems 4.3 and 4.5. If $k < d$, then we successively compute $u(1), u'(1), \ldots, u^{(k-1)}(1)/(k-1)!$ at a cost of $O(kd)$ with classical arithmetic. We can compute the products $a \cdot Eu$ and $b \cdot u$ modulo x^k and add them with $O(\min\{k, n\} \cdot \min\{k, d\})$ and $O(\mathsf{M}(k))$ arithmetic operations, respectively. $\square$

Lemma 9.10. *Let $L = aD + b$ be a linear differential operator, with $a, b \in F[x]$ not both zero, $u \in F[x]$, $n = \deg L$, $d = \deg u$, and $k \leq n + d$. Then we can compute the lowest k coefficients of Lu using $O(\min\{k, n\} \cdot \min\{k, d\})$ arithmetic operations in F with classical arithmetic and $O(\mathsf{M}(k))$ with fast arithmetic.*

Proof. We can compute the products $a \cdot u'$ and $b \cdot u$ modulo x^k and add them with $O(\min\{k, n\} \cdot \min\{k, d\})$ and $O(\mathsf{M}(k))$ arithmetic operations, respectively. $\square$

In the following sections we present several algorithms for solving first order linear differential and difference equations with polynomial coefficients, in historical order.

Let L be a first order linear differential or difference operator with coefficients in $F[x]$, $n = \deg L$, and $c \in F[x]$. Many known algorithms to solve the equation $Lu = c$ in characteristic zero determine first an upper bound d on the degree of a possible solution u, via Corollary 9.6. Then they set up a system of $n + d + 1$ linear equations in the $d + 1$ unknown coefficients of u with respect to some suitable basis which is equivalent to the original equation. The most commonly used basis is the monomial basis $1, x, x^2, \ldots$. This is known as the *method of undetermined coefficients* and first appears in Proposition VII "Ex aequatione quantitates duas fluentes vel solas vel una cum earum fluixionibus involvente quantitatem alterutram in series convergente extrahere." of Newton (1691/92). The coefficient matrix of the linear system is triangular with at most one nonzero diagonal entry, and the standard linear algebra approach takes $O(nd + d^2)$ field operations. We discuss this method in the following section. In later sections, we will present asymptotically fast algorithms with only $O^\sim(n + d)$ operations.

If $\delta_L \in \mathbb{N}$, then either $\ker L = \{0\}$ or it is a one-dimensional F-subspace of $F[x]$, generated by a unique monic polynomial of degree δ_L, by Lemma 9.1 and Corollary 9.6. Then the solution u of the equation $Lu = c$ may not be unique, i.e., there may be one degree of freedom. There are two natural ways to handle this. The degree of freedom may either correspond to the choice of the coefficient of x^{δ_L} in u, or to the value $u(s)$ at some suitably chosen point $s \in F$, as in the proof of Lemma 9.1.

In those algorithms presented below that pursue the first approach, there is at most one polynomial $w \in F[x]$ of degree at most d satisfying the upper $d + 1$ linear equations of the system equivalent to $Lw = c$ and such that the coefficient of x^{δ_L} in w vanishes. If no such w exists, then $Lu = c$ is unsolvable. Similarly, there is a unique monic polynomial $h \in F[x]$ of degree δ_L satisfying the upper $\delta_L + 1$ linear equations of the system equivalent to $Lh = 0$. The algorithms compute these polynomials w and h, and then try to determine κ, the coefficient of x^{δ_L} in u, in such a way that $u = w + \kappa h$ satisfies the lower n linear equations of the system equivalent to $Lu = c$. Then $\ker L$ is Fh if $Lh = 0$ and $\{0\}$ otherwise, and the solution space $L^{-1}c$ of the inhomogeneous equation is $u + \ker L$, if a suitable κ exists, and empty otherwise.

The algorithms choosing the second possibility usually first perform a Taylor shift $x \longmapsto x + s$ on L and c, thus transforming the evaluation point s into the origin. Next, they compute the unique solutions w with $w(0) = 0$ and h with $h(0) = 1$ of the lower $d + 1$ and $\delta_L + 1$ linear equations equivalent to $Lw = c$ and $Lh = 0$, respectively. As in the first case, such a polynomial w need not exist, in which case $Lu = c$ is unsolvable. Then they try to determine κ, the constant coefficient of u, in such a way that $u + \kappa h$ also satisfies the upper n equations of the linear system equivalent to $Lu = c$. Finally, they transform u and h via the Taylor shift $x \longmapsto x - s$ to obtain a solution of the original equation. As above, $\ker L$ is Fh if $Lh = 0$ and $\{0\}$ otherwise, and the solution space $L^{-1}c$ of the inhomogeneous equation is $u + \ker L$, if a suitable κ exists, and empty otherwise. The first approach has the advantage that no Taylor shifts are necessary.

Remark 9.11. *We note that we can employ the techniques from the previous chapter for computing all polynomial solutions of a homogeneous first order equation $Lu = 0$. If $L = aE + b$ or $L = aD + b$, respectively, with $a, b \in F[x]$, then we compute a Gosper-Petkovšek form or a GP'-form, respectively, $(f, g, u) \in F[x]^3$ of the rational function $-b/a$, as described in Chap. 8. If f and g are constant, then $\ker L = Fu$, and otherwise $\ker L = \{0\}$. (See also Corollaries 8.19 and 8.43.) In contrast to the algorithms presented in this chapter, these algorithms have the* *advantage that they yield a representation of u of polynomial size in the input size, if the coefficients of u are not explicitly needed. For that reason, we concentrate on the inhomogeneous case in what follows, i.e., on determining $L^{-1}c$ for a nonzero polynomial $c \in F[x]$.*

9.1 The Method of Undetermined Coefficients

In the spirit of Newton (1691/92), we do not state this method as an algorithm working on matrices, but rather as a recursive algorithm working with polynomials. This view is also taken, e.g., by Abramov (1971) and Rothstein (1976, 1977). So let F be a field of characteristic zero, L be a nonzero linear first order differential or difference operator with coefficients in $F[x]$, $n = \deg L$, and $c \in F[x]$. Given the degree bound $d \in \mathbb{N}$ from Corollary 9.6 on a possible solution $u \in F[x]$ of the equation $Lu = c$, the idea is to equate the coefficients of x^{n+d} in Lu and c in order to determine the coefficient u_d of x^d in u and then proceed recursively with $c - L(u_d x^d)$ instead of c. The following algorithm finds a polynomial $u \in F[x]$ such that $\deg(Lu - c) < n$, if one exists.

Algorithm 9.12 (Method of undetermined coefficients: part I).
Input: A linear differential operator $L = aD + b$ or linear difference operator $L = aE + b$, with $a, b \in F[x]$ not both zero and $n = \deg L$, a polynomial $c \in F[x]$, where F is a field of characteristic zero, an integer $d \in \mathbb{Z}$, and the polynomials Lx^i for $0 \leq i \leq d$.
Output: Polynomials $u, r \in F[x]$ with $\deg u \leq d$, $\deg r < n$, and $r = Lu - c$, or otherwise "unsolvable".

1. **if** $\deg c < n$ **then return** 0 and $-c$
2. **if** $d < \deg c - n$ or $\delta_L = \deg c - n$ **then return** "unsolvable"
3. **if** $d = \delta_L$ **then** $u_d \longleftarrow 0$
   else let $l \in F$ be the coefficient of x^{n+d} in Lx^d and $u_d \in F$ be the coefficient of x^{n+d} in c/l
4. **call** the algorithm recursively with input L, $c - u_d Lx^d$, and $d - 1$, to obtain $v, r \in F[x]$ with $\deg v \leq d - 1$, $\deg r < n$, and $r = L(u_d x^d + v) - c$
5. **if** the recursive call returns "unsolvable" **then return** "unsolvable"
   else return $u_d x^d + v$ and r

In practice, one would divide a, b, c by their greatest common divisor before applying any algorithm to solve the first order differential or difference equation. Paule (1995) remarks that we may even achieve that a, b, c are *pairwise* coprime in the difference case. For example, if $g = \gcd(a, c)$ is nonconstant and coprime to b and u satisfies $a\,Eu + bu = c$, then necessarily g divides u, and $u^* = u/g$ is a solution of the linear difference equation

$$\frac{a\,Eg}{g} \cdot Eu^* + bu^* = \frac{c}{g},$$

whose right hand side has smaller degree than the right hand side of the original equation. Similar simplifications work in the difference case if $\gcd(b, c)$ is nonconstant and coprime to a, and in the differential case.

The following lemma investigates the solution space determined by the output specification of Algorithm 9.12.

Lemma 9.13. *Let $L = aD + b$ or $L = a\Delta + b$ be a linear differential or difference operator, with $a, b \in F[x]$ not both zero, $n = \deg L$, and $c \in F[x] \setminus \{0\}$ of degree at most $n + d$. Consider the sets $S_0 = \{u \in F[x]: \deg u \leq d \text{ and } \deg(Lu) < n\}$ and $S_c = \{u \in F[x]: \deg u \leq d \text{ and } \deg(Lu - c) < n\}$.*

- *(i) S_0 is an F-linear subspace of $F[x]$. We have $S_0 = \{0\}$ if and only if $\delta_L \notin \{0, \ldots, d\}$, and otherwise S_0 is one-dimensional and generated by a unique monic polynomial of degree δ_L.*
- *(ii) If $\delta_L \notin \{0, \ldots, d\}$, then $S_c = \{u\}$ for a unique polynomial $u \in F[x]$ of degree $\deg c - n$. Otherwise, either $S_c = \emptyset$ or there exists a unique nonzero polynomial $u \in S_c$ of degree at most d such that the coefficient of x^{δ_L} in u vanishes, and $S_c = u + S_0$.*

Proof. By Corollary 9.5, we have $\deg(Lu) \leq n + \deg u \leq n + d$ for all $u \in F[x]$ of degree at most d.

- (i) We have $\deg Lu < n$ if and only if the coefficients of $x^{n+d}, \ldots, x^{n+1}, x^n$ in Lu vanish. If we regard the coefficients of u as indeterminates, then we can rephrase this condition as a homogeneous system of $d + 1$ linear equations in these $d + 1$ indeterminates. This proves that S_0 is a linear subspace. Let $R = (r_{ij})_{0 \leq i, j \leq d}$ be the $(d + 1) \times (d + 1)$ coefficient matrix of this system, where $r_{ij} \in F$ is the coefficient of x^{n+i} in Lx^j for all i, j. Then Lemma 9.3 implies that R is lower triangular, such that $r_{ij} = 0$ if $i < j$, and the diagonal entries are $r_{ii} = a_{n+1}j + b_n$, where $a_{n+1}, b_n \in F$ are the coefficients of x^{n+1} in a and of x^n in b, respectively. In the differential case, R comprises the upper $d + 1$ rows of the matrix in Fig. 9.1 below.
  By definition of δ_L, we have $r_{ii} = a_{n+1}(\delta_L - i)$ if $\deg a = n + 1$, and $r_{ii} = b_n \neq 0$ and $\delta_L = \infty$ if $\deg a < n + 1$. Thus R is regular and $S_0 = \{0\}$ if and only if $\delta_L \notin \{0, \ldots, d\}$. If this is not the case, then R has rank d and S_0 is one-dimensional. We have one degree of freedom in choosing the coefficient of x^{δ_L} in u, so that S_0 contains a unique monic polynomial $u \in F[x]$ of degree δ_L and $S_0 = F \cdot u$.
- (ii) If $\deg c < n$, then $S_c = S_0$, and we may assume that $\deg c \geq n$. Then $\deg(Lu - c) < n$ implies that $\deg(Lu) = \deg c$, and Corollary 9.5 shows that $\deg u = \deg c - n$ if $\delta_L \notin \{0, \ldots, d\}$. If we let v be the vector with the coefficients of $x^{n+d}, \ldots, x^{n+1}, x^n$ in c and identify $u \in F[x]$ with its coefficient vector in F^{d+1}, then $u \in S_c$ if and only if $Ru = c$, and the claims follow from (i). $\square$

Theorem 9.14. *Algorithm 9.12 works correctly as specified and uses $O(nd + d^2)$ arithmetic operations in F.*

Proof. By similar arguments as in the proof of Corollary 9.6, the algorithm returns the correct result in steps 1 or 2. So we may assume that $\delta_L \neq \deg c - n$ and $d \geq \deg c - n \geq 0$. If $d \neq \delta_L$, then $\deg(Lx^d) = n + d$, by Corollary 9.5, and hence $l \neq 0$ and u_d is well-defined. Since the algorithm terminates in step 2 if $d < 0$, we may conclude by induction on d that the recursive call returns the correct result. If the algorithm does not return "unsolvable", then we have $\deg(u_d x^d + v) \leq d$ and $\deg(L(u_d x^d + v) - c) < n$, and the result returned is correct. Conversely, assume that there exists a polynomial $u \in F[x]$ of degree at most d satisfying $\deg(Lu - c) < n$. By Lemma 9.13, we may assume that $\deg u < d$ if $d = \delta_L$. Thus the coefficient of x^d in u is equal to u_d in the latter case. Otherwise, this follows from comparing the coefficients of x^{n+d} in Lu and c. In any case, $\deg(u - u_d x^d) \leq d - 1$ and $\deg(L(u - u_d x^d) - (c - u_d Lx^d)) = \deg(Lu - c) < n$, and hence $\deg v \leq d - 1$ and $\deg(Lv - (c - u_d Lux^d)) < n$ has the solution $v = u - u_d x^d$. We conclude that the output of the algorithm is correct if the recursive call returns "unsolvable".

Step 3 costs one division in F. If the algorithm does not stop in step 2, then $\deg c \leq n + d$, and Corollary 9.5 shows that $\deg(Lx^d) \leq n + d$ as well. Thus the cost for computing $c - u_d Lx^d$ is at most $n + d$ additions and the same number of multiplications in F. Hence the total cost is at most

$$\sum_{0 \leq i \leq d} (1 + 2(n + i)) = (d + 1)(2n + 1 + d/2) \in O(nd + d^2) . \quad \Box$$

The following algorithm uses Corollary 9.6 to determine a degree bound d, computes all Lx^i required by Algorithm 9.12, calls this algorithm, and then essentially checks whether the returned solution u in fact satisfies $Lu = c$. Things are a bit more involved if $\delta_L \in \mathbb{N}$, since then the solution may not be unique.

Algorithm 9.15 (Method of undetermined coefficients: part II).
Input: A linear differential operator $L = aD + b$ or linear difference operator $L = aE + b$, with $a, b \in F[x]$ not both zero and $n = \deg L$, and a polynomial $c \in F[x]$, where F is a field of characteristic zero.
Output: A nonzero polynomial $u \in F[x]$ such that $Lu = c$, or otherwise "unsolvable".

0. use Corollary 9.6 to determine a degree bound d for u
   if $d < 0$ **then return** "unsolvable"
1. compute Lx^i for $0 \leq i \leq d$
2. **call** Algorithm 9.12 to compute $w, w^* \in F[x]$ with $\deg w \leq d$, $\deg w^* < n$, and $w^* = Lw - c$
   if the algorithm returns "unsolvable" **then return** "unsolvable"
3. **if** $\delta_L \notin \mathbb{N}$ **then** $h \longleftarrow 0$, $\quad h^* \longleftarrow 0$
   else
       call Algorithm 9.12 to compute $v, h^* \in F[x]$ with $\deg v \leq \delta_L - 1$,
       $\deg h^* < n$, and $h^* = Lv + Lx^{\delta_L}$
       $h \longleftarrow x^{\delta_L} + v$
4. **if** there exists some $\kappa \in F$ such that $w + \kappa h \neq 0$ and $w^* + \kappa h^* = 0$
   then return $w + \kappa h$ **else return** "unsolvable"

If $\delta_L \in \mathbb{N}$, then there are several possibilities to choose the constant κ, the coefficient of x^{δ_L}. Natural choices are, e.g., to set κ to zero, or to determine κ in such a way that the degree of $w + \kappa h$ is minimal if $d = \delta_L$.

Theorem 9.16. *Algorithm 9.15 works correctly as specified and uses $O(nd + d^2)$ arithmetic operations in F.*

Proof. It is clear that the result returned is correct if the algorithm returns "unsolvable" in step 0 or 2. Otherwise, if the algorithm does not return "unsolvable" in step 4, then $L(w + \kappa h) - c = w^* + \kappa h^* = 0$, and the result returned is correct. Conversely, suppose that $u \in F[x] \setminus \{0\}$ satisfies $Lu = c$. Then Corollary 9.6 implies that $\deg u \leq d$. If $\delta_L \notin \mathbb{N}$, then this u is unique, Lemma 9.13 implies that $w = u$ is computed in step 2, $w^* = 0$, and the algorithm does not return "unsolvable" in step 4. Now suppose that $\delta_L \in \mathbb{N}$. Then Lemma 9.13 implies that u belongs to the space S_c, so that the algorithm does not return "unsolvable" in step 2, and $u - w \in S_0$. The lemma also implies that Algorithm 9.12 does not return "unsolvable" in step 3 of Algorithm 9.15 and $S_0 = \{\kappa h: \kappa \in F\}$. Thus there exists a suitable κ in step 4, and the algorithm does not return "unsolvable".

In the differential case, computing $Lx^i = iax^{i-1} + bx^i$ takes $O(n)$ operations for each i, in total $O(nd)$. In the difference case, we compute $a \cdot (x + 1)$, ..., $a \cdot (x + 1)^d$ by iterated multiplication with $x + 1$, and then add bx^i to obtain Lx^i. This takes $O(n + d)$ operations for each i, in total $O(nd + d^2)$. Since $\delta_L \leq d$ if $\delta_L \in \mathbb{N}$, the cost for steps 2 and 3 is $O(nd + d^2)$ operations, by Theorem 9.14. Finally, finding a suitable κ in step 4 takes $O(n)$ operations. $\square$

In Sect. 9.5 and 9.6 below, we discuss asymptotically fast variants of the method of undetermined coefficients.

9.2 Brent and Kung's Algorithm for Linear Differential Equations

The algorithm for solving first order differential equations that we describe in this section is due to Brent & Kung (1978). In their paper, the algorithm is stated for power series, and also a generalization of it for higher order equations. When counting only arithmetic operations, it is the fastest currently known algorithm, with a running time of $O(\mathsf{M}(n + d))$ field operations when the input has degree n and the solution has degree d. It is based on the following facts, which we quote from Brent & Kung's article without proofs.

Definition 9.17. *Let F be a field of characteristic zero. The differential operator D and the integral operator I are defined for a formal power series $f = \sum_{i \geq 0} f_i x^i \in F[[x]]$ by*

$$Df = \sum_{i \geq 0} (i + 1)f_{i+1} x^i, \qquad If = \sum_{i \geq 1} \frac{f_{i-1}}{i} x^i .$$

If $f_0 = 0$, then the exponential *of f is* $\exp(f) = \sum_{i \geq 0} f^i/i!$, *and the* logarithm *of* $1 - f$ *is* $\log(1 - f) = -\sum_{i \geq 1} f^i/i$.

Fact 9.18. *Let F be a field of characteristic zero, $f = \sum_{i \geq 0} f_i x^i$, $g = \sum_{i \geq 0} g_i x^i$ in $F[[x]]$ two formal power series, with $f \equiv g$ mod x^n for some $n \in \mathbb{N}_{\geq 1}$.*

(i) $DIf = f$, $IDf = f - f_0$, $Df \equiv Dg$ mod x^{n-1}, and $If \equiv Ig$ mod x^{n+1}.
(ii) If $f_0 = 0$, then

$$\exp(\log(1 - f)) = 1 - f, \qquad \log(\exp(f)) = f,$$

$$D(\exp(f)) = (Df) \cdot \exp(f), \quad D(\log(1 - f)) = -(Df) \cdot (1 - f)^{-1}.$$

Moreover, $\exp(f) \equiv \exp(g)$ mod x^n and $\log(1 - f) \equiv \log(1 - g)$ mod x^n.

Fact 9.19. *Let F be a field of characteristic zero, $f = \sum_{i \geq 0} f_i x^i$, $g \in F[[x]]$ two formal power series, and $n \in \mathbb{N}_{\geq 1}$.*

(i) We can compute $f + g$ mod x^n from f mod x^n and g mod x^n with n additions in F.
(ii) We can compute Df mod x^n from f mod x^{n+1} with $n - 1$ multiplications in F.
(iii) We can compute If mod x^n from f mod x^{n-1} with $n - 2$ divisions by positive integers below n in F.
(iv) We can compute $f \cdot g$ mod x^n from f mod x^n and g mod x^n with $\mathsf{M}(n)$ additions and multiplications in F.
(v) If $f_0 = 1$, then we can compute f^{-1} mod x^n from f mod x^n with $O(\mathsf{M}(n))$ additions and multiplications in F.
(vi) If $f_0 = 0$, then we can compute

$$\log(1 - f) \text{ mod } x^n = -I((Df) \cdot (1 - f)^{-1}) \text{ mod } x^n$$

from f mod x^n with $n - 2$ divisions by positive integers below n plus $O(\mathsf{M}(n))$ additions and multiplications in F.
(vii) If $f_0 = 0$, then we can compute $g = \exp(f)$ mod x^n from f mod x^n by the following Newton iteration for the equation $(\log g) - f = 0$:

$$g_0 = 1, \quad g_{i+1} = g_i - g_i \left(I((Dg_i) \cdot g_i^{-1}) - f \right) \text{ rem } x^{\lceil 2^{i+1-k} n \rceil}$$

for $0 \leq i < k = \lceil \log_2 n \rceil$, and finally $g = g_k$ mod x^n. This takes $O(n)$ divisions by positive integers below n and $O(\mathsf{M}(n))$ additions and multiplications in F.

The following algorithm implements the well-known method of "variation of the constant" for solving inhomogeneous linear differential equations.

Algorithm 9.20 (Brent & Kung's algorithm). ▬▬▬▬▬▬
Input: Polynomials $a, b, c = \sum_i c_i x^i \in F[x]$, where F is a field of characteristic zero, with $a(0) \neq 0$.
Output: A nonzero polynomial $u \in F[x]$ solving $au' + bu = c$, or otherwise "unsolvable".

    0. use Corollary 9.6 to determine a degree bound d for u
       if $d < 0$ **then return** "unsolvable"
    1. $f \longleftarrow ba^{-1} \text{ rem } x^d, \quad g \longleftarrow ca^{-1} \text{ rem } x^d$
    2. $h \longleftarrow \exp(-If) \text{ rem } x^{d+1}, \quad w \longleftarrow h \cdot I(gh^{-1}) \text{ rem } x^{d+1}$
    3. **if** there exists $\kappa \in F$ such that $w + \kappa h \neq 0$ and $a \cdot (w' + \kappa h') + b \cdot (w + \kappa h) = c$
       then return $w + \kappa h$ **else return** "unsolvable"

The constant κ is the constant coefficient of u. The following theorem is also proven by Brent & Kung (1978).

Theorem 9.21. *Algorithm 9.20 works correctly and uses* $O(\mathsf{M}(n) + \mathsf{M}(d))$ *arithmetic operations in* F.

Proof. We first note that $au' + bu \equiv c \bmod x^d$ if and only if $u' + fu \equiv g \bmod x^d$ holds, for all $u \in F[x]$ of degree at most d, and similarly for the corresponding homogeneous equations. Now

$$h' = Dh \equiv D(\exp(-If)) = -(DIf) \cdot \exp(-If) = -hf \bmod x^d \ ,$$

and hence $ah' + bh \equiv 0 \bmod x^d$. Similarly,

$$w' = Dw \equiv D(h \cdot I(gh^{-1})) = (Dh) \cdot I(gh^{-1}) + h \cdot DI(gh^{-1}) = -fw + g \bmod x^d \ ,$$

so that $aw' + bw \equiv c \bmod x^d$. By construction, we have $h(0) = 1$ and $w(0) = 0$. Let

$$H \quad = \quad \{u \in F[x] : \deg u \le d \text{ and } au' + bu \equiv 0 \bmod x^d\} \ ,$$
$$W \quad = \quad \{u \in F[x] : \deg u \le d \text{ and } au' + bu \equiv c \bmod x^d\} \ .$$

Then clearly $h \in H \setminus \{0\}$ and $W = w + H$. The $d \times (d + 1)$ coefficient matrix of the linear system $au' + bu \equiv 0 \bmod x^d$ for $\deg u \le d$ with respect to the monomial basis $1, x, x^2, \ldots$ is triangular, with diagonal entries equal to $a(0)$ (see the matrix R_* in Fig. 9.1 below), and hence it has full rank d. Thus its kernel H is one-dimensional, and we find that $H = \{\kappa h : \kappa \in F\}$.

By Corollary 9.6, the answer "unsolvable" is correct in step 0 if $d < 0$, and it is clear that the algorithm answers correctly if it does not return "unsolvable" in step 3. Conversely, suppose that there exists an $u \in F[x]$, which has degree at most d, by Corollary 9.6, such that $au' + bu = c$. However, such a u clearly belongs to W, so that $u = w + \kappa h$ for $\kappa = u(0)$, and hence the algorithm does not return "unsolvable" in step 3.

Steps 1 and 2 take $O(\mathsf{M}(d))$ field operations, by Fact 9.19. Since the lower d coefficients of $ah' + bh$ and $aw' + bw - c$ are zero, by construction, it is sufficient to compute the upper $n + 1$ coefficients in step 3. By a similar proof as for Lemma 9.10, this takes $O(\mathsf{M}(n))$ field operations, and finding a suitable κ or proving its nonexistence takes $O(n)$ operations. $\square$

With classical arithmetic, Theorem 9.21 yields a cost estimate of $O(n^2 + d^2)$ arithmetic operations. However, we recommend to use the simpler Algorithm 9.15 for classical arithmetic.

Van der Hoeven (1997) discusses an algorithm for lazy multiplication of formal power series modulo x^d with running time $O(\mathsf{M}(d) \log d)$. This algorithm can be used to find power series solutions of linear first order differential equations within the same time bound. This bound is slower by a factor of $\log d$ than the bound for Brent & Kung's algorithm. It is not clear whether there is an efficient analog of Algorithm 9.20 for or whether the lazy multiplication technique can be applied to first order difference equations.

The restriction that $a(0)$ be nonzero can be easily removed by shifting everything by a suitable field element.

Corollary 9.22. *Let F be a field of characteristic zero, $L = aD + b$ a linear differential operator, with $a, b \in F[x]$, $a \neq 0$, and $n = \deg L$, let $c \in F[x]$, and assume that the degree bound $d = \max(\{\deg c - n, \delta_L\} \cap \mathbb{Z})$ from Corollary 9.6 is nonnegative and $\#F \geq 2n + 2$. Using Algorithm 9.20, we can decide whether the differential equation $au' + bu = c$ has a solution $u \in F[x]$, and if so, compute one, using an expected number of $O(\mathsf{M}(n + d))$ operations in F with fast arithmetic.*

Proof. We choose some $s \in F$ such that $a(s) \neq 0$. Then $u \in F[x]$ solves the original difference equation if and only if $a(x+s)u(x+s)' + b(x+s)u(x+s) = c(x+s)$, since $u'(x + s) = u(x + s)'$. Now $a(x + s)$ does not have 0 as a root, and we can apply Algorithm 9.20. If we choose s uniformly at random from a finite subset of F of cardinality at least $2n + 2$, then we will find a suitable s after an expected number of at most 2 trials, at an expected cost of $O(n)$ additions and multiplications. The degrees of a, b, c, u are at most $n + d$, and hence the additional cost for computing the Taylor shifts is $O(\mathsf{M}(n + d))$ with fast arithmetic, by Theorem 4.5. $\square$

9.3 Rothstein's SPDE Algorithm

The algorithm that we discuss in this section is due to Rothstein (1976, 1977) (see also Bronstein 1990). In fact, Rothstein gives an algorithm for the general case of elementary differential fields, and we will only use the base case of rational functions.

The idea is as follows. Let F be a field and $L = aD + b$ be a linear differential operator, with $a, b \in F[x]$ not both zero, $c, u \in F[x]$ such that $Lu = c$, and assume that $d = \deg u \geq \deg a > 0$. Rothstein's algorithm successively computes the a-adic expansion of u. We can write $u = qa + r$ with unique polynomials $q, r \in F[x]$ such that $\deg r < \deg a$. Plugging this into the differential equation, we find

$$
\begin{aligned}
c &= Lu = a(q'a + qa' + r') + b(qa + r) = a(aq' + (b + a')q + r') + br \\
  &= az + br,
\end{aligned}
$$

where $z = aq' + (b + a')q + r'$. The polynomial q has degree at most $d - \deg a$ and satisfies the first order differential equation $Mq = aq' + (b + a')q = z - r'$, where $M = L + a'$. Conversely, if a and b are coprime, then there exist polynomials $z, r \in F[x]$ satisfying $az + br = c$ and $\deg r < \deg a$. Such polynomials can be obtained from the Extended Euclidean Algorithm, which computes $s, t \in F[x]$ with $as + bt = 1$, and then we may take $r = tc \operatorname{rem} a$ and $z = (c - bz)/a$. If z_1, r_1 is another pair of polynomials with $az_1 + br_1 = c$ and $\deg r_1 < \deg a$, then $a(z - z_1) = b(r_1 - r)$, the coprimality of a and b implies that $a \mid (r_1 - r)$, and $\deg(r_1 - r) < \deg a$ yields $r_1 = r$ and $z_1 = z$. Thus the polynomials z and r are unique. If there is a solution $q \in F[x]$ of degree at most $d - \deg a$ of the equation $Mq = z - r'$, then a similar calculation as above shows that $u = qa + r$ is a solution of degree at most d of the original equation $Lu = c$. Thus, to find u, we may compute z and r as described above, solve for q recursively, and finally put $u = qa + r$.

For simplicity, we only analyze the algorithm for the case where $\deg a - 1 \geq \deg b$, since otherwise it may happen that $\deg a = 0$ and $b \neq 0$ during the recursion, and a different algorithm would have to be called to handle this case. (Rothstein then uses the method of indeterminate coefficients.)

Algorithm 9.23 (Rothstein's SPDE algorithm).

Input: Polynomials $a, b, c \in F[x]$, where F is a field of characteristic zero, such that $a \neq 0$ and $\deg a - 1 \geq \deg b$, and $d \in \mathbb{Z}$.

Output: A polynomial $u \in \mathbb{Z}[x]$ of degree at most d solving $au' + bu = c$, or otherwise "unsolvable".

1. **if** $c = 0$ and $d < 0$ **then return** 0
2. **if** $d < 0$ or $d + \deg a - 1 < \deg c$ **then return** "unsolvable"
3. **call** the Extended Euclidean Algorithm to compute $s, t, g \in F[x]$ such that $as + bt = g = \gcd(a, b)$, $\deg s < \deg b - \deg g$, and $\deg t < \deg a - \deg g$
4. **if** $g \nmid c$ **then return** "unsolvable"
5. $A \longleftarrow a/g, \quad B \longleftarrow b/g, \quad C \longleftarrow c/g$
6. **if** $\deg A = 0$ **then return** $\int A^{-1}C$
7. $r \longleftarrow tC \operatorname{rem} A, \quad z \longleftarrow (C - Br)/A$
8. **call** the algorithm recursively with input $A, B + A', z - r'$, and $d - \deg A$ to compute $q \in F[x]$ of degree at most $d - \deg A$ satisfying $Aq' + (B + A')q = z - r'$
9. **if** the recursive call returns "unsolvable" **then return** "unsolvable"
   else return $qA + r$

The only choice in the algorithm is for the constant coefficient of $\int A^{-1}C$ in step 6. In the homogeneous case, where $c = 0$, we may obtain a nonzero solution of $au' + bu = 0$ if this constant coefficient is nonzero.

Theorem 9.24. *If Algorithm 9.23 returns "unsolvable", then the differential equation $au' + bu = c$ has no polynomial solution of degree at most d. Otherwise, the polynomial returned is a solution of degree at most d of this equation. If $n = \deg a$,* *then the cost for the algorithm with classical arithmetic is $O(n^2 + d^2)$ arithmetic operations in F.*

Proof. It is clear that the result returned is correct if the algorithm returns in steps 1, 2, or 4, by Corollary 9.5. So we may assume that $c \neq 0$, $d \geq 0$, and $g \mid c$. Clearly, the two equations $au' + bu = c$ and $Au' + Bu = C$ are equivalent, and we have $As + Bt = 1$. If $\deg A = 0$, then $\deg a - 1 \geq \deg b$ implies that $-1 = \deg A - 1 \geq \deg B$, and hence $b = B = 0$. Then the differential equation degenerates to $Au' = C$, or equivalently, $u' = A^{-1}C$, and the result returned in step 6 is correct. We now assume that $\deg A \geq 1$. Then $A \neq 0$ and $\deg(B + A') \leq \max\{\deg B, \deg A - 1\} \leq \deg A - 1$, and by induction, the recursive call in step 8 returns the correct result. If it does not return "unsolvable", then the discussion preceding the algorithm implies that $u = qA + r$ has degree at most d and solves the differential equation $au' + bu = c$. Similarly, if the recursive call returns "unsolvable", then the discussion preceding the algorithm shows that the equation $au' + bu = c$ has no polynomial solution of degree at most d.

Let $m = \deg c$ and $k = \deg A$. Steps 3 through 5 take $O(n^2 + mn)$ field operations. The cost for step 6 is $O(m)$, and step 7 takes $O(k(m + k))$ operations. We have

$$\deg z \leq \max\{\deg B + \deg r, \deg C - \deg A\} \in O(m + n) \,,$$

so that computing $B + A'$ and $z - r'$ in step 8 takes $O(m+n)$ operations. Finally, the cost for step 9 is $O(kd)$. Using $m \leq n + d - 1$ when the algorithm does not return in step 2, we find that the overall cost for the algorithm is $O(n^2 + nd)$ plus the cost for the recursive call, with the parameters n, d replaced by $k, d - k$. If the recursion stops in steps 1, 2, 4, or 6, then the cost is $O(n^2 + nd)$ as well. Inductively, we find that the total cost of the algorithm is $O((n + d)^2)$ or $O(n^2 + d^2)$ field operations. $\square$

We do not state a cost estimate for fast arithmetic, since then the cost is still $\Omega(d^2)$ in the worst case, e.g., when $\deg a = 1$ throughout the recursive process. It is not clear whether divide & conquer techniques can be applied to obtain an asymptotically fast variant.

However, there is a variant of Algorithm 9.23 for difference equations, which seems to be new. We denote by Σ the formal inverse of the difference operator Δ, such that $\Delta\Sigma f = f$ and $\Sigma\Delta f - f$ is a constant, for all polynomials f.

Algorithm 9.25 (Rothstein's SPDE algorithm for difference equations).
Input: Polynomials $a, b, c \in F[x]$, where F is a field of characteristic zero, such that $a \neq 0$ and $\deg a - 1 \geq \deg(a + b)$, and $d \in \mathbb{Z}$.
Output: A polynomial $u \in \mathbb{Z}[x]$ of degree at most d solving $a \cdot Eu + b \cdot u = c$, or otherwise "unsolvable".

1. **if** $c = 0$ and $d < 0$ **then return** 0
2. **if** $d < 0$ or $d + \deg a - 1 < \deg c$ **then return** "unsolvable"
3. **call** the Extended Euclidean Algorithm to compute $s, t, g \in F[x]$ such that $as + bt = g = \gcd(a, b)$, $\deg s < \deg b - \deg g$, and $\deg t < \deg a - \deg g$

4. **if** $g \nmid c$ **then return** "unsolvable"
5. $A \longleftarrow a/g, \quad B \longleftarrow b/g, \quad C \longleftarrow c/g$
6. **if** $\deg A = 0$ **then return** $\Sigma A^{-1} C$
7. $r \longleftarrow tC \text{ rem } A, \quad z \longleftarrow (C - Br)/A$
8. **call** the algorithm recursively with input $EA, B, z - Er$, and $d - \deg A$ to compute $q \in F[x]$ of degree at most $d - \deg A$ satisfying $EA \cdot Eq + B \cdot q = z - Er$
9. **if** the recursive call returns "unsolvable" **then return** "unsolvable"
   else return $qA + r$

Theorem 9.26. *If Algorithm 9.25 returns "unsolvable", then the difference equation* $a \cdot Eu + bu = c$ *has no polynomial solution of degree at most d. Otherwise, the polynomial returned is a solution of degree at most d of this equation. If $n = \deg a$, then the cost for the algorithm with classical arithmetic is $O(n^2 + d^2)$ arithmetic operations in F.*

Proof. It is clear that the result returned is correct if the algorithm returns in steps 1, 2, or 4, by Corollary 9.5. So we may assume that $c \neq 0$, $d \geq 0$, and $g \mid c$. The two equations $a \cdot Eu + bu = c$ and $A \cdot Eu + Bu = C$ are equivalent, and we have $As + Bt = 1$. If $\deg A = 0$, then $\deg a - 1 \geq \deg(a + b)$ implies that $-1 = \deg A - 1 \geq \deg(A + B)$, and hence $a + b = A + B = 0$. Then the differential equation degenerates to $A(Eu - u) = C$, or equivalently, $\Delta u = A^{-1}C$, and the result returned in step 6 is correct. We now assume that $\deg A \geq 1$. Then $A \neq 0$ and $\deg(B + A) = \deg(b + a) - \deg g \leq \deg a - 1 - \deg g = \deg A - 1$, and by induction, the recursive call in step 8 returns the correct result. If it does not return "unsolvable", then $u = qA + r$ has degree at most d and

$$A \cdot Eu + Bu = A(EA \cdot Eq + Bq + Er) + Br = Az + Br = c. \qquad (9.2)$$

Conversely, if $A \cdot Eu + Bu = C$ has a polynomial solution of degree at most d and we divide u by A with remainder, yielding $q, r \in F[x]$ with $\deg r < \deg A$ and $u = qA + r$, then (9.2) implies that $EA \cdot Eq + bq = z - Er$. Thus the equation $A \cdot Eu + Bu = C$ has no polynomial solution of degree at most d if the recursive call returns "unsolvable".

Let $m = \deg c$ and $k = \deg A$. Steps 3 through 5 take $O(n^2 + mn)$ field operations. In step 6, we convert $A^{-1}C$ into the falling factorial basis, use $\Sigma x^{\underline{i}} = x^{\underline{i+1}}/(i + 1)$ to compute the antidifference, and convert back to the usual monomial basis. This takes $O(m^2)$ field operations, by Corollary 4.14. Step 7 costs $O(k(m + k))$ operations. As in the proof of Theorem 9.24, we have $\deg z \in O(m + n)$, and hence computing EA and $z - r'$ in step 8 takes $O(k^2 + m + n)$ operations, by Theorem 4.3. Finally, the cost for step 9 is $O(kd)$. Using $m \leq n + d - 1$ when the algorithm does not return in step 2, we find that the overall cost for the algorithm is $O(n^2 + nd)$ plus the cost for the recursive call, with the parameters n, d replaced by $k, d - k$. If the recursion stops in steps 1, 2, 4, or 6, then the cost is $O((n + d)^2)$. Inductively, we find that the total cost of the algorithm is $O((n + d)^2)$ or $O(n^2 + d^2)$ field operations. $\square$

9.4 The ABP Algorithm

In this section, we present and analyze the special case of an algorithm due to Abramov, Bronstein & Petkovšek (1995) for linear difference and differential equations with polynomial coefficients, namely the case of first order equations. The key idea is what Graham, Knuth & Patashnik (1994), §1.3, call the "repertoire method". Suppose that L is a linear differential or difference operator with polynomial coefficients and we want to solve $Lu = c$. Compute $r^{(j)} = Lu^{(j)}$ for several "simple" polynomials $u^{(j)}$ (for example, $u^{(j)} = x^j$ is suitable in the differential case), and then try to represent c as a linear combination of the $r^{(j)}$. If the $u^{(j)}$ are suitably chosen, then this transforms the linear system corresponding to the original equation into an equivalent one whose coefficient matrix is banded triangular, and therefore the transformed system is particularly easy to solve. Since it is easier, we start to describe the algorithm in the differential case, which is closely related to the method of undetermined coefficients.

Algorithm 9.27 (ABP algorithm for first order differential equations). ▬▬▬
Input: Polynomials $a, b, c = \sum_i c_i x^i \in F[x]$, where F is a field of characteristic 0, with $a(0) \neq 0$.
Output: A nonzero polynomial $u \in F[x]$ solving $au' + bu = c$, or otherwise "unsolvable".

0. use Corollary 9.6 to determine a degree bound d for u
    if $d < 0$ **then return** "unsolvable"
1. **for** $0 \leq j \leq d$ compute $r^{(j)} = \sum_i r_{ij} x^i = ajx^{j-1} + bx^j$
2. $w_0 \longleftarrow 0, \quad h_0 \longleftarrow 1$
    for $1 \leq i \leq d$ **do**
    $$w_i \longleftarrow (c_{i-1} - \textstyle\sum_{j<i} r_{ij} w_j)/ia(0), \quad h_i \longleftarrow -\textstyle\sum_{j<i} r_{ij} h_j/ia(0)$$
3. $h \longleftarrow \sum_{0 \leq i \leq d} h_i x^i, \quad w \longleftarrow \sum_{0 \leq i \leq d} w_i x^i$
    if there is a $\kappa \in F$ such that $w + \kappa h \neq 0$ and $a \cdot (w' + \kappa h') + b \cdot (w + \kappa h) = c$
    then return $w + \kappa h$ **else return** "unsolvable" ▬▬▬▬▬▬

The constant κ is the constant coefficient of u.

Theorem 9.28. *Algorithm 9.27 works correctly as specified. If $L = aD + b$ has degree n, then the algorithm takes $O(nd)$ additions and multiplications and $O(d)$ divisions in F.*

Proof. If the algorithm returns "unsolvable" in step 0, then this is correct, by Corollary 9.6. If the algorithm does not return "unsolvable", then it clearly returns a solution of the differential equation. It remains to show that the algorithm returns a solution if there exists one. We let $u = \sum_{0 \leq i \leq d} u_i x^i \in F[x]$, possibly with $u_d = 0$, be a solution of the differential equation $Lu = c$. Since L is linear, we have

$$\sum_{0 \leq j \leq d} u_j r^{(j)} = \sum_{0 \leq j \leq d} u_j Lx^j = Lu = c .$$

Let $e = d + \deg L$, $R = (r_{ij}) \in F^{(e+1)\times(d+1)}$, identify u and c with their coefficient vectors in F^{d+1} and F^{e+1}, respectively, and let R_*, c_* and R^*, c^* be the lower d rows and the upper $e - d + 1$ rows of R and c, respectively. This is illustrated in Fig. 9.1. Then R is the coefficient matrix of the linear system $Lu = c$ with respect to the monomial basis $x^e, x^{e-1}, \ldots, x, 1$. Thus $Ru = c$, and hence also $R_* u = c_*$. The matrix R is banded triangular, with $r_{i,i+1} = (i+1)a(0) \neq 0$ and $r_{i,j} = 0$ for $0 \le i < d$ and $i + 2 \le j \le d$, and the rank of R_* is equal to d.

$$
\begin{array}{c}
R^* \left\{ \begin{array}{c} \\ \\ \\ \\ \end{array} \right. \\
R_* \left\{ \begin{array}{c} \\ \\ \\ \\ \\ \end{array} \right.
\end{array}
\left(\begin{array}{cccccc}
r_{e,d} & 0 & \cdots & \cdots & \cdots & 0 \\
r_{e-1,d} & r_{e-1,d-1} & 0 & \cdots & \cdots & 0 \\
\vdots & \ddots & \ddots & \ddots & & \vdots \\
r_{d,d} & r_{d,d-1} & \cdots & & & \\
\hline
r_{d-1,d} & r_{d-1,d-1} & \cdots & & & \\
0 & r_{d-2,d-1} & \cdots & & & \\
\vdots & \ddots & \ddots & \ddots & \ddots & \vdots \\
0 & \cdots & 0 & r_{1,2} & r_{1,1} & r_{1,0} \\
0 & \cdots & \cdots & 0 & r_{0,1} & r_{0,0}
\end{array}\right)
\begin{pmatrix} u_d \\ \vdots \\ \vdots \\ \vdots \\ u_0 \end{pmatrix}
=
\begin{pmatrix} c_e \\ c_{e-1} \\ \vdots \\ c_d \\ c_{d-1} \\ c_{d-2} \\ \vdots \\ c_1 \\ c_0 \end{pmatrix}
\begin{array}{c}
\left. \begin{array}{c} \\ \\ \\ \\ \end{array} \right\} c^* \\
\left. \begin{array}{c} \\ \\ \\ \\ \\ \end{array} \right\} c_*
\end{array}
$$

Fig. 9.1. The linear system $Lu = c$ with respect to the monomial basis

If we identify h and w with their coefficient vectors in F^{d+1}, then we see that $R_* h = 0$ and $R_* w = c_*$. Since R_* has rank d, its kernel is one-dimensional and consists of all scalar multiples of h, and $w + Fh$ is the solution space of the linear system $R_* y = c_*$. In particular, since $R_* u = c_*$, we find that $u = w + u(0)h$. Thus there exists a $\kappa \in F$ such that $L(w + \kappa h) = c$, and the algorithm correctly returns a solution of the differential equation.

The cost for step 0 is negligible. Computing $r^{(j)}$ in step 1 takes $O(n)$ additions and multiplications in F for each j, in total $O(nd)$. Since the bandwidth of the matrix is $e - d + 2 = n + 1$, the cost for the ith iteration of the loop in step 2 is $O(n)$ additions and multiplications in F plus one division, in total $O(nd)$ additions and multiplications and $O(d)$ divisions. Finally, in step 3 we only need to find a $\kappa \in F$ such that $\kappa R^* h = c^* - R^* w$. Computing $R^* h$ and $c^* - R^* w$ amounts to another $O(nd)$ additions and multiplications, and we can determine a suitable κ (or its non-existence) and compute $w + \kappa h$ from these two vectors with one division and $O(n)$ additions and multiplications. □

Corollary 9.29. *Let F be a field of characteristic zero, $L = aD + b$ a linear differential operator, with $a, b \in F[x]$, $a \neq 0$, and $n = \deg L$, let $c \in F[x]$, and assume that the degree bound $d = \max(\{\deg c - n, \delta_L\} \cap \mathbb{Z})$ from Corollary 9.6 is nonnegative. Using Algorithm 9.27, we can decide whether the differential equation $au' + bu = c$ has a solution $u \in F[x]$, and if so, compute one, using an expected number of $O(n^2 + d^2)$ operations in F with classical arithmetic and $O(nd + \mathsf{M}(n + d))$ with fast arithmetic.*

Proof. The proof for fast arithmetic is completely analogous to the proof of Corollary 9.22. The cost for the Taylor shifts is $O(n^2 + d^2)$ operations with classical arithmetic, by Theorem 4.3. $\square$

In order to describe the analog of the ABP algorithm for difference equations, we assume that our input polynomials a, b, c are given in the falling factorial basis and we want to find the solution u in the same basis as well. The conversions between this and the usual monomial basis can be easily achieved with the algorithms from Chap. 4. Before presenting the algorithm, we need a technical lemma from Abramov, Bronstein & Petkovšek (1995).

Lemma 9.30. *Let* $j \in \mathbb{N}$.

(i) For all $l \in \mathbb{N}$*, we have*

$$x^{\underline{l}}x^{\underline{j}} = \sum_{\max\{l,j\} \leq i \leq l+j} \binom{l}{i-j}\binom{j}{i-l}(l+j-i)!\, x^{\underline{i}}\,.$$

(ii) If $f = \sum_{0 \leq l \leq n} f_l x^{\underline{l}}$*, then the falling factorial coefficient* r_{ij} *of* $x^{\underline{i}}$ *in* $f \cdot x^{\underline{j}}$ *is*

$$r_{ij} = \sum_{i-j \leq l \leq i} f_l \binom{l}{i-j}\binom{j}{i-l}(l+j-i)!\,, \tag{9.3}$$

for $0 \leq i \leq n + j$*. In particular,* $r_{ij} = 0$ *if* $i < j$ *and* $r_{ii} = f(j)$*.*

Proof. (i) We start with the Vandermonde identity (see, e.g., Exercise 23.9 in von zur Gathen & Gerhard 1999), rewritten as the binomial convolution formula for the falling factorials: if y is another indeterminate, then

$$(x + y)^{\underline{l}} = \sum_{0 \leq i \leq l} \binom{l}{i}x^{\underline{i}}y^{\underline{l-i}}\,.$$

Substituting $x = x - j$ and $y = j$, multiplying both sides by $x^{\underline{j}}$, and noting that $j^{\underline{l-i}} = 0$ if $j < l - i$, we obtain

$$
\begin{aligned}
x^{\underline{l}}x^{\underline{j}} &= \sum_{\max\{l-j,0\} \leq i \leq l} \binom{l}{i}x^{\underline{i}}(x-j)^{\underline{i}}j^{\underline{l-i}} = \sum_{\max\{l-j,0\} \leq i \leq l} \binom{l}{i}j^{\underline{l-i}}x^{\underline{i+j}} \\
&= \sum_{\max\{l,j\} \leq i \leq l+j} \binom{l}{i-j}\binom{j}{i-l}(l+j-i)!\, x^{\underline{i}}\,.
\end{aligned}
$$

(ii) This follows by plugging (i) into $\sum_i r_{ij}x^{\underline{i}} = \sum_{0 \leq l \leq n} f_l x^{\underline{l}}x^{\underline{j}}$, exchanging the two summation signs, and comparing coefficients. The sum in (9.3) is empty if $i < j$, and for $i = j$ it has the value

$$\sum_{0 \leq l \leq i} f_l \binom{i}{i-l}l! = \sum_{0 \leq l \leq i} f_l i^{\underline{l}} = f(i)\,. \quad \square$$

Algorithm 9.31 (ABP algorithm for first order difference equations). ▬▬▬
Input: Polynomials $a, b, c = \sum_i c_i x^{\underline{i}} \in F[x]$ in the falling factorial basis, where
F is a field of characteristic zero, and the degree bound d for $L = a\Delta + b$ from
Corollary 9.6, such that $d \in \mathbb{N}$ and $a(0), \ldots, a(d-1) \neq 0$.
Output: A nonzero polynomial $u \in F[x]$ solving $a \, \Delta u - bu = c$, or otherwise
"unsolvable".

   1. **for** $0 \leq j \leq d$ compute $r^{(j)} = \sum_i r_{ij} x^{\underline{i}} = ajx^{\underline{j-1}} + bx^{\underline{j}}$
   2. $w_0 \longleftarrow 0, \quad h_0 \longleftarrow 1$
      for $1 \leq i \leq d$ **do**
         $w_i \longleftarrow (c_{i-1} - \sum_{j<i} r_{ij} w_j)/ia(i-1)$
         $h_i \longleftarrow -\sum_{j<i} r_{ij} h_j / ia(i-1)$
   3. $h \longleftarrow \sum_{0 \leq i \leq d} h_i x^{\underline{i}}, \quad w \longleftarrow \sum_{0 \leq i \leq d} w_i x^{\underline{i}}$
     if there is a $\kappa \in F$ such that $w + \kappa h \neq 0$ and $a \cdot (\Delta w + \kappa \, \Delta h) + b \cdot (w + \kappa h) = c$
     then return $w + \kappa h$ **else return** "unsolvable" ▬▬▬

Theorem 9.32. *Algorithm 9.31 works correctly as specified. If* $\deg L = n$*, then the
algorithm takes* $O(nd)$ *additions and multiplications and* $O(d)$ *divisions in* F.

Proof. The proof parallels the proof of Theorem 9.28. If the algorithm does not
return "unsolvable", then it clearly returns a solution of the difference equation, and
it remains to show that the algorithm returns a solution if there exists one. We let
$u = \sum_{0 \leq i \leq d} u_i x^{\underline{i}} \in F[x]$, possibly with $u_d = 0$, be a solution of the difference
equation $Lu = c$. Since L is linear, we have

$$\sum_{0 \leq j \leq d} u_j r^{(j)} = \sum_{0 \leq j \leq d} u_j L x^{\underline{j}} = Lu = c .$$

Let $e = d + \deg L$, $R = (r_{ij}) \in F^{(e+1) \times (d+1)}$, identify u and c with their coefficient
vectors in F^{d+1} and F^{e+1} with respect to the falling factorial basis $\mathcal{F}$, respectively,
and let R_*, c_* and R^*, c^* be the lower d rows and the upper $e - d + 1$ rows of R
and c, respectively. Then R is the coefficient matrix of the linear system $Lu = c$
with respect to the falling factorial basis $x^{\underline{e}}, x^{\underline{e-1}}, \ldots, x^{\underline{1}}, 1$, and we have $Ru = c$
and $R_* u = c_*$. Lemma 9.30 implies that the matrix R is banded triangular, with
$r_{i,i+1} = (i+1)a(i) \neq 0$ and $r_{i,j} = 0$ for $0 \leq i < d$ and $i + 2 \leq j \leq d$, and the
rank of R_* is equal to d. Thus we have the same picture as in Fig. 9.1.

    If we identify h and w with their coefficient vectors in F^{d+1} with respect to $\mathcal{F}$,
then we find that $R_* h = 0$ and $R_* w = c_*$. As in the proof of Theorem 9.28, we
conclude that $u = w + u(0)h$ and the algorithm correctly returns a solution of the
difference equation.

    In step 1, we recursively compute $ax^{\underline{j}} = (x - j + 1) \cdot ax^{\underline{j-1}}$ and $bx^{\underline{j+1}} =
(x - j) \cdot bx^{\underline{j}}$, and calculate $r^{(j+1)}$ from these polynomials, for $0 \leq j < d$. Using
Lemma 9.30 with $j = 1$, this takes $O(n)$ additions and multiplications in F for
each j, in total $O(nd)$. The cost for steps 2 and 3 is $O(nd)$ additions and multiplications plus $O(d)$ divisions, as in the proof of Theorem 9.28. $\square$

Corollary 9.33. *Let F be a field of characteristic zero, $L = a\Delta + b$ a linear difference operator, with $a, b \in F[x]$, $a \neq 0$, and $n = \deg L$, let $c \in F[x]$, and assume that the degree bound $d = \max(\{\deg c - n, \delta_L\} \cap \mathbb{Z})$ of Corollary 9.6 is nonnegative. Using Algorithm 9.31, we can decide whether the difference equation $a \Delta u + bu = c$ has a solution $u \in F[x]$, and if so, compute one, using an expected number of $O(n^2 + d^2)$ operations in F with classical arithmetic and $O(nd + \mathsf{M}(n + d) \log(n + d))$ with fast arithmetic, if F has cardinality at least $2(n + 1)d$.*

Proof. First, we find a point $s \in F$ such that the values $a(s), a(s - 1), \ldots, a(s - d + 1)$ are all nonzero. If we choose s uniformly at random from a finite subset $S \subseteq F$ of cardinality at least $2n + 2$ such that $\{s, s - 1, \ldots, s - d + 1\} \cap \{s^*, s^* - 1, \ldots, s^* - d + 1\} = \emptyset$ for all distinct $s, s^* \in S$, then we can expect to find a suitable s with at most two trials. For each trial, we can check the condition by evaluating a using $O(nd)$ additions and multiplications in F, as in the proof of Theorem 9.32. Since the shift by s is an automorphism of the polynomial ring $F[x]$, we have $\Delta(u(x + s)) = (\Delta u)(x + s)$ for all $u \in F[x]$, and u is a solution of the original difference equation if and only if

$$a(x + s) \Delta u(x + s) + b(x + s)u(x + s) = c(x + s) .$$

Thus we shift a, b, c by s and convert the shifted polynomials to the falling factorial basis. Then, since $a(x + s)$ does not vanish at $x = 0, \ldots, d - 1$, we can apply Algorithm 9.31, and finally we convert u back to the monomial basis. The polynomials a, b, c, u have degrees at most $n + d$, whence the conversions take $O(n^2 + d^2)$ operations with classical arithmetic, by Theorems 4.3 and 4.14, and $O(\mathsf{M}(n + d) \log(n + d))$ with fast arithmetic, by Theorems 4.5, 4.15, and 4.16, and the estimate follows from Theorem 9.28. $\square$

9.5 A Divide-and-Conquer Algorithm: Generic Case

In this section and the following one, we discuss an algorithm by von zur Gathen & Gerhard (1997) for solving linear first order difference and differential equations. It works by writing the equation as a linear system with respect to the monomial basis; no change of basis is necessary. The algorithm is essentially a divide & conquer version of the method of undetermined coefficients.

Let $a, b, c \in F[x]$ be polynomials, and let $L = aD + b$ or $L = aE + b$. The main observation for the algorithm is that the highest coefficients of a possible solution $u \in F[x]$ of $Lu = c$ depend only on the highest coefficients of a, b, and c. This is due to the fact that the coefficient matrix is lower triangular, as illustrated in Fig. 9.1 for the differential case. The idea is now, similarly to the asymptotically fast Euclidean Algorithm (see Aho, Hopcroft & Ullman 1974; Strassen 1983), to write $u = Ux^m + V$, with $m = \lceil (\deg u)/2 \rceil$ and polynomials $U, V \in F[x]$ such that $\deg V < m$. Then we compute first the upper half U, using only the upper halfs of a, b, and c, plug the partial solution obtained into the original equation, yielding $LV = c - L(Ux^m)$, and recursively solve the latter equation for the lower half V. For $m = \deg u$, this idea is precisely Algorithm 9.12.

We first illustrate this in an example. For a polynomial $f \in F[x]$ and an integer k, we denote by $f \upharpoonright k$ the polynomial part of $x^{-k}f$. Thus $f \upharpoonright k$ is equal to the quotient of f on division by x^k if $k \geq 0$, and to $x^{-k}f$ if $k < 0$, and $\mathrm{lc}(f \upharpoonright k) = \mathrm{lc}(f)$ if both polynomials are nonzero. We have $\deg(f \upharpoonright k) = \deg f - k$ if $f \upharpoonright k \neq 0$, and $\deg(x^k(f \upharpoonright k) - f) < k$.

Example 9.34. *We study the generic example of a linear first order difference equation with* $\deg a = \deg b = 3$ *and* $\deg c = 6$. *So let* $a = \sum_{0 \leq i \leq 3} a_i x^i$, $b = \sum_{0 \leq i \leq 3} b_i x^i$, $c = \sum_{0 \leq i \leq 6} c_i x^i$ *and* $u = \sum_{0 \leq i \leq 3} u_i x^i$ *in* $F[x]$ *such that* $Lu = a \, Eu + bu = c$. *Comparing coefficients on both sides yields the linear system*

$$
\begin{aligned}
c_6 &= b_3^* u_3 , \\
c_5 &= b_3^* u_2 + (3a_3 + b_2^*)u_3 , \\
c_4 &= b_3^* u_1 + (2a_3 + b_2^*)u_2 + (3a_3 + 3a_2 + b_1^*)u_3 , \\
c_3 &= b_3^* u_0 + (a_3 + b_2^*)u_1 + (a_3 + 2a_2 + b_1^*)u_2 \\
    &\quad + (a_3 + 3a_2 + 3a_1 + b_0^*)u_3 , \\
c_2 &= b_2^* u_0 + (a_2 + b_1^*)u_1 + (a_2 + 2a_1 + b_0^*)u_2 \\
    &\quad + (a_2 + 3a_1 + 3a_0)u_3 , \\
c_1 &= b_1^* u_0 + (a_1 + b_0^*)u_1 + (a_1 + 2a_0)u_2 + (a_1 + 3a_0)u_3 , \\
c_0 &= b_0^* u_0 + a_0 u_1 + a_0 u_2 + a_0 u_3 ,
\end{aligned}
$$

where $b_i^* = a_i + b_i$ *for* $0 \leq i \leq 3$. *If we have* $b_3^* \neq 0$, *then the first four equations uniquely determine* $u_0, \ldots, u_3$, *and* u_2 *and* u_3 *can already be computed from the first two equations. Let* $U = u \upharpoonright 2 = u_3 x + u_2$, $A = (x+1)^2 a \upharpoonright 4 = a_3 x + 2a_3 + a_2$, $B = x^2 b \upharpoonright 4 = b_3 x + b_2$, *and* $C = c \upharpoonright 4 = c_6 x^2 + c_5 x + c_4$. *Then*

$$
\begin{aligned}
A \cdot EU + B \cdot U &= b_3^* u_3 x^2 + \left(b_3^* u_2 + (3a_3 + b_2^*)u_3 \right) x + \cdots \\
&= c_6 x^2 + c_5 x + \cdots ,
\end{aligned}
$$

i.e., U *is the unique polynomial satisfying* $\deg(A \cdot EU + B \cdot U - C) \leq \deg C - 2$, *or equivalently,* $\deg(a \cdot E(Ux^2) + b \cdot Ux^2 - c) \leq \deg c - 2$.

If we have determined U, *we write* $u = Ux^2 + V$, *with* $V = u_1 x + u_0$, *and plug this into* $Lu = c$:

$$
\begin{aligned}
a \cdot EV + b \cdot V &= a(EU \cdot (x+1)^2 + EV) + b(Ux^2 + V) \\
&\quad - ((x+1)^2 a \cdot EU + x^2 b \cdot U) \\
&= c - ((x+1)^2 a \cdot EU + x^2 b \cdot U) .
\end{aligned}
$$

This is again a linear first order difference equation for V *which can then be solved. In general, the degrees of* U *and* V *are about half the degree of* u, *and they can be determined recursively.*

In general, the operator L may not be injective. Lemma 9.1 shows that its kernel is one-dimensional in that case, and hence the solution space of $Lu = c$ is either empty or one-dimensional as well. Corollary 9.6 implies that any nonzero polynomial in the kernel of L has degree δ_L, and hence $\ker L = \{0\}$ if $\delta_L \notin \mathbb{N}$. If the last condition is true, then we call this the *generic case*. The following algorithm works in the generic difference case and computes $u \in F[x]$ such that $\deg(Lu - c) < \deg L$.

Algorithm 9.35 (Difference divide & conquer: generic case). ▬▬▬▬▬
Input: A linear difference operator $L = aE + b = a\Delta + (a + b)$, where $a, b \in F[x]$
  satisfy $a + b \neq 0$ and F is a field of characteristic zero, and $c \in F[x]$ with
  $\deg a - 1 \leq \deg(a + b) = \deg L = n$, $d = \deg c - n$, and $\delta_L \notin \{0, \dots, d\}$.
Output: A polynomial $u \in F[x]$ with $u = 0$ or $\deg u = d$ if $d \geq 0$, such that
  $\deg(Lu - c) < n$.

1. **if** $d < 0$ **then return** $u = 0$
2. **if** $d = 0$ **then return** $u = \mathrm{lc}(c)/\mathrm{lc}(a + b)$
3. $m \longleftarrow \lceil d/2 \rceil$, $t \longleftarrow n - (d - m)$
   $A_1 \longleftarrow (x+1)^m a \upharpoonright (t+m)$, $B_1 \longleftarrow x^m b \upharpoonright (t+m)$, $C_1 \longleftarrow c \upharpoonright (t+m)$
4. **call** the algorithm recursively with input $L_1 = A_1 E + B_1$ and C_1 to obtain
   $U \in F[x]$ such that $\deg(L_1 U - C_1) < d - m$
5. $A_2 \longleftarrow a \upharpoonright (n - m)$, $B_2 \longleftarrow b \upharpoonright (n - m)$,
   $C_2 \longleftarrow (c - a(x + 1)^m EU - bx^m U) \upharpoonright (n - m)$
6. **call** the algorithm recursively with input $L_2 = A_2 E + B_2$ and C_2, yielding
   $V \in F[x]$ with $\deg(L_2 V - C_2) < m$
7. **return** $Ux^m + V$ ▬▬▬▬▬▬▬▬▬▬▬▬▬▬▬▬▬▬▬▬▬▬▬▬

We note that, by definition, $\delta_L \neq 0$ already implies that $a+b \neq 0$ and $\deg a - 1 \leq \deg(a + b)$.

Theorem 9.36. *Algorithm 9.35 works correctly and uses $O(\mathsf{M}(d) \log d)$ arithmetic operations in F with fast arithmetic.*

Proof. We prove correctness by induction on $d = \deg c - \deg L$. The output is clearly correct of $d \leq 0$, and we may assume that $d > 0$. Then $0 \leq d - m = \lfloor d/2 \rfloor \leq \lceil d/2 \rceil = m \leq d$, and at least one of the last two inequalities is strict.

In step 3, we find that for $i \geq t + m$, the coefficients of x^i in $x^{t+m}(A_1 + B_1)$ and $a \cdot (x + 1)^m + bx^m = Lx^m$ coincide. Since $\delta_L \neq m$, Corollary 9.5 implies that $\deg Lx^m = \deg L + \deg x^m = n + m$, and since $n + m \geq t + m$, we find that $\deg(x^{t+m}(A_1 + B_1)) = n + m$ and $\deg(A_1 + B_1) = n - t = d - m \geq 0$. In particular, $A_1 + B_1$ is nonzero. Now

$$\deg A_1 - 1 \leq \deg a - 1 - t \leq n - t = d - m = \deg(A_1 + B_1) = \deg L_1 .$$

We claim that $\delta_{L_1} \notin \{0, \dots, d - m\}$. This is clear if $A_1 = 0$ or $\deg a - 1 < \deg L = n$, since then $\deg A_1 - 1 < \deg L_1$ and $\delta_{L_1} = \infty$. Otherwise, we have

$$\mathrm{lc}(A_1 + B_1) = \mathrm{lc}(Lx^m) = \mathrm{lc}(a)(m - \delta_L) = -\mathrm{lc}(A_1)(\delta_L - m) ,$$

by Lemma 9.3. Thus $\delta_{L_1} = \delta_L - m$, and the claim is proved. Now

$$\begin{aligned}
\deg C_1 &= \deg c - (t+m) = n - t + d - m = 2(d-m)\,, \\
\deg C_1 - \deg L_1 &= 2(d-m) - (d-m) = d - m < d\,,
\end{aligned}$$

and by induction, the output of the recursive call in step 4 is correct, i.e., $\deg U = \deg C_1 - \deg L_1 = d - m$ and $\deg(L_1U - C_1) < \deg L_1 = d - m$ (we note that $U \neq 0$ since otherwise $d - m = \deg C_1 = \deg(L_1U - C_1) < d - m$).

In step 5, we have

$$\begin{aligned}
\deg A_2 - 1 &\leq \deg a - 1 - (n-m) \leq m\,, \\
\deg(A_2 + B_2) &= \deg(a+b) - (n-m) = m = \deg L_2\,.
\end{aligned}$$

Similarly as above, we find that $\delta_{L_2} = \infty$ if $A_2 = 0$ or $\deg a - 1 < n$ and $\delta_{L_2} = \delta_L$ otherwise, so that $\delta_{L_2} \notin \{0, \ldots, m-1\}$. Let $c^* = c - L(Ux^m)$, so that $C_2 = c^* \upharpoonright (n-m)$. By the definition of $\upharpoonright$, the degrees of $(x+1)^m a - x^{t+m} A_1$, $x^m b - x^{t+m} B_1$, and $c - x^{t+m} C_1$ are less than $t+m$, and we conclude that

$$\begin{aligned}
\deg c^* &= \deg(L(Ux^m) - c) = \deg((x+1)^m a\, EU + x^m bU - c) \\
&= \deg\Big(x^{t+m}(L_1U - C_1) + ((x+1)^m a - x^{t+m}A_1)EU \\
&\qquad + (x^m b - x^{t+m}B_1)U - (c - x^{t+m}C_1) \Big) \\
&< t + m + d - m = n + m\,.
\end{aligned}$$

Hence

$$\deg C_2 - \deg L_2 \leq \deg c^* - (n-m) - m < m \leq d\,,$$

and by induction, the output of the recursive call in step 6 is correct as well, i.e., $\deg V \leq \deg C_2 - \deg L_2 < m$ and $\deg(L_2V - C_2) < \deg L_2 = m$.

Finally,

$$\begin{aligned}
\deg u &= \deg(Ux^m + V) = \deg U + m = d\,, \\
Lu - c &= L(Ux^m) + LV - (c^* + L(Ux^m)) = LV - c^*\,,
\end{aligned}$$

and

$$\begin{aligned}
\deg(Lu - c) &= \deg(LV - c^*) \\
&= \deg\Big(x^{n-m}(L_2V - C_2) + (a - A_2 x^{n-m})EV \\
&\qquad + (b - B_2 x^{n-m})V - (c^* - C_2 x^{n-m}) \Big) \\
&< n - m + m = n\,,
\end{aligned}$$

where we have used that the degrees of $a - A_2 x^{n-m}$, $b - B_2 x^{n-m}$, and $c - C_2 x^{n-m}$ are less than $n - m$.

We denote the cost of the algorithm for inputs with $\deg c - \deg L = d$ by $T(d)$. The cost of steps 1 and 2 is $O(1)$. In step 3, we first compute $(x+1)^m$ with repeated squaring, and then compute the leading $d - m + 1$ coefficients of $(x+1)^m a$, at a cost of $O(\mathsf{M}(d))$ operations. This is the total cost for step 3, since the computation of B_1 and C_1 requires no arithmetic operations. In step 4, we have $\deg C_1 - \deg L_1 = d - m = \lfloor d/2 \rfloor$, and hence the cost for the recursive call is at most $T(\lfloor d/2 \rfloor)$. In step 5, we compute the coefficients of $x^{n-m}, \ldots, x^{n+d}$ in c^*, similarly to step 3, at a cost of $O(\mathsf{M}(d + m))$ operations. The computation of A_2 and B_2 is for free. Finally, in step 6 we have $\deg C_2 - \deg L_2 \leq m - 1 \leq \lfloor d/2 \rfloor$, and the cost for the recursive call is at most $T(\lfloor d/2 \rfloor)$. Step 7 is essentially free. Summarizing, we have

$$T(0) \in O(1) \text{ and } T(d) \leq 2T(\lfloor d/2 \rfloor) + O(\mathsf{M}(d)) \text{ if } d \geq 1 .$$

The running time bound now follows from unraveling the recursion. $\square$

With classical arithmetic, similar arguments show that the running time is $O(d^2)$. However, we recommend to use the simpler Algorithm 9.12 for classical arithmetic.

Corollary 9.37. *Let F be a field of characteristic zero, $L = aE + b = a\Delta + (a+b)$ be a linear difference operator, with $a, b \in F[x]$, $a + b \neq 0$, and $\deg a - 1 \leq \deg(a + b) = n$, and $c \in F[x]$ with $d = \deg c - n \geq 0$ and $\delta_L \notin \mathbb{N}$. Then we have an algorithm that either computes the unique polynomial $u \in F[x]$ satisfying $Lu = c$, which has degree d, or correctly asserts that no solution exists. It uses $O(\mathsf{M}(d) \log d + \mathsf{M}(n))$ arithmetic operations in F.*

Proof. We apply algorithm 9.35 and check its output by computing $a\,Eu + bu - c$. For the check, we need only compute the lower n coefficients of Lu, and Theorem 9.36 and Lemma 9.9 yield the claim. $\square$

Here is the analog of Algorithm 9.35 in the differential case. If $L = aD + b$ is a linear differential operator, then $\delta_L \neq 0$ implies that $b \neq 0$ and $\deg a - 1 \leq \deg b$, as in the difference case.

Algorithm 9.38 (Differential divide & conquer: generic case). ▬▬▬▬
Input: A linear differential operator $L = aD + b$, where $a, b \in F[x]$ with $b \neq 0$ for a field F of characteristic zero, and $c \in F[x]$ with $\deg a - 1 \leq \deg b = n = \deg L$, $d = \deg c - n$, and $\delta_L \notin \{0, \ldots, d\}$.
Output: A polynomial $u \in F[x]$ with $u = 0$ or $\deg u = d$ if $d \geq 0$, such that $\deg(Lu - c) < n$.

1. **if** $d < 0$ **then return** $u = 0$
2. **if** $d = 0$ **then return** $u = \mathrm{lc}(c)/\mathrm{lc}(b)$
3. $m \longleftarrow \lceil d/2 \rceil, \quad t \longleftarrow n - (d - m)$
   $A_1 \longleftarrow x^m a \upharpoonright (t + m), \quad B_1 \longleftarrow (mx^{m-1}a + x^m b) \upharpoonright (t + m)$
   $C_1 \longleftarrow c \upharpoonright (t + m)$
4. **call** the algorithm recursively with input $L_1 = A_1 D + B_1$ and C_1 to obtain $U \in F[x]$ such that $\deg(L_1 U - C_1) < d - m$
5. $A_2 \longleftarrow a \upharpoonright (n - m), \quad B_2 \longleftarrow b \upharpoonright (n - m),$
   $C_2 \longleftarrow (c - ax^m U' - (amx^{m-1} + bx^m)U) \upharpoonright (n - m)$

6. **call** the algorithm recursively with input $L_2 = A_2 D + B_2$ and C_2, yielding $V \in F[x]$ with $\deg(L_2 V - C_2) < m$
7. **return** $U x^m + V$

The proofs of the following theorem and corollary are completely analogous to the proofs of Theorem 9.36 and Corollary 9.37 and therefore omitted.

Theorem 9.39. *Algorithm 9.38 works correctly and uses $O(\mathsf{M}(d) \log d)$ arithmetic operations in F with fast arithmetic.*

Corollary 9.40. *Let F be a field of characteristic zero, $L = aD + b$ be a linear difference operator, with $a, b \in F[x]$, $b \neq 0$, and $\deg a - 1 \leq \deg b = n$, and $c \in F[x]$ with $d = \deg c - n \geq 0$ and $\delta_L \notin \mathbb{N}$. Then we have an algorithm that either computes the unique polynomial $u \in F[x]$ satisfying $Lu = c$, which has degree d, or correctly asserts that no solution exists. It uses $O(\mathsf{M}(d) \log d + \mathsf{M}(n))$ arithmetic operations in F.*

9.6 A Divide-and-Conquer Algorithm: General Case

Let

$$a = \sum_{0 \leq i \leq n+1} a_i x^i, \quad b = \sum_{0 \leq i \leq n+1} b_i x^i, \quad u = \sum_{0 \leq i \leq d} u_i x^i, \text{ and } c = \sum_{0 \leq i \leq k} c_i x^i,$$

with $n + 1 = \deg a = \deg b$, $d = \deg u$, $k \in \mathbb{N}$, and assume that $(\deg a) - 1 \geq \deg(a + b)$ (i.e., $a_{n+1} + b_{n+1} = 0$). The coefficient matrix of the linear system in the coefficients of u equivalent to the linear difference equation $Lu = a \, Eu + bu = a \, \Delta u + (a + b)u = c$ is triangular, with the coefficient of u_i in the equation corresponding to x^{n+i} equal to $a_{n+1}i + a_n + b_n = a_{n+1}(i - \delta_L)$, by Lemma 9.3. For example, if $a_3 = b_3$ in Example 9.34, then $n = \deg L = \deg a - 1 = 2 \geq \deg(a + b)$, $\delta_L = -(a_3 + b_3)/a_3$, and the linear system reads

$$
\begin{aligned}
c_6 &= 0, \\
c_5 &= (3a_3 + b_2^*)u_3, \\
c_4 &= (2a_3 + b_2^*)u_2 + (3a_3 + 3a_2 + b_1^*)u_3, \\
c_3 &= (a_3 + b_2^*)u_1 + (a_3 + 2a_2 + b_1^*)u_2 \\
&\quad + (a_3 + 3a_2 + 3a_1 + b_0^*)u_3, \\
c_2 &= b_2^* u_0 + (a_2 + b_1^*)u_1 + (a_2 + 2a_1 + b_0^*)u_2 \\
&\quad + (a_2 + 3a_1 + 3a_0)u_3, \\
c_1 &= b_1^* u_0 + (a_1 + b_0^*)u_1 + (a_1 + 2a_0)u_2 + (a_1 + 3a_0)u_3, \\
c_0 &= b_0^* u_0 + a_0 u_1 + a_0 u_2 + a_0 u_3,
\end{aligned}
$$

where again $b_i^* = a_i + b_i$ for $0 \leq i \leq 3$. In general, at most one of the subdiagonal entries $a_n(i - \delta_L)$ vanishes, and this can only happen if δ_L is a nonnegative integer.

Then there may be a degree of freedom in the choice of u_{δ_L} (corresponding to a nonzero solution of the homogeneous equation $a \cdot Eu + b \cdot u = 0$), in which case we might simply choose $u_{\delta_L} = 0$. However, it may happen that $\delta_L \in \mathbb{N}$ and $\ker L = \{0\}$, and in that case u_{δ_L} has to be chosen consistently with the other equations.

We now use Algorithm 9.35 to solve the difference equation $Lu = a\,Eu + bu = a\,\Delta u + (a+b)u = c$ in the case where $\delta_L \in \mathbb{N}$. We first compute the coefficients of x^i in u for $i > \delta_L$ and then take care of a possible freedom in choosing the coefficient of x^{δ_L}. The idea is to compute polynomials $U, V, W \in F[x]$ such that $\deg U < d - \delta_L$, $\deg V, \deg W < \delta_L$, and the set of all solutions of $\deg(Lu - c) < \deg L$ is $\{Ux^{\delta_L+1} + V + \kappa(x^{\delta_L} + W): \kappa \in F\}$, using Algorithm 9.35, and then to check whether some (or every) $\kappa \in F$ gives a solution of $Lu = c$ (this corresponds to the checking in Corollary 9.37).

Algorithm 9.41 (Difference divide & conquer: general case). ▬▬▬▬▬
Input: A linear difference operator $L = aE + b = a\Delta + (a+b)$, where $a, b \in F[x]$
  and $a \neq 0$ for a field F of characteristic zero, and $c \in F[x]$ with $\deg(a+b) \le$
  $\deg a - 1 = n = \deg L$, $\delta_L \in \mathbb{N}$, and $d = \max\{\deg c - n, \delta_L\}$.
Output: A polynomial $u \in F[x]$ of degree at most d such that $Lu = c$, or otherwise
  "unsolvable".

1. **if** $d = \delta_L$ **then** $U \longleftarrow 0$, **goto** 4
2. $m \longleftarrow \delta_L + 1$, $t \longleftarrow n - (d - m)$
   $A \longleftarrow (x+1)^m a \upharpoonright (t+m)$, $B \longleftarrow x^m b \upharpoonright (t+m)$, $C \longleftarrow c \upharpoonright (t+m)$
3. **call** Algorithm 9.35 with input $L_1 = AE + B$ and C to obtain $U \in F[x]$ such
   that $\deg(L_1 U - C) < d - m$
4. $c^* \longleftarrow c - L(Ux^m)$, **if** $\delta_L = \deg c^* - n$ **then return** "unsolvable"
5. **call** Algorithm 9.35 with input L and c^* to obtain $V \in F[x]$ such that
   $\deg(LV - c^*) < n$
6. $h^* \longleftarrow LV - c^*$, **if** $h^* = 0$ **then return** $Ux^m + V$
7. $c^{**} \longleftarrow Lx^{\delta_L}$
   call Algorithm 9.35 with input L and $-c^{**}$ to obtain $W \in F[x]$ such that
   $\deg(LW + c^{**}) < n$
8. $h^{**} \longleftarrow LW + c^{**}$
   if there exists some $\kappa \in F$ such that $h^* + \kappa h^{**} = 0$ in $F[x]$
   then return $Ux^m + V + \kappa(x^{\delta_L} + W)$ **else return** "unsolvable" ▬▬▬▬▬

We note that by definition, $\delta_L \in \mathbb{N}$ already implies that $a \neq 0$ and $\deg a - 1 \ge \deg(a+b)$. The constant κ is the coefficient of x^{δ_L} in u.

Theorem 9.42. *Algorithm 9.41 works correctly, using* $O(\mathsf{M}(d)\log d + \mathsf{M}(n+d))$ *arithmetic operations in* F.

Proof. If the algorithm returns $Ux^m + V$ in step 6, then $L(Ux^m + V) - c = LV - c^* = h^* = 0$, and $Ux^m + V$ solves the difference equation. If the algorithm returns $u = Ux^m + V + \kappa(x^{\delta_L} + W)$ in step 8, then

$$Lu - c = LV + \kappa(Lx^{\delta_L} + LW) - c^* = h^* + \kappa(c^{**} + LW) = h^* + \kappa h^{**} = 0 \,,$$

and the output is correct as well.

Conversely, suppose that there exists a polynomial $u \in F[x]$ with $\deg u \leq d$ such that $Lu = c$. If $\delta_L < d$, then we have $\delta_L < m \leq d$ in step 2. As in the proof of Theorem 9.36, the coefficients of x^i in $x^{t+m}(A+B)$ and $Lx^m \bmod x^{t+m}$ coincide for $i \geq t + m$, Corollary 9.5 shows that $\deg(Lx^m) = n + m$, and $n \geq t$ implies that $\deg(A + B) = d - m \geq 0$. Now

$$
\begin{aligned}
\deg A - 1 &= \deg a - 1 - t = n - t = d - m = \deg L_1 \,, \\
\deg C &= \deg c - (t + m) = n - t + d - m = 2(d - m) \,, \\
\deg C - \deg L_1 &= 2(d - m) - d - m = d - m \,.
\end{aligned}
$$

As in the proof of Theorem 9.36, Lemma 9.3 implies that $\delta_{L_1} = \delta_L - m = -1$. Thus Algorithm 9.35 returns the correct result in step 3, so that we have $\deg U = \deg C - \deg L_1 = d - m$ and $\deg(L_1 U - C) < \deg L_1 = d - m$. Since this U is unique, by Lemma 9.13, and $U^* = u \restriction m$ satisfies $\deg(L_1 U^* - C) < d - m$, we find that $U = U^*$. If $\delta_L = d$ in step 2, then $U = U^* = 0$.

In step 4, we have $L(u - Ux^m) = c^*$. As in the proof of Theorem 9.35, we get $\deg c^* < n + m$, and hence $\delta_L \geq d^* = \deg c^* - n$. Corollary 9.6 shows that $\delta_L \neq d^*$, and therefore $\delta_L > d^*$. Then Algorithm 9.35 correctly computes $V \in F[x]$ of degree d^* such that $\deg(LV - c^*) < \deg L = n$. We find

$$L(Ux^m + V) - c = LV + L(Ux^m - u) = LV - c^* = h^*$$

in step 6. If $h^* = 0$, then the algorithm correctly returns the solution $Ux^m + V$.

Otherwise, if $h^* \neq 0$, then Corollary 9.5 shows that $\deg c^{**} < n + \delta_L$, and Algorithm 9.35 correctly computes $W \in F[x]$ of degree $\deg c^{**} - n < \delta_L$ such that $\deg(L(x^{\delta_L} + W)) < \deg L = n$ in step 7. In step 8, we then have $h^{**} = L(x^{\delta_L} + W)$. If $h^{**} = 0$, then $\ker L = F \cdot (x^\delta + W)$, by Lemma 9.1, and otherwise $\ker L = \{0\}$. Since $\deg c^* < n + \delta_L$, Lemma 9.13 implies that the space S_{c^*} of all polynomials $y \in F[x]$ of degree at most δ_L satisfying $\deg(Ly - c^*) < n$ is one-dimensional, and we find $S_{c^*} = V + F \cdot (x^{\delta_L} + W)$. Now $u - Ux^m \in S_{c^*}$, and we conclude that $u - Ux^m = V + \sigma \cdot (x^{\delta_L} + W)$, where $\sigma \in F$ is the coefficient of x^{δ_L} in u. Then

$$h^* + \sigma h^{**} = L(V + \sigma(x^{\delta_L} + W)) - c^* = L(u - Ux^m) - c^* = 0 \,,$$

and there exists a suitable κ in step 8. If $\ker L = \{0\}$, then this κ is unique, namely $\kappa = \sigma$. Otherwise $x^{\delta_L} + W$ generates $\ker L$ and $Ux^m + V$ is a solution of the difference equation $Ly = c$. However, the algorithm would have already stopped in step 6 in that case.

As in the proof of Theorem 9.36, the cost for step 2 is $O(\mathsf{M}(d))$ arithmetic operations. Step 3 takes $O(\mathsf{M}(d - \delta_L)\log(d - \delta_L))$, by Theorem 9.36. Computing $L(Ux^m)$ and c^* in step 4, h^* in step 6, Lx^{δ_L} in step 7, and h^{**} in step 8 takes $O(\mathsf{M}(n+d))$ operations, by Lemma 9.9. The cost for the two calls to Algorithm 9.35

in steps 5 and 7 is $O(M(\delta_L) \log \delta_L)$. Finally, finding a suitable κ in step 8 takes $O(n)$ operations. The cost estimate now follows by adding up costs. $\square$

In the nonunique case, where $\ker L \neq \{0\}$, Algorithm 9.41 stops in step 6 if $Lu = c$ is solvable, but may need to proceed until step 8 to find out that the difference equation is unsolvable (see Example 9.43 (ii) below). If we know somehow in advance that $\ker L \neq \{0\}$, then we may already return "unsolvable" in step 6 if $h^* \neq 0$, since then $\ker L = F \cdot (x^\delta + W)$ and $h^{**} = 0$ in step 8. This may be useful in Gosper's (1978) algorithm, since Lisoněk, Paule & Strehl (1993) have shown that the condition $\ker L \neq \{0\}$ corresponds to the case where Gosper's algorithm is used to compute the antidifference of a rational function.

Example 9.43. *In the following examples, we use Algorithms 9.35 and 9.41 to solve $Lu = c$ for the linear difference operator $L = aE + b = a\Delta + a + b$, and write $n = \deg L$ and $\delta = \delta_L$ for short.*

(i) *Let $a = x^2$ and $b = -x^2 + x/2 - 1$. Then $n = 1$ and $\delta = -1/2 \notin \mathbb{N}$. If we let $c = x - 2$, then we obtain $u = 2$ in step 1 of Algorithm 9.35. We check that*

$$Lu = x^2 \cdot 2 + \left(-x^2 + \frac{1}{2}x - 1\right) \cdot 2 = x - 2,$$

and u solves the difference equation. On the other hand, for $c = x$, we also get $u = 2$ in step 1, but now $Lu = c - 2 \neq c$, and $Lu = c$ is unsolvable.

(ii) *Let $a = x^2$ and $b = -x^2 - 2x - 1$, so that $n = 1$ and $\delta = 2$. For $c = x^2 + x$, we have $d = \max\{\deg c - n, \delta\} = \delta = 2$. Thus $U = 0$ in step 4 of Algorithm 9.41, $c^* = c$ and $\deg c^* - n = 1 < \delta$, and in step 5 we obtain $V = -x$. Then*

$$LV = x^2(-x - 1) + (-x^2 - 2x - 1)(-x) = x^2 + x,$$

whence $h^* = 0$ in step 6, and $u = -x$ solves the difference equation.
On the other hand, for $c = x^2 + x + 1$ we have d, U, c^*, and V as before, but now $h^* = -1$ in step 6. Then

$$c^{**} = Lx^\delta = x^2(x + 1)^2 + (-x^2 - 2x - 1)x^2 = 0$$

and $W = 0$ in step 7, and $h^{**} = 0$ in step 8. Thus $h^* + \kappa h^{**} = -1 \neq 0$ for all $\kappa \in F$, and $Lu = c$ is unsolvable.
In fact, $\ker L = F \cdot (x^\delta + W) = F \cdot x^2$, and the set of all solutions of $Lu = x^2 + x$ is

$$Ux^{\delta+1} + V + \ker L = \{\kappa x^2 - x : \kappa \in F\}.$$

(iii) *Let $a = x^2$ and $b = -x^2 - x - 1/4$. Then $n = 1$ and $\delta = 1$, and for $c = 4x^3 + 3x^2 + x$, we have $d = \max\{\deg c - n, \delta\} = 2 > \delta$. Thus $m = 2$, $t = 1$, $A = (x+1)^2 a \upharpoonright 3 = x + 2$, $B = x^2 b \upharpoonright 3 = -x - 1$, and $C = c \upharpoonright 3 = 4$ in step 2, and in step 3 we obtain $U = 4$. Next, we have*

$$
\begin{aligned}
c^* &= c - L(Ux^m) \\
&= 4x^3 + 3x^2 + x - x^2 \cdot 4(x+1)^2 - \left(-x^2 - x - \frac{1}{4}\right) \cdot 4x^2 \\
&= x,
\end{aligned}
$$

and $\deg c^* - n = 0 < \delta$. *Now we compute* $V = -1$ *in step 5, and*

$$h^* = LV - c^* = -x^2 - \left(-x^2 - x - \frac{1}{4}\right) - x = \frac{1}{4} \neq 0$$

in step 6. In step 7, we get

$$c^{**} = Lx^\delta = x^2(x+1) + \left(-x^2 - x - \frac{1}{4}\right)x = -\frac{1}{4}x ,$$

and $W = -1/4$. *Finally,*

$$h^{**} = LW + c^{**} = -\frac{1}{4}x^2 - \frac{1}{4}\left(-x^2 - x - \frac{1}{4}\right) - \frac{1}{4}x = \frac{1}{16}$$

in step 8, and $h^* + \kappa h^{**} = 1/4 + \kappa/16 = 0$ *if and only if* $\kappa = -4$. *Then* $u = 4x^2 - 1 - 4 \cdot (x - 1/4) = 4x(x - 1)$, *and we check that*

$$
\begin{aligned}
Lu &= x^2 \cdot 4(x+1)x + \left(-x^2 - x - \frac{1}{4}\right) \cdot 4x(x-1) \\
&= 4x^3 + 3x^2 + x ,
\end{aligned}
$$

and u *solves* $Lu = c$.

If we take $c = 4x^3 + 4x^2 + x$, *then* d *and* U *are as before, but now* $c^* = x^2 + x$, $\deg c^* - n = 1 = \delta$, *and the algorithm returns "unsolvable" in step 4. In fact, the homogeneous equation* $Lu = 0$ *only has the trivial solution* $u = 0$.

Here is the analog of Algorithm 9.35 in the differential case.

Algorithm 9.44 (Differential divide & conquer: general case).
Input: A linear differential operator $L = aD + b$, where $a, b \in F[x]$ and $a \neq 0$ for a field F of characteristic zero, and $c \in F[x]$ with $\deg b \leq \deg a - 1 = n = \deg L$, $\delta_L \in \mathbb{N}$, and $d = \max\{\deg c - n, \delta_L\}$.
Output: A polynomial $u \in F[x]$ of degree at most d such that $Lu = c$, or otherwise "unsolvable".

1. **if** $d = \delta_L$ **then** $U \longleftarrow 0$,   **goto** 4
2. $m \longleftarrow \delta_L + 1$, $t \longleftarrow n - (d - m)$
   $A \longleftarrow x^m a \upharpoonright (t + m)$, $B \longleftarrow (m x^{m-1} a + x^m b) \upharpoonright (t + m)$
   $C \longleftarrow c \upharpoonright (t + m)$
3. **call** Algorithm 9.38 with input $L_1 = AD + B$ and C to obtain $U \in F[x]$ such that $\deg(L_1 U - C) < d - m$
4. $c^* \longleftarrow c - L(Ux^m)$,   **if** $\delta_L = \deg c^* - n$ **then return** "unsolvable"
5. **call** Algorithm 9.38 with input L and c^* to obtain $V \in F[x]$ such that $\deg(LV - c^*) < n$
6. $h^* \longleftarrow LV - c^*$,   **if** $h^* = 0$ **then return** $Ux^m + V$
7. $c^{**} \longleftarrow Lx^{\delta_L}$
   call Algorithm 9.38 with input L and $-c^{**}$ to obtain $W \in F[x]$ such that $\deg(LW + c^{**}) < n$

8. $h^{**} \longleftarrow LW + c^{**}$
   if there exists some $\kappa \in F$ such that $h^* + \kappa h^{**} = 0$ in $F[x]$
   then return $Ux^m + V + \kappa(x^{\delta_L} + W)$ **else return** "unsolvable"

As in the difference case, $\delta_L \in \mathbb{N}$ implies that $a \neq 0$ and $\deg a - 1 \geq \deg b$. The proof of the following theorem is analogous to the proof of Theorem 9.42.

Theorem 9.45. *Algorithm 9.44 works correctly, using* $O(\mathsf{M}(d) \log d + \mathsf{M}(n + d))$ *arithmetic operations in F.*

9.7 Barkatou's Algorithm for Linear Difference Equations

The algorithm that we will discuss in this section is due to Barkatou (1999), who used it for the solution of linear first order matrix difference equations. The idea is to evaluate the coefficients of the difference equation at the points $0, 1, 2, \ldots$. This gives a linear system for the values of the solution at these points, and we can recover the solution by interpolation. The algorithm bears strong similarities to the ABP algorithm: both transform the original linear system into an equivalent one which is nicely structured, by performing a suitable change of basis. The coefficient matrix of the transformed system in Barkatou's algorithm is even simpler than for the ABP algorithm: it is banded triangular, but the bandwidth is only two.

Algorithm 9.46 (Barkatou's algorithm).
Input: Polynomials $a, b, c \in F[x]$, where F is a field of characteristic zero, and the
   degree bound d from Corollary 9.6 for $L = aE + b = a\Delta + (a + b)$, such that
   $d \in \mathbb{N}$ and $a(0), \ldots, a(d - 1) \neq 0$.
Output: A nonzero polynomial $u \in F[x]$ solving $a\,Eu + bu = c$, or otherwise
   "unsolvable".

1. **for** $0 \leq i < d$ **do** $\alpha_i \longleftarrow a(i), \quad \beta_i \longleftarrow b(i), \quad \gamma_i \longleftarrow c(i)$
2. $\omega_0 \longleftarrow 0, \quad \eta_0 \longleftarrow 1$
   for $0 \leq i < d$ **do** $\omega_{i+1} \longleftarrow (\gamma_i - \beta_i \omega_i)/\alpha_i, \quad \eta_{i+1} \longleftarrow -\beta_i \eta_i / \alpha_i$
3. compute by interpolation polynomials $w, h \in F[x]$ of degrees at most d such
   that $w(i) = \omega_i$ and $h(i) = \eta_i$ for $0 \leq i \leq d$
4. **if** there is a $\kappa \in F$ such that $w + \kappa h \neq 0$ and $a \cdot (Ew + \kappa\,Eh) + b \cdot (w + \kappa h) = c$
   then return $w + \kappa h$ **else return** "unsolvable"

The constant κ is the constant coefficient of u.

Theorem 9.47. *Algorithm 9.46 works correctly as specified. If $\deg L = n$, then it takes $O(nd + d^2)$ field operations with classical arithmetic and $O(\mathsf{M}(d) \log d + \mathsf{M}(n + d))$ with fast arithmetic.*

Proof. It is clear that the algorithm returns a solution of the difference equation if it does not return "unsolvable". So we assume that $u \in F[x]$ of degree at most d satisfies $Lu = c$. Then

$$\alpha_i u(i + 1) + \beta_i u(i) = a(i)u(i + 1) + b(i)u(i) = c(i) = \gamma_i$$

for $0 \leq i < d$. This is a linear system of d equations for the $d + 1$ values $u(0), \ldots, u(d)$. By assumption, we have $\alpha_i \neq 0$ for all i, and hence the solution space of this linear system is one-dimensional. By construction, the vector $\omega = (\omega_0, \ldots, \omega_d)$ belongs to this solution space, and $\eta = (\eta_0, \ldots, \eta_d)$ is a nonzero solution of the corresponding homogeneous system, where all γ_i are zero. Thus the solution space is precisely $\omega + F\eta$, and hence $u(i) = \omega_i + u(0)\eta_i$ for $0 \leq i \leq d$. Since interpolation is a linear operation in the coefficients and $\deg u \leq d$, we find that $u = w + u(0)h$, and the algorithm returns a solution of the difference equation.

We have $\deg a, \deg b \leq n + 1$ and $\deg c \leq n + d$. With classical arithmetic, we can evaluate a, b, c at all points $0, \ldots, d - 1$ in step 1 using $O(nd + d^2)$ arithmetic operations in F. With fast arithmetic, we first compute c rem $x^{\underline{d}}$, taking $O(\mathsf{M}(d) \log d + \mathsf{M}(n))$ operations, by Lemma 3.15, and then evaluate this remainder and a, b at all points using $O(\mathsf{M}(d) \log d)$ operations. Computing the ω_i, η_i in step 2 takes $O(d)$ field operations. The cost for the interpolation in step 3 is $O(d^2)$ field operations with classical arithmetic and $O(\mathsf{M}(d) \log d)$ with fast arithmetic. In step 4, we check whether $c - a\,Ew - bw$ is proportional to $a\,Eh + bh$. This takes $O(nd + d^2)$ with classical arithmetic and $O(\mathsf{M}(n + d))$ with fast arithmetic, by Lemma 9.9. Finally, we can find a proportionality factor κ (or prove its nonexistence) and compute $w + \kappa h$ with another $O(n + d)$ operations. $\square$

As in Sect. 9.4, the restriction that $a(0), \ldots, a(d - 1)$ be nonzero is not severe.

Corollary 9.48. *Let F be a field of characteristic zero, $L = aE + b = a\Delta + a + b$ a linear difference operator, with $a, b \in F[x]$, $a \neq 0$, and $n = \deg L$, let $c \in F[x]$, and assume that the degree bound $d = \max(\{\deg c - n, \delta_L\} \cap \mathbb{Z})$ of Corollary 9.6 is nonnegative. Using Algorithm 9.46, we can decide whether the difference equation $a\,Eu + bu = c$ has a solution $u \in F[x]$, and if so, compute one, using an expected number of $O(n^2 + d^2)$ operations in F with classical arithmetic and $O(\mathsf{M}(d) \log d + \mathsf{M}(n + d))$ with fast arithmetic.*

The proof is analogous to the proof of Corollary 9.33; the additional conversion cost is only $O(\mathsf{M}(n + d))$ with fast arithmetic since no conversion to the falling factorial basis is required.

An analog of Barkatou's algorithm for linear differential equations would be to evaluate the coefficients of the equation and their derivatives at zero. This gives a system of linear equations for the derivatives of the solution at zero. However, this is essentially the same as writing the differential equation as a system of linear equations with respect to the monomial basis.

9.8 Modular Algorithms

In this section, we present new modular algorithms for finding polynomial solutions of linear first order differential or difference equations with coefficients in $\mathbb{Z}[x]$. The idea is to reduce the inputs modulo several single precision primes, perform one of the algorithms from the previous sections modulo each prime, and reconstruct the result in $\mathbb{Q}[x]$ via Chinese remaindering and rational number reconstruction. If p is a prime greater than the degrees of the input and the output polynomials, then the algorithms from the previous sections, which were specified for characteristic zero, work literally modulo p. However, it may still happen that p is an unlucky prime, namely if the equation in $\mathbb{Q}[x]$ has a unique solution, and the equation taken modulo p has several solutions.

Let L be a linear first order differential or difference operator with coefficients in $\mathbb{Z}[x]$, and let $c \in \mathbb{Z}[x]$. As we have already noted, the equation $Lu = c$ for $u \in \mathbb{Q}[x]$ with $\deg u \leq d$ is equivalent to a certain system of linear equations in the $d + 1$ unknown coefficients of u. Thus the coefficients of u can be expressed as quotients of certain minors of the coefficient matrix of this linear system, by Cramer's rule. Since there may be several possible solutions u, we introduce below the notion of a *normalized* solution and show that it is unique. Later we show that if u is the normalized solution of $Lu = c$ and p is a "lucky" prime, then $u \bmod p$ is the normalized solution of $Lu \equiv c \bmod p$, which allows us to reconstruct u correctly from its modular images.

Definition 9.49. *Let R be a ring, $f \in R[x]$, and $n \in \mathbb{N}$. We denote the coefficient of x^n in f by $[x^n]f$.*

Definition 9.50. *Let F be a field, $a, b \in F[x]$ not both zero, $L = aD + b$ or $L = a\Delta + b$ a linear differential or difference operator, respectively, $u \in F[x]$, and $c = Lu$. We say that u is a* normalized *solution of the equation $Lu = c$ if $[x^{\deg v}]u = 0$ for all nonzero $v \in \ker L$.*

Lemma 9.51. *With the notation as in Definition 9.50, there is at most one normalized solution of the equation $Lu = c$.*

Proof. Let $v \in F[x]$ be another normalized solution. Then $u - v \in \ker L$. Suppose that $u \neq v$ and let $d = \deg(u - v)$. By assumption, the coefficient of x^d in u and v vanishes, and hence so does the coefficient of x^d in $u - v$. This contradiction proves that $u = v$. $\square$

In particular, $u = 0$ is the normalized solution of the homogeneous equation $Lu = 0$.

Problem 9.52 (Polynomial solutions of first order equations). *Given a first order linear difference or differential operator L with coefficients in $\mathbb{Z}[x]$ and a nonzero polynomial $c \in \mathbb{Z}[x]$, decide if $L^{-1}c$ is non-empty, and if so, compute the unique normalized solution $u \in \mathbb{Q}[x]$.*

See Remark 9.11 for the homogeneous case.

The following theorem is the basis for our modular algorithm. It says when the modular image of a normalized solution of the differential or difference equation is a normalized solution of the modular equation.

Theorem 9.53. *Let Z be an integral domain of characteristic zero with field of fractions Q, $a, b \in Z[x]$ not both zero, $L = aD + b$ or $L = a\Delta + b$ a linear differential or* *difference operator, respectively, and $c \in Z[x]$. Let $d \in \mathbb{N}$ be the bound from Corollary 9.6 on the degree of a polynomial solution $u \in Q[x]$ of $Lu = c$. Moreover, let $I \subseteq Z$ be a maximal ideal coprime to $\mathrm{lc}(abc)$, denote reduction modulo I by a bar and the residue class field Z/I by $\overline{Z}$, and suppose that $\overline{Z}$ has positive characteristic $p > d$. Finally, let $V = \{v \in \overline{Z}[x] : \deg v < p\}$.*

(i) *If $\dim \ker L = \dim(V \cap \ker \overline{L}) = 0$ and $u \in Q[x]$ satisfies $Lu = c$, then u is normalized, the denominators of u are invertible modulo I, and $v = \overline{u}$ is the unique normalized solution of the equation $\overline{L}v = \overline{c}$.*

(ii) *If $\dim \ker L = 1$, then $\delta_L \in \{0, \ldots, d\}$ and $\dim(V \cap \ker \overline{L}) = 1$. If the equation $Lu = c$ is solvable, then it has a unique normalized solution $u \in Q[x]$, the denominators of u are invertible modulo I, and $v = \overline{u}$ is the unique normalized solution of the equation $\overline{L}v = \overline{c}$.*

Proof. Let $e = \deg L + d$ and $R = (r_{ij})$ be the $(e+1) \times (d+1)$ coefficient matrix of the linear system $Lu = c$ for $\deg u \leq d$ with respect to the usual monomial basis $1, x, x^2, \ldots$, as illustrated in Fig. 9.1 on p. 166 for the differential case. Then $\mathrm{rank}\, R = d + 1 - \dim \ker L \in \{d, d+1\}$, by Lemma 9.1 and Corollary 9.6. Similarly, $\overline{R} = R \bmod I$ is the coefficient matrix of the linear system $\overline{L}v = \overline{c}$ when $\deg v \leq d$, and Lemma 9.1 implies that $\dim(V \cap \ker \overline{L}) \leq 1$ and $\mathrm{rank}\, \overline{R} \in \{d, d+1\}$.

(i) We first note that $\deg u \leq d$, by Corollary 9.6, and that both R and $\overline{R}$ have full column rank $d + 1$. Since $\ker L = \{0\}$, u is clearly normalized. Let A be any $(d+1) \times (d+1)$ submatrix of R such that $\det A \not\equiv 0 \bmod I$, comprising rows $0 \leq i_0 < i_1 < \cdots < i_d \leq e$ of R, and let $w \in Z^{d+1}$ be the column vector with the coefficients of $x^{i_0}, x^{i_1}, \ldots, x^{i_d}$ in c. If we identify u with its coefficient vector in Q^{d+1}, then $Au = w$, and Cramer's rule implies that $(\det A)u \in Z^{d+1}$. Since $\det A$ is a unit modulo I, so are the denominators of u. Thus $\overline{u}$ is well defined, and taking the equation $Lu = c$ modulo I yields $\overline{L}\overline{u} = \overline{c}$. If $w \in \ker \overline{L}$ is nonzero, then $\dim(V \cap \ker \overline{L}) = 0$ implies that $\deg w \geq p > \deg \overline{u}$, so that the coefficient of $x^{\deg w}$ in $\overline{u}$ is zero. Thus $v = \overline{u}$ is a normalized solution of $\overline{L}v = \overline{c}$, and uniqueness follows from Lemma 9.51.

(ii) The first assertion was shown in Corollary 9.6. Since $\dim \ker L = 1$, we have $\mathrm{rank}\, R = d$, and all $(d+1) \times (d+1)$ minors of R, and hence also of $\overline{R}$, vanish. Thus $\mathrm{rank}\, \overline{R} = d$ and $\dim(V \cap \ker \overline{L}) = d + 1 - \mathrm{rank}\, \overline{R} = 1$.

All nonzero polynomials in $\ker L$ and $V \cap \ker \overline{L}$ have degree δ_L, by Corollaries 9.6 and 9.7, respectively. Thus, if we augment the matrix R by the row $(0, \ldots, 0, 1, 0, \ldots, 0)$, whose only nonzero entry is a 1 at the position corresponding to x^{δ_L}, and denote the resulting $(e+2) \times (d+1)$ matrix by R^*, then both R^* and $\overline{R}^*$ have full column rank $d + 1$.

By adding a suitable element of $\ker L$ to u if necessary, we may assume that u is normalized. As in (i), let A be any nonsingular $(d+1) \times (d+1)$ submatrix of R^* such that $\det A \not\equiv 0 \bmod I$, and let $w \in Z^{d+1}$ be the corresponding subvector of the coefficient vector of c augmented by a single entry 0. Then the remaining claims follow from similar arguments as in (i). $\square$

Definition 9.54. *Let L be a linear first order difference or differential operator with coefficients in $\mathbb{Z}[x]$, and let $c \in \mathbb{Z}[x]$ be nonzero. For a prime $p \in \mathbb{N}$, let $V = \{u \in \mathbb{F}_p[x] \colon \deg u < p\}$ and denote reduction modulo p by a bar. Then p is unlucky with respect to problem 9.52 if $\dim(V \cap \ker \overline{L}) > \dim \ker L$ or both $L^{-1}c = \{0\}$ and $V \cap \overline{L}^{-1}\overline{c} \neq \{0\}$. Otherwise, p is lucky.*

The following corollary gives a condition when a prime is lucky, and expresses the coefficients of a normalized solution of the differential or difference equation as quotients of determinants.

Corollary 9.55. *With the assumptions as in Theorem 9.53, let $Z = \mathbb{Z}$, $Q = \mathbb{Q}$, and $I = \langle p \rangle$ for a prime $p > d$. Moreover, let $e = \deg L + d$ and R be the $(e+1) \times (d+1)$ coefficient matrix of the linear system $Lu = c$ with respect to the monomial basis, and let S be the matrix R augmented by the column vector in $\mathbb{Z}^{e+1}$ with the coefficients of c. We define two matrices R^* and S^*, as follows. If $\ker L = \{0\}$, then $R^* = R$ and $S^* = S$. Otherwise, if $\dim \ker L = 1$, then we obtain R^* and S^* from R and S by augmenting a row vector $(0, \ldots, 0, 1, 0, \ldots, 0)$ in $\mathbb{Z}^{d+1}$ and $\mathbb{Z}^{d+2}$, respectively, with the entry 1 at the position corresponding to the coefficient of x^{δ_L} in u. Then $\operatorname{rank} S^* \geq \operatorname{rank} R^* = d + 1$. Finally, we let A be any nonsingular $(d+1) \times (d+1)$ submatrix of R^* and $w \in \mathbb{Z}^{d+1}$ be the corresponding subvector of the coefficient vector of c, and if $\operatorname{rank} S^* > \operatorname{rank} R^*$, we let B be any nonsingular $(d+2) \times (d+2)$ submatrix of S^*.*

(i) *If $Lu = c$ is solvable, then the coefficient of x^i in the normalized solution u is $(\det A_i)/\det A$, where A_i is the matrix A with the ith column replaced by w for $0 \leq i \leq d$. If furthermore $p \nmid \det A$, then p is a lucky prime and we are in one of the two cases of Theorem 9.53.*

(ii) *If $Lu = c$ has no solution, then $\operatorname{rank} S^* = d + 2 > \operatorname{rank} R^*$, and if furthermore $p \nmid \det B$, then p is a lucky prime and $Lu \equiv c \bmod p$ has no solution of degree at most d.*

Proof. (i) follows from the proof of Theorem 9.55. The first statement of (ii) follows by a standard linear algebra argument. If $p \nmid \det B$, then $\operatorname{rank} \overline{S^*} = \operatorname{rank} S^* > \operatorname{rank} R^* \geq \operatorname{rank} \overline{R^*}$, and the second claim follows by the same linear algebra argument. $\square$

We start with a modular algorithm for the differential case. We assume that the input polynomials have degree less than $2^{\omega-2}$ and the degree bound on the output polynomial is less than $2^{\omega-1}$, where ω is the word size of our processor. This is sufficient in practice, since otherwise the input or the output would be so large that it probably would not fit into main memory in a dense representation.

Algorithm 9.56 (Modular Brent & Kung / ABP algorithm). ▬▬▬▬▬▬▬
Input: Polynomials $a, b, c \in \mathbb{Z}[x]$, with a, b not both zero, $\|a\|_\infty, \|b\|_\infty < 2^\lambda$, $\|c\|_\infty < 2^\gamma$, $n = \max\{\deg a - 1, \deg b\} < 2^{\omega-2}$, and $m = \deg c$.
Output: The normalized polynomial $u \in \mathbb{Q}[x]$ solving $au' + bu = c$, or otherwise "unsolvable".

0. use Corollary 9.6 to determine a degree bound d for u
   if $d < 0$ **then return** "unsolvable"
   if $d \geq 2^{\omega-1}$ **then return** "output too large"

1. $\alpha \longleftarrow \lceil (\log_2((d+1)^{3/2}2^\lambda)^{d+1})/(\omega-1) \rceil$
   $\beta \longleftarrow \lceil (\log_2 2((d+2)^{3/2}2^\lambda)^{d+2}2^\gamma)/(\omega-1) \rceil$
   $\eta \longleftarrow \lceil (2\lambda+\gamma)/(\omega-1) \rceil, \quad r \longleftarrow \alpha + 2\beta + \eta$
   choose single precision primes $2^{\omega-1} < p_1 < \ldots < p_r$
   $S \longleftarrow \{p_i : 1 \leq i \leq r \text{ and } p_i \nmid \mathrm{lc}(abc)\}$

2. **for all** $p \in S$ **do**

3.   **repeat**
           choose $s \in \{0, \ldots, 2n+1\}$ uniformly at random
       until $a(s) \not\equiv 0 \bmod p$

4. $M \longleftarrow a(x+s)D + b(x+s)$
       $a_p \longleftarrow a(x+s) \bmod p, \quad b_p \longleftarrow b(x+s) \bmod p$
       $c_p \longleftarrow c(x+s) \bmod p$

5.   **call** steps 1 and 2 of Brent & Kung's algorithm 9.20 or steps 1 and 2 of
       the ABP algorithm 9.27 for $F = \mathbb{F}_p$ with input a_p, b_p, c_p and the degree
       bound d, yielding $w_p, h_p \in \mathbb{Z}[x]$ with $h_p \not\equiv 0 \bmod p$ and the following
       properties:
       - $Mh_p \equiv 0 \bmod p$ if this homogeneous equation has a nontrivial solu-
         tion modulo p of degree at most d,
       - the inhomogeneous equation $Mv \equiv c(x+s) \bmod p$ has a solution
         $v \in \mathbb{Z}[x]$ of degree at most d if and only if there exists $\kappa \in \mathbb{Z}$ such
         that $M(w_p + \kappa h_p) \equiv c(x+s) \bmod p$

6.   **if** $Mh_p \equiv 0 \bmod p$ **then**
               $r_p \longleftarrow 1$
               if $\deg h_p \neq \delta_M$ **then** remove p from S
               else if $Mw_p \equiv c(x+s) \bmod p$
                   then choose $\kappa \in \{0, \ldots, p-1\}$ such that the coefficient of
                       x^{δ_M} in $w_p(x-s) + \kappa h_p(x-s)$ vanishes,
                       $u_p \longleftarrow w_p(x-s) + \kappa h_p(x-s)$
                   else $u_p \longleftarrow$ "unsolvable"

7.   **else**
               $r_p \longleftarrow 0$
               if there exists $\kappa \in \{0, \ldots, p-1\}$ such that $M(w_p + \kappa h_p) \equiv$
               $c(x+s) \bmod p$
               then $u_p \longleftarrow w_p(x-s) + \kappa h_p(x-s)$ **else** $u_p \longleftarrow$ "unsolvable"

8. **if** $u_p =$ "unsolvable" for less than $\alpha + \beta$ primes $p \in S$
   then remove all such primes from S **else return** "unsolvable"

9. **if** there exist primes $p \in S$ with $r_p = 0$
   then remove all p with $r_p = 1$ from S

10. use Chinese remaindering and rational number reconstruction to compute $u \in$
    $\mathbb{Q}[x]$ with $u \equiv u_p \bmod p$ for all $p \in S$

11. **return** u

For classical arithmetic, we use the ABP algorithm in step 5, and Brent & Kung's algorithm for fast arithmetic.

Theorem 9.57 (Correctness of Algorithm 9.56). *If $d < 2^{\omega-1}$, then Algorithm 9.56 correctly solves Problem 9.52 for linear first order differential equations.*

Proof. Let $d \geq 0$ be as computed in step 0 and $L = aD + b$. Then $(Lv)(x + s) = M(v(x + s))$ for all $v \in \mathbb{Q}[x]$ and $s \in \mathbb{Z}$. Now let $p \in S$ and s be as in step 3. Then $p > 2^{\omega-1} > d$. We consider the reductions $\overline{L}$ and $\overline{M}$ of the linear differential operators L and M, respectively, modulo p and their actions on polynomials in $\mathbb{F}_p[x]$. If we let $V = \{v \in \mathbb{F}_p[x] \colon \deg v < p\}$, then Lemma 9.1 implies that

- $\dim(V \cap \ker \overline{L}) = \dim(V \cap \ker \overline{M}) \leq 1$,
- all nonzero polynomials $h \in V$ with $\overline{L}h = 0$ or with $\overline{M}h = 0$ have degree $\delta_L \bmod p = \delta_M \bmod p$,
- the differential equation $\overline{M}v = c(x + s) \bmod p$ has a solution $v \in V$ if and only if the equation $\overline{L}v = c \bmod p$ has.

Let A, A_i, B be as in Corollary 9.55. The corollary implies that all primes that are unlucky with respect to Problem 9.52 divide $\det A$ or $\det B$ if $L^{-1}c \neq \emptyset$ or $L^{-1}c = \emptyset$, respectively. The coefficients of A and of all but one column of A_i and B are coefficients of $ajx^{j-1} + bx^j$ with $0 \leq j \leq d$. Therefore they are of absolute value less than $(d + 1)2^\lambda$, and the coefficients of the remaining column are absolutely bounded by 2^γ. Using Hadamard's inequality (Lemma 3.1 (i)), we find $|\det A| \leq ((d + 1)^{3/2}2^\lambda)^{d+1} \leq 2^{(\omega-1)\alpha}$ and $|\det A_i|, |\det B| \leq (d+2)^{(d+2)/2}((d+1)2^\lambda)^{d+1}2^\gamma \leq 2^{(\omega-1)\beta-1}$. Thus the number of single precision primes between $2^{\omega-1}$ and 2^ω dividing $\det A$ or $\det B$ is at most α or β, respectively. In particular, since $\alpha \leq \beta$, the number of unlucky single precision primes is at most β. Moreover, we have $|\mathrm{lc}(abc)| \leq 2^{2\lambda+\gamma} \leq 2^{(\omega-1)\eta}$, so that at most η single precision primes divide $\mathrm{lc}(abc)$. Since we start with $r = \alpha + 2\beta + \eta$ primes, the set S contains at least $\alpha + 2\beta$ of them after step 2. We claim that S contains at least $\alpha + \beta$ primes in step 10, and that all of them are lucky.

If $\deg h_p \neq \delta_M$ in step 6, then Corollaries 9.6 and 9.7 imply that $\delta_L = \delta_M$ is not in $\mathbb{N}$, $\dim \ker L = \dim \ker M = 0$, and $\dim(V \cap \ker \overline{L}) = \dim(V \cap \ker \overline{M}) = r_p = 1$. Thus only unlucky primes are removed from S in step 6.

If $L^{-1}c = \emptyset$, then all primes p such that $Lv \equiv c \bmod p$ has a solution v of degree at most d are unlucky. There are at most β unlucky single precision primes. Thus for at least $\alpha+\beta$ primes in S, neither $Lv \equiv c \bmod p$ nor $Mv \equiv c(x+s) \bmod p$ has a solution of degree at most d, and the algorithm returns the correct result in step 8.

Now assume that $L^{-1}c \neq \emptyset$ and $u \in L^{-1}c$ is the normalized solution. We have $r_p = \dim(V \cap \ker \overline{L}) = \dim(V \cap \ker \overline{M})$ after step 7. The primes for which $Lu \equiv c \bmod p$ is unsolvable or $\dim \ker L < r_p$ are removed from S in steps 8 or 9, respectively, and only the lucky primes survive.

We have shown that the primes that were removed from S in steps 6, 8, or 9 are unlucky. Since there are at most β unlucky primes, there remain at least $\alpha+\beta$ primes in S after step 9, all lucky. For each such prime p, u_p is the normalized solution of $Lv \equiv c \bmod p$, and Theorem 9.53 implies that u_p is the modular image of u.

By Corollary 9.55, the coefficients of u are of the form $(\det A_i)/\det A$. The product all primes in S in step 10 exceeds $2^{(\alpha+\beta)(\omega-1)} > 2|\det A_i| \cdot |\det A|$ for all i, and hence the numerators and denominators of the coefficients of u can be uniquely reconstructed from their modular images in step 10, by Fact 3.18. $\square$

If $c = 0$, then Algorithm 9.56 always returns $u = 0$. To obtain a nonzero solution of the homogeneous equation, if one exists, one would assume that h_p is monic in step 5 and reconstruct $u \equiv h_p \bmod p$ for all $p \in S$ in step 10, if all $p \in S$ have $r_p = 1$, and otherwise there is no nonzero solution. See also Remark 9.11.

Theorem 9.58 (Cost of Algorithm 9.56). *If we assume that there exist sufficiently many single precision primes in step 1 and ignore the cost for finding them, then Algorithm 9.56 uses $O((n\lambda + m\gamma + n^2 + dr)r)$ word operations with classical arithmetic and $O((n + d)\mathsf{M}(r) \log r + \mathsf{M}(n+d)r)$ with fast arithmetic, where $r \in \Theta(d(\lambda + \log d) + \gamma)$.*

Proof. The cost for reducing $\mathrm{lc}(abc)$ modulo p_i in step 2 is $O(\eta)$ word operations for each i, together $O(r\eta)$ word operations with classical arithmetic, and $O(\mathsf{M}(r) \log r)$ with fast arithmetic. This is dominated by the cost for the other steps.

In step 3, we first reduce the polynomials a, b, c modulo p, taking $O(n\lambda + m\gamma)$ word operations for each p, together $O((n\lambda + m\gamma)r)$ with classical arithmetic, and $O((n+m)\mathsf{M}(r) \log r)$ for all primes with fast arithmetic. Computing $a(s) \bmod p$ by Horner's rule takes $O(n)$ arithmetic operations in $\mathbb{F}_p$. We have $2n + 2 \leq 2^{\omega-1} < p$, so that the elements $0, 1, \ldots, 2n + 1$ are distinct modulo p, and since a has at most $n + 1$ roots modulo p, the expected number of iterations of the loop in step 3 is at most two. Thus the expected cost of the repeat loop is $O(n)$ word operations per prime.

The cost for the three Taylor shifts in step 4 is $O(n^2 + m^2)$ word operations with classical arithmetic and $O(\mathsf{M}(n) + \mathsf{M}(m))$ with fast arithmetic per prime, by Theorems 4.3 and 4.5. With classical arithmetic, we use the ABP algorithm in step 5, at a cost of $O(nd)$ word operations per prime, by Theorem 9.28. With fast arithmetic, we call Brent & Kung's algorithm, taking $O(\mathsf{M}(n + d))$ word operations per prime, by Theorem 9.21. In step 6, we compute $Mh_p \bmod p$ and $Mw_p \bmod p$, taking $O(nd)$ word operations with classical arithmetic and $O(\mathsf{M}(n + d))$ with fast arithmetic for each prime, by Lemma 9.10. The cost for computing the Taylor shifts by $-s$ in steps 6 and 7 is $O(d^2)$ and $O(\mathsf{M}(d))$ word operations per prime with classical and fast arithmetic, respectively, and the cost for finding κ is $O(n + d)$ for each prime. Thus the overall cost for steps 2 through 7 is $O((n\lambda + m\gamma + n^2 + m^2 + d^2)r)$ word operations with classical arithmetic and $O((n + m)\mathsf{M}(r) \log r + (\mathsf{M}(n + d) + \mathsf{M}(m))r)$ with fast arithmetic.

Finally, the cost for Chinese remaindering and rational number reconstruction in step 10 is $O(r^2)$ and $O(\mathsf{M}(r) \log r)$ word operations per coefficient with classical and fast arithmetic, respectively, together $O(dr^2)$ and $O(d\,\mathsf{M}(r) \log r)$ word operations, respectively. Now $m \leq d + n$, and the claims follow by adding up costs. $\square$

Corollary 9.59. *With the assumptions of Theorem 9.58 and the additional assumptions* $m, d \in O(n)$ *and* $\gamma \in O(\lambda)$, *Algorithm 9.56 takes* $O(n^3(\lambda^2 + \log^2 n))$ *word operations with classical arithmetic and* $O(n\,\mathsf{M}(n(\lambda + \log n))\log(n\lambda))$ *or* $O^\sim(n^2\lambda)$ *with fast arithmetic.*

Here is an analog of Algorithm 9.56 in the difference case, which uses Barkatou's algorithm 9.46 for the modular computation.

Algorithm 9.60 (Modular Barkatou algorithm). ▬▬▬▬▬
Input: Polynomials $a, b, c \in \mathbb{Z}[x]$, with a, b not both zero, $\|a\|_\infty, \|b\|_\infty < 2^\lambda$, $\|c\|_\infty < 2^\gamma$, $n = \max\{\deg a - 1, \deg(a + b)\}$, and $m = \deg c$.
Output: The normalized polynomial $u \in \mathbb{Q}[x]$ solving $a\,Eu + bu = c$, or otherwise "unsolvable".

0. use Corollary 9.6 to determine a degree bound d for u
   if $d < 0$ **then return** "unsolvable"
   if $2(n+1)d \geq 2^{\omega-1}$ **then return** "input or output too large"
1. $\alpha \longleftarrow \lceil (\lambda + d + \log_2(d+1)/2 + 1)(d+1)/(\omega-1) \rceil$
   $\beta \longleftarrow \lceil ((\lambda + d + \log_2(d+2)/2 + 1)(d+2) + \gamma + 1)/(\omega-1) \rceil$
   $\eta \longleftarrow \lceil (2\lambda + \gamma + 1)/(\omega-1) \rceil, \quad r \longleftarrow \alpha + 2\beta + \eta$
   choose single precision primes $2^{\omega-1} < p_1 < \ldots < p_r$
   $S \longleftarrow \{p_i \colon 1 \leq i \leq r \text{ and } p_i \nmid \mathrm{lc}(a(a+b)c)\}$
2. **for all** $p \in S$ **do**
3.   **repeat**
          choose $s \in \{d-1, 2d-1, 3d-1, \ldots, 2(n+1)d-1\}$ uniformly
          at random
       until $a(s), a(s-1), \ldots, a(s-d+1)$ are all nonzero modulo p
4. $M \longleftarrow a(x+s)E + b(x+s)$
       $a_p \longleftarrow a(x+s) \bmod p, \quad b_p \longleftarrow b(x+s) \bmod p$
       $c_p \longleftarrow c(x+s) \bmod p$
5.   **call** steps 1-3 of Barkatou's algorithm 9.46 for $F = \mathbb{F}_p$ with input a_p, b_p, c_p and the degree bound d, yielding $w_p, h_p \in \mathbb{Z}[x]$ with $h_p \not\equiv 0 \bmod p$ and the following properties:
       • $Mh_p \equiv 0 \bmod p$ if this homogeneous equation has a nontrivial solution modulo p of degree at most d,
       • the inhomogeneous equation $Mv \equiv c(x+s) \bmod p$ has a solution $v \in \mathbb{Z}[x]$ of degree at most d if and only if there exists $\kappa \in \mathbb{Z}$ such that $M(w_p + \kappa h_p) \equiv c(x+s) \bmod p$
6.   **if** $Mh_p \equiv 0 \bmod p$ **then**
          $r_p \longleftarrow 1$
          if $\deg h_p \neq \delta_M$ **then** remove p from S
          else if $Mw_p \equiv c(x+s) \bmod p$
             then choose $\kappa \in \{0, \ldots, p-1\}$ such that the coefficient of
             x^{δ_M} in $w_p(x-s) + \kappa h_p(x-s)$ vanishes,
             $u_p \longleftarrow w_p(x-s) + \kappa h_p(x-s)$
             else $u_p \longleftarrow$ "unsolvable"

7.   **else**
$$r_p \longleftarrow 0$$
if there exists $\kappa \in \{0, \ldots, p-1\}$ such that $M(w_p + \kappa h_p) \equiv c(x+s) \bmod p$
then $u_p \longleftarrow w_p(x-s) + \kappa h_p(x-s)$ **else** $u_p \longleftarrow$ "unsolvable"

8. **if** $u_p =$ "unsolvable" for less than $\alpha + \beta$ primes $p \in S$
   then remove all such primes from S **else return** "unsolvable"
9. **if** there exist primes $p \in S$ with $r_p = 0$
   then remove all p with $r_p = 1$ from S
10. use Chinese remaindering and rational number reconstruction to compute $u \in \mathbb{Q}[x]$ with $u \equiv u_p \bmod p$ for all $p \in S$
11. **return** u

Theorem 9.61 (Correctness of Algorithm 9.60). *If* $2(n+1)d < 2^{\omega-1}$, *then Algorithm 9.60 correctly solves Problem 9.52 for linear first order difference equations.*

Proof. Let $L = aE + b$. The proof is completely analogous to the proof of Theorem 9.57, with the following exceptions. Let A, A_i, B be as in Corollary 9.55. Each column of A contains some coefficients of $Lx^j = a \cdot (x+1)^j + bx^j$ for some $j \leq d$. If $a = \sum_{0 \leq i \leq n+1} a_i x^i$, then the coefficient of x^k in $a \cdot (x+1)^j$ has absolute value

$$\left| \sum_{i+l=k} a_i \binom{j}{l} \right| < 2^\lambda \sum_{i+l=k} \binom{j}{l} \leq 2^{\lambda+j} .$$

Thus all entries of A are of absolute value less than $2^{\lambda+d+1}$. The same is true for all but one column of A_i and B, and the entries in the remaining column are absolutely bounded by 2^γ. Thus Hadamard's inequality (Lemma 3.1 (i)) yields $|\det A| \leq (d+1)^{(d+1)/2} 2^{(\lambda+d+1)(d+1)} \leq 2^{(\omega-1)\alpha}$ and $|\det A_i|, |\det B| \leq (d+2)^{(d+2)/2} 2^{(\lambda+d+1)d+\gamma} \leq 2^{(\omega-1)\beta-1}$. $\square$

Theorem 9.62 (Cost of Algorithm 9.60). *If we assume that there exist sufficiently many single precision primes in step 1 and ignore the cost for finding them, then Algorithm 9.60 uses* $O((n\lambda + m\gamma + n^2 + dr)r)$ *word operations with classical arithmetic and* $O((n+d)M(r) \log r + (M(d) \log d + M(n+d))r)$ *with fast arithmetic, where* $r \in \Theta(d(\lambda + d) + \gamma)$.

Proof. We proceed as in the proof of Theorem 9.58. As in Algorithm 9.56, the cost for reducing $\mathrm{lc}(a(a+b)c)$ modulo all p_i in step 1 is dominated by the cost for the other steps.

The cost for reducing a, b, c modulo all primes p in step 3 is the same as in the proof of Theorem 9.58. In addition, computing $a(s), a(s-1), \ldots, a(s-d+1)$ modulo p in step 3 takes $O(nd)$ word operations per prime with classical arithmetic and $O(M(n) + M(d) \log d)$ with fast arithmetic. Since $2(n+1)d < 2^{\omega-1} < p$, the elements $0, 1, \ldots, 2(n+1)d - 1$ are distinct modulo p. Now a has at most $n+1$ roots modulo p, and hence at least $n+1$ of the sets $\{jd, jd+1, \ldots, (j+1)d-1\}$ for $0 \leq j \leq 2n+1$ contain no root of a modulo p. Thus the expected number of iterations of the repeat loop is at most two, and the expected cost is $O(ndr)$ word operations for all primes with classical arithmetic and $O((M(n) + M(d) \log d)r)$ with fast arithmetic.

The cost for step 5 is $O(nd + d^2)$ word operations per prime with classical arithmetic and $O(M(d) \log d + M(n + d))$ with fast arithmetic, by Theorem 9.47. In steps 6 and 7, we compute $M h_p \bmod p$ and $M w_p \bmod p$, taking $O(nd + d^2)$ word operations with classical arithmetic and $O(M(n + d))$ with fast arithmetic for each prime, by Lemma 9.9. The cost for step 4 and for the Taylor shifts and for finding κ in steps 6 and 7 is the same as in the proof of Theorem 9.58. Thus the overall cost for steps 3 through 7 is $O((n\lambda + m\gamma + n^2 + m^2 + d^2)r)$ word operations with classical arithmetic and

$$O((n + m)M(r) \log r + (M(d) \log d + M(n + d) + M(m))r)$$

with fast arithmetic.

The cost for step 10 is $O(dr^2)$ and $O(d\, M(r) \log r)$ word operations with classical and fast arithmetic, respectively, as in the proof of Theorem 9.58, and the claims follow from $m \leq n + d$ by adding up costs. $\square$

Corollary 9.63. *With the assumptions of Theorem 9.62 and the additional assumptions $m, d \in O(n)$ and $\gamma \in O(\lambda)$, Algorithm 9.60 takes $O(n^3(\lambda^2 + n^2))$ word operations with classical arithmetic and*

$$O(n\, M(n(\lambda + n)) \log(n\lambda)) \text{ or } O^\sim(n^2(\lambda + n))$$

with fast arithmetic.

Under the assumptions of Corollaries 9.59 and 9.63, the output size of Algorithms 9.56 and 9.60 is $O(n^2(\lambda + \log n))$ and $O(n^2(\lambda + n))$ words, respectively. Hence both algorithms with fast arithmetic are – up to logarithmic factors – asymptotically optimal for those inputs where these upper bounds are achieved.

We now discuss modular versions of the method of undetermined coefficients and its asymptotically fast variants, the divide-and-conquer algorithms from Sect. 9.5 and 9.6. We start with the difference case.

Algorithm 9.64 (Modular difference divide & conquer).
Input: Polynomials $a, b, c \in \mathbb{Z}[x]$, with a, b not both zero, $\|a\|_\infty, \|b\|_\infty < 2^\lambda$, $\|c\|_\infty < 2^\gamma$, $n = \max\{\deg a - 1, \deg(a + b)\}$, and $m = \deg c$.
Output: The normalized polynomial $u \in \mathbb{Q}[x]$ solving $Lu = c$, where $L = aE + b = a\Delta + a + b$, or otherwise "unsolvable".

0. use Corollary 9.6 to determine a degree bound d for u
   if $d < 0$ **then return** "unsolvable"
   if $d \geq 2^{\omega-1}$ **then return** "output too large"
1. $\alpha \longleftarrow \lceil (\lambda + d + \log_2(d + 1)/2 + 1)(d + 1)/(\omega - 1) \rceil$
   $\beta \longleftarrow \lceil ((\lambda + d + \log_2(d + 2)/2 + 1)(d + 2) + \gamma + 1)/(\omega - 1) \rceil$
   $\eta \longleftarrow \lceil (2\lambda + \gamma + 1)/(\omega - 1) \rceil, \quad r \longleftarrow \alpha + 2\beta + \eta$
   choose single precision primes $2^{\omega-1} < p_1 < \ldots < p_r$
   $S \longleftarrow \{p_i : 1 \leq i \leq r \text{ and } p_i \nmid \mathrm{lc}(a(a + b)c)\}$

2. **for** all $p \in S$ **do**
3.   **call** Algorithm 9.15, 9.35, or 9.41 for $F = \mathbb{F}_p$ with input $a \bmod p$, $b \bmod p, c \bmod p$ and the degree bound d, yielding the dimension $r_p \in \{0, 1\}$ of the kernel of $L \bmod p$ on polynomials of degree at most d and $u_p \in \mathbb{Z}[x]$ of degree at most d such that u_p is the normalized solution of $Lv \equiv c \bmod p$, or otherwise $u_p =$"unsolvable"
4. **if** $u_p =$"unsolvable" for less than $\alpha + \beta$ primes $p \in S$
   then remove all such primes from S **else return** "unsolvable"
5. **if** there exist primes $p \in S$ with $r_p = 0$
   then remove all p with $r_p = 1$ from S
6. use Chinese remaindering and rational number reconstruction to compute $u \in \mathbb{Q}[x]$ with $u \equiv u_p \bmod p$ for all $p \in S$
7. **return** u

The algorithm that is called in step 3 has to be modified to meet the required specifications, as follows. Algorithm 9.15, which is called when classical arithmetic is used, need not compute d in step 0 and should only return "unsolvable" in that step when $c \neq 0$. The rank r_p is one if $h^* = 0$ and zero otherwise. Finally, $\kappa = 0$ should be chosen in step 4 if possible. If fast arithmetic is used, then one of the two other algorithms is called. Algorithm 9.35 is only called when $\deg a - 1 < \deg(a + b)$, and then $\delta_L = \infty$ and the solution is unique, so that $r_p = 0$. The algorithm should check whether $L(Ux^m + V) = 0$, and return "unsolvable" otherwise. Algorithm 9.41 should in addition return $r_p = 1$ if $h^{**} = 0$, and $r_p = 0$ otherwise. The value of κ in Algorithm 9.41 should be chosen to be zero if possible.

If $c = 0$ and a nonzero solution of the homogeneous equation $Lu = 0$ is desired, then Algorithm 9.64 can be modified as follows. There is no such solution if $\delta_L \notin \mathbb{N}$. Otherwise, the call to Algorithm 9.15 in step 3 should return h if $h^* = 0$ and "unsolvable" otherwise. Similarly, the call to Algorithm 9.41 in step 3 should return $x^{\delta_L} + W$ if $h^{**} = 0$ and "unsolvable" otherwise. See also Remark 9.11.

Theorem 9.65. *If $d < 2^{\omega-1}$ then Algorithm 9.64 correctly solves Problem 9.52 for linear first order difference equations. If we assume that there exist sufficiently many single precision primes in step 1 and ignore the cost for finding them, then the algorithm uses $O((n\lambda + m\gamma + nd + dr)r)$ word operations with classical arithmetic and $O((n + d)\mathsf{M}(r) \log r + (\mathsf{M}(d) \log d + \mathsf{M}(n + d))r)$ with fast arithmetic, where $r \in \Theta(d(\lambda + d) + \gamma)$.*

Proof. Correctness follows as in the proof of Theorem 9.61. For the cost analysis, we proceed as in the proof of Theorem 9.62. As in the proof of Theorem 9.61, the cost for reducing $\mathrm{lc}(a(a + b)c)$ modulo all p_i in step 1 is negligible.

Reducing a, b, c modulo all primes $p \in S$ in step 3 takes $O((n\lambda + m\gamma)r)$ word operations with classical arithmetic and $O((n+m)\mathsf{M}(r) \log r)$ with fast arithmetic, as in the proof of Theorem 9.61. For each prime, computing r_p and u_p takes $O(nd + d^2)$ word operations with classical arithmetic, by Theorem 9.16, and $O(\mathsf{M}(d) \log d + \mathsf{M}(n + d))$ word operations with fast arithmetic, by Corollary 9.37 and Theorem 9.42. Thus the overall cost for steps 2 and 3 is $O((n\lambda + m\gamma + nd + d^2)r)$

with classical arithmetic and $O((n + m)\mathsf{M}(r)\log r + (\mathsf{M}(d)\log d + \mathsf{M}(n + d))r)$ with fast arithmetic. Finally, step 6 takes $O(dr^2)$ with classical and $O(d\,\mathsf{M}(r)\log r)$ with fast arithmetic, as in the proof of Theorem 9.61, and the claim follows from $m \leq n + d$. $\square$

Corollary 9.66. *With the assumptions of Theorem 9.65 and the additional assumptions $m, d \in O(n)$ and $\gamma \in O(\lambda)$, Algorithm 9.64 takes $O(n^3(\lambda^2 + n^3))$ word operations with classical arithmetic and*

$$O(n\,\mathsf{M}(n(\lambda + n))\log(n\lambda)) \ or \ O^\sim(n^2(\lambda + n))$$

with fast arithmetic.

Here is the analogous algorithm in the differential case.

Algorithm 9.67 (Modular differential divide & conquer). ▬▬▬▬▬▬▬
Input: Polynomials $a, b, c \in \mathbb{Z}[x]$, with a, b not both zero, $\|a\|_\infty, \|b\|_\infty < 2^\lambda$, $\|c\|_\infty < 2^\gamma$, $n = \max\{\deg a - 1, \deg b\} < 2^{\omega-2}$, and $m = \deg c$.
Output: The normalized polynomial $u \in \mathbb{Q}[x]$ solving $Lu = c$, where $L = aD + b$, or otherwise "unsolvable".

0. use Corollary 9.6 to determine a degree bound d for u
   if $d < 0$ **then return** "unsolvable"
   if $d \geq 2^{\omega-1}$ **then return** "output too large"
1. $\alpha \longleftarrow \lceil (\log_2((d + 1)^{3/2}2^\lambda)^{d+1})/(\omega - 1) \rceil$
   $\beta \longleftarrow \lceil (\log_2 2((d + 2)^{3/2}2^\lambda)^{d+2}2^\gamma)/(\omega - 1) \rceil$
   $\eta \longleftarrow \lceil (2\lambda + \gamma)/(\omega - 1) \rceil, \quad r \longleftarrow \alpha + 2\beta + \eta$
   choose single precision primes $2^{\omega-1} < p_1 < \ldots < p_r$
   $S \longleftarrow \{p_i : 1 \leq i \leq r \text{ and } p_i \nmid \mathrm{lc}(abc)\}$
2. **for** all $p \in S$ **do**
3.   **call** Algorithm 9.15, 9.38, or 9.44 for $F = \mathbb{F}_p$ with input $a \bmod p$, $b \bmod p, c \bmod p$ and the degree bound d, yielding the dimension $r_p \in \{0, 1\}$ of the kernel of $L \bmod p$ on polynomials of degree at most d and $u_p \in \mathbb{Z}[x]$ of degree at most d such that u_p is the normalized solution of $Lv \equiv c \bmod p$, or otherwise $u_p =$"unsolvable"
4. **if** $u_p =$"unsolvable" for less than $\alpha + \beta$ primes $p \in S$
   then remove all such primes from S **else return** "unsolvable"
5. **if** there exist primes $p \in S$ with $r_p = 0$
   then remove all p with $r_p = 1$ from S
6. use Chinese remaindering and rational number reconstruction to compute $u \in \mathbb{Q}[x]$ with $u \equiv u_p \bmod p$ for all $p \in S$
7. **return** u ▬▬▬▬▬▬▬▬▬▬▬

We omit the proof of the following theorem, since it is analogous to the proofs of Theorems 9.57 and 9.65.

Theorem 9.68. *If $d < 2^{\omega-1}$, then Algorithm 9.67 correctly solves Problem 9.52 for linear first order differential equations. If we assume that there exist sufficiently* *many single precision primes in step 1 and ignore the cost for finding them, then the algorithm uses $O((n\lambda + m\gamma + nd + dr)r)$ word operations with classical arithmetic and $O((n+d)\mathsf{M}(r)\log r + (\mathsf{M}(d)\log d + \mathsf{M}(n+d))r)$ with fast arithmetic, where $r \in \Theta(d(\lambda + \log d) + \gamma)$.*

Corollary 9.69. *With the assumptions of Theorem 9.68 and the additional assumptions $m, d \in O(n)$ and $\gamma \in O(\lambda)$, Algorithm 9.67 takes $O(n^3(\lambda^2 + \log^2 n))$ word operations with classical arithmetic and $O(n\,\mathsf{M}(n(\lambda + \log n))\log(n\lambda))$ or $O^\sim(n^2\lambda)$ with fast arithmetic.*

Table 9.2. Cost estimates in field operations for various algorithms computing polynomial solutions of linear first order differential or difference equations with polynomial coefficients

Algorithm	Classical	Fast
method of undetermined coefficients, 9.15	$nd + d^2$	
Brent & Kung, 9.20 (differential)	$n^2 + d^2$	$\mathsf{M}(n) + \mathsf{M}(d)$
Rothstein, 9.23, 9.25	$n^2 + d^2$	
ABP, 9.27 (differential),	$n^2 + d^2$	$nd + \mathsf{M}(n+d)$
9.31 (difference)		$nd + \mathsf{M}(n+d)\log(n+d)$
divide & conquer, 9.41, 9.44		$\mathsf{M}(d)\log d + \mathsf{M}(n+d)$
Barkatou, 9.46 (difference)	$n^2 + d^2$	$\mathsf{M}(d)\log d + \mathsf{M}(n+d)$

Table 9.2 gives a survey of the arithmetic cost estimates for the algorithms from Sect. 9.1 through 9.7, both for classical and for fast arithmetic. All entries are upper bounds in the O-sense. A missing entry indicates that we did not analyze the corresponding variant.

Table 9.3. Cost estimates in word operations for various modular algorithms computing polynomial solutions of linear first order differential or difference equations with coefficients in $\mathbb{Z}[x]$

Algorithm	Classical	Fast
9.56 (differential)	$(n\lambda + m\gamma + n^2 + dr)r$	$(n+d)\,\mathsf{M}(r)\log r +$ $\mathsf{M}(n+d)r$
9.60 (difference)	$(n\lambda + m\gamma + n^2 + dr)r$	$(n+d)\,\mathsf{M}(r)\log r +$ $(\mathsf{M}(d)\log d + \mathsf{M}(n+d))r$
9.67 (differential)	$(n\lambda + m\gamma + nd + dr)r$	$(n+d)\,\mathsf{M}(r)\log r +$ $(\mathsf{M}(d)\log d + \mathsf{M}(n+d))r$
9.64 (difference)	$(n\lambda + m\gamma + nd + dr)r$	$(n+d)\,\mathsf{M}(r)\log r +$ $(\mathsf{M}(d)\log d + \mathsf{M}(n+d))r$

Table 9.3 gives an overview of the cost estimates in word operations for the modular algorithms from this section, both for classical and for fast arithmetic. All entries are upper bounds in the O-sense. We have $r \in \Theta(d(\lambda + \log d) + \gamma)$ in the

differential case and $r \in \Theta(d(\lambda + d) + \gamma)$ in the difference case. For the special case where $m, d \in O(n)$ and $\gamma \in O(\lambda)$, the dominant cost in all modular algorithms is for the rational number reconstruction. Then the cost estimates for all algorithms are $O(n^3(\lambda^2 + \log^2 n))$ with classical arithmetic and $O(n \, \mathsf{M}(n(\lambda + \log n)) \log(n\lambda))$ with fast arithmetic in the differential case, and $O(n^3(\lambda^2 + n^2))$ with classical arithmetic and $O(n \, \mathsf{M}(n(\lambda + n)) \log(n\lambda))$ in the difference case.

10. Modular Gosper and Almkvist & Zeilberger Algorithms

Putting the results from Chap. 8 and 9 together, we obtain the following cost analysis for our new modular variant of Gosper's algorithm.

Algorithm 10.1 (Modular Gosper algorithm).
Input: Nonzero polynomials $f, g \in \mathbb{Z}[x]$ of degree at most n and max-norm less than 2^{λ}.
Output: Polynomials $w \in \mathbb{Q}[x]$ and $v \in \mathbb{Z}[x]$ solving $\dfrac{f}{g} \cdot \dfrac{Ew}{Ev} - \dfrac{w}{v} = 1$, or otherwise "unsolvable".

1. **call** Algorithm 8.2 to compute a Gosper-Petkovšek form a, b, v of f/g
2. **call** Algorithm 9.64 (or Algorithm 9.60) with input $a, -E^{-1}b, v$ to compute $u \in \mathbb{Q}[x]$ with $a \cdot Eu - E^{-1}b \cdot u = v$
   if it returns "unsolvable" **then return** "unsolvable"
3. **return** $w = E^{-1}b \cdot u$ and v

Correctness of the algorithm has been shown by Gosper (1978).

Theorem 10.2. *Let $M = fE - g = f\Delta + f - g$. If $e = \mathrm{dis}(f,g)$ is the dispersion of f and g and $\delta = \max(\{0, \delta_M\} \cap \mathbb{N})$, then $e < 2^{\lambda+2}$, $\delta < 2^{\lambda+1}$, and Algorithm 10.1 takes*

$$O(e^5 n^5 + e^3 n^3 \lambda^2 + \delta^5 + \delta^3 \lambda^2)$$

word operations with classical arithmetic, and the cost with fast arithmetic is

$$O((en + \delta)\mathsf{M}((en + \delta)(en + \delta + \lambda))\log(e\delta n\lambda))$$

or

$$O^{\sim}(e^3 n^3 + e^2 n^2 \lambda + \delta^3 + \delta^2 \lambda) \,.$$

Proof. The bound on e follows from Theorem 8.18. If $\delta_M \neq \infty$, then

$$\delta_M = \frac{g_{\deg M} - f_{\deg M}}{\mathrm{lc}(f)} \leq |g_{\deg M}| + |f_{\deg M}| < 2^{\lambda+1} \,.$$

Theorem 8.18 says that $\deg v \leq en$ and $\log_2 \|v\|_{\infty} \in O(e(n + \lambda))$ and that the cost for step 1 is $O(n^4 + n^2\lambda^2 + e^3(n^3 + n\lambda^2))$ word operations with classical arithmetic and

J. Gerhard: Modular Algorithms, LNCS 3218, pp. 195-205, 2004.
© Springer-Verlag Berlin Heidelberg 2004

$$O\Big((n^2\, \mathsf{M}(n+\lambda) + n\lambda\, \mathsf{M}(n) + \mathsf{M}(n^2))\log(n+\lambda)$$
$$+(en\, \mathsf{M}(e(n+\lambda)) + e\lambda\, \mathsf{M}(en))\log(e(n+\lambda))\Big)$$

with fast arithmetic.

Now let $L = aE - E^{-1}b$. Then

$$M\left(\frac{E^{-1}b}{v}u\right) = g\frac{a \cdot Ev}{b \cdot v} \cdot \frac{b}{Ev}Eu - g\frac{E^{-1}b}{v}u$$
$$= \frac{g}{v}(a \cdot Eu - E^{-1}b \cdot u) = \frac{g}{v}Lu$$

for all $u \in F[x]$, and Lemma 9.8, which holds more generally when z is a rational function or even a formal Laurent series in x^{-1}, implies that $\deg L = \deg M - (\deg g - \deg b) \le n$ and $\delta_L = \delta_M - \deg b + \deg v$.

Since a divides f and b divides g, their degrees are at most n and their coefficients are of word size $O(n + \lambda)$, by Mignotte's bound (Fact 3.3), and the same is true for the coefficients of $E^{-1}b$, by Lemma 4.2. Corollary 9.6 yields the degree bound $d = \deg v - \deg L = \deg v - \deg b - (\deg M - \deg g)$ for u if $\delta_M \notin \mathbb{Z}$ and

$$d = \max\{\delta_L, \deg v - \deg L\} = \deg v - \deg b + \max\{\delta_M, -(\deg M - \deg g)\}$$

if $\delta_M \in \mathbb{Z}$, and hence $d \in O(en + \delta)$. Now Theorems 9.65 and 9.62 with λ replaced by $O(n + \lambda)$, $m = \deg v \le en$, and $\gamma = 1 + \lfloor \log_2 \|v\|_\infty \rfloor \in O(e(n + \lambda))$ imply that the cost for step 2 is $O((en + \delta)^3(en + \delta + \lambda)^2)$ word operations with classical arithmetic and

$$O((en + \delta)\mathsf{M}((en + \delta)(en + \delta + \lambda))\log(e\delta n\lambda))$$

with fast arithmetic. Step 3 can be incorporated in the modular computation in step 2. The additional cost for this is negligible, and the claims follow by adding up costs. □

The dominant step for Algorithm 10.1 is the rational reconstruction in the last step of the modular algorithm for computing u. The output size of Algorithm 10.1 is essentially the size of u, namely $O((en + \delta)^2(en + \delta + \lambda))$ words, and hence the algorithm with fast arithmetic is – up to logarithmic factors – asymptotically optimal for those inputs where the output size is close to the upper bound.

Corollary 10.3. *If $e, \delta \in O(n)$, then Algorithm 10.1 takes $O(n^{10} + n^6\lambda^2)$ word operations with classical arithmetic, and the cost with fast arithmetic is*

$$O(n^2\, \mathsf{M}(n^2(n^2 + \lambda))\log(n\lambda)) \text{ or } O^\sim(n^6 + n^4\lambda).$$

Here is the analog of Algorithm 10.1 in the differential case.

Algorithm 10.4 (Modular Almkvist & Zeilberger algorithm). ▬▬▬
Input: Nonzero coprime polynomials $f, g \in \mathbb{Z}[x]$ of degree at most n and max-norm less than 2^λ.
Output: Polynomials $w \in \mathbb{Q}[x]$ and $v \in \mathbb{Z}[x]$ solving $\left(\dfrac{w}{v}\right)' + \dfrac{f}{g} \cdot \dfrac{w}{v} = 1$, or otherwise "unsolvable".

1. **call** Algorithm 8.28 to compute a GP'-form a, b, v of f/g
2. **call** Algorithm 9.67 (or Algorithm 9.56) with input $b, a + b', v$ to compute $u \in \mathbb{Q}[x]$ such that $bu' + (a + b')u = v$
   if it returns "unsolvable" **then return** "unsolvable"
3. **return** $w = bu$ and v

Correctness of the algorithm is proved in Almkvist & Zeilberger (1990).

Theorem 10.5. *Let* $M = gD + f$. *If* $e = \varepsilon(f, g)$ *and* $\delta = \max(\{0, \delta_M\} \cap \mathbb{N})$, *then* $e \leq (n + 1)^n 2^{2n\lambda}$, $\delta \leq 2^\lambda$, *and Algorithm 10.4 takes*

$$O((e^3 n^5 + \delta^3 n^2)(\lambda^2 + \log^2 n))$$

word operations with classical arithmetic, and the cost with fast arithmetic is

$$O\Big((en + \delta)\, \mathsf{M}((en + \delta)n(\lambda + \log n))\log(e\delta n\lambda) + \mathsf{M}(n(\lambda + \log n))\, \mathsf{M}(n)\log n\Big)$$

or

$$O^\sim(e^2 n^3 \lambda + \delta^2 n\lambda) .$$

Proof. The bound on e follows from Theorem 8.42, and the bound on δ is immediate from the definition of δ_M. Moreover, Theorem 8.42 says that $\deg v \in O(en)$, $\log_2 \|v\|_\infty \in O(e(n + \lambda))$, and that the cost for step 1 is

$$O(n^4(\lambda^2 + \log^2 n) + e^3(n^3 + n\lambda^2))$$

word operations with classical arithmetic and

$$O\Big(n^2(\mathsf{M}(n(\lambda + \log n)) + (\lambda + \log n)\mathsf{M}(n)\log n)$$
$$+ \mathsf{M}(n(\lambda + \log n))(\mathsf{M}(n)\log n + n\log \lambda)$$
$$+ (en\, \mathsf{M}(e(n + \lambda)) + e\lambda\, \mathsf{M}(en))\log(e(n + \lambda))\Big)$$

with fast arithmetic. Let $L = bD + a + b'$. Then

$$
\begin{aligned}
M(bu/v) &= g\left(\frac{bu}{v}\right)' + f\frac{ub}{v} \\
&= g\left(\frac{bu'}{v} + \frac{b'u}{v} - \frac{buv'}{v^2} + \left(\frac{a}{b} + \frac{v'}{v}\right)\frac{bu}{v}\right) \\
&= \frac{g}{v}\left(bu' + (a + b')u\right) = \frac{g}{v}Lu
\end{aligned}
$$

for all $u \in F[x]$. As in the proof of Theorem 10.2, Lemma 9.8 implies that $\deg L = \deg M - (\deg g - \deg b)$ and $\delta_L = \delta_M - \deg b + \deg v$.

By the remark following the proof of Theorem 8.36, the coefficients of a are of word size $O(n(\lambda + \log n))$. Since b divides g, Mignotte's bound (Fact 3.3) implies that the coefficients of b are of word size $O(n + \lambda)$, and hence the coefficients of b

and $a+b'$ are of word size $O(n(\lambda+\log n))$ as well. As in the proof of Theorem 10.2, the degree bound d for u is $d = \deg v - \deg L = \deg v - \deg b - (\deg M - \deg g)$ if $\delta_M \notin \mathbb{Z}$ and $d = \deg v - \deg b + \max\{\delta_M, -(\deg M - \deg g)\}$ otherwise, by Corollary 9.6. Thus $d \in O(en + \delta)$, and Theorems 9.68 and 9.58 with λ replaced by $O(n(\lambda + \log n))$, $m = \deg v \le en$, and $\gamma = 1 + \lfloor \log_2 \|v\|_\infty \rfloor \in O(e(n + \lambda))$ imply that step 2 takes $O((en+\delta)^3 n^2(\lambda^2 + \log^2 n))$ word operations with classical arithmetic and

$$O\Big((en + \delta)\,\mathsf{M}((en + \delta)n(\lambda + \log n))\log(e\delta n\lambda)\Big)$$

with fast arithmetic, respectively. As in the proof of Theorem 10.2, the cost for step 3 is negligible, and the claims follow by adding up costs. $\square$

As for Algorithm 10.1, the dominant step is the rational reconstruction in the last step of the modular algorithm for computing u. The output size of Algorithm 10.4 is essentially the size of u, namely $O((en+\delta)^2 n(\lambda+\log n))$ words, and hence the algorithm with fast arithmetic is – up to logarithmic factors – asymptotically optimal for those inputs where the output size is close to the upper bound.

Corollary 10.6. *If $e, \delta \in O(n)$, then Algorithm 10.4 takes $O(n^8(\lambda^2 + \log^2 n))$ word operations with classical arithmetic, and the cost with fast arithmetic is*

$$O(n^2\,\mathsf{M}(n^3(\lambda + \log n))\log(n\lambda) + \mathsf{M}(n(\lambda + \log n))\,\mathsf{M}(n)\log n) \text{ or } O^{\sim}(n^5\lambda) \text{ .}$$

10.1 High Degree Examples

In this section, we report on a particular class of solutions of the *key equation*

$$Lu = a \cdot Eu - (E^{-1}b)u = a \cdot \Delta u + (a - E^{-1}b)u = v \qquad (10.1)$$

in Gosper's algorithm 10.1, where $L = aE - E^{-1}b = a\Delta + a - E^{-1}b$. Corollary 9.6 shows that there are two essentially different possibilities for the degree of such a solution u: it can be expressed either solely in terms of the degrees of a, b, v, or solely in terms of certain *coefficients* of a and b. Up to now, there seem to be known only few nontrivial examples (where a, b, v result from a proper, i.e., nonrational hypergeometric summation problem) leading to the second case; Lisoněk, Paule & Strehl (1993) give two. Since the degree of u affects the running time of Gosper's algorithm, it is interesting to know whether the second case is of real impact or merely a curiosity.

The main contribution of this section is that we exhibit a certain class of proper hypergeometric summation problems leading to a key equation with degrees of a, b, v at most two and a unique solution u of arbitrarily high degree (Theorem 10.7). We also consider similar classes for the differential case (Theorem 10.12) and give examples in both cases. We also give classes of examples where the denominators v in Gosper's and Almkvist & Zeilberger's algorithm have arbitrarily high degree. The material is based on Gerhard (1998).

Lisoněk, Paule & Strehl (1993) have shown that the key equation (10.1) has at most one solution in every case when it originates from the summation of a non-rational hypergeometric term. The case where $\deg u = \delta_L$ in Corollary 9.6 is surprising at first sight, since then the degree of the solution u to (10.1) does not depend on the degree of a, b, v but only on the coefficients of a and b. For the case where Gosper's algorithm is used to compute the antidifference of a rational function, Lisoněk, Paule & Strehl have shown that the solution space of (10.1) is either one-dimensional or empty (see also Lemma 9.1). In the former case, there is a unique solution u of degree $\deg v - \deg L$, and all other solutions are of the form $u + \kappa h$ for a constant $\kappa \in F$, where h is a nonzero solution of degree δ_L of the corresponding homogeneous difference equation $Lh = 0$, or equivalently,

$$\frac{E^{-1}b}{a} = \frac{Eh}{h} , \qquad (10.2)$$

which then is the Gosper-Petkovšek representation of $(E^{-1}b)/a$. Lisoněk, Paule & Strehl conclude that in the rational case, the case $\deg u = \delta_L$ need not be considered.

In the non-rational hypergeometric case, it is not clear at first sight whether the key equation admits solutions of degree δ_L at all. Examples for this situation have been given by Lisoněk, Paule & Strehl (1993). We will now exhibit a class of hypergeometric expressions for which we show that (10.1) always has a unique solution of degree δ_L, which may be arbitrary large.

Let F be a field of characteristic zero and

$$\frac{f}{g} = \frac{x^2 + f_1 x + f_0}{x^2 + (f_1 + d)x + g_0} , \qquad (10.3)$$

with $f_1, f_0, g_0 \in F$ and $d \in \mathbb{N}$ such that $d \geq 2$. We say that f and g are *shift-coprime* if $\gcd(f, E^i g) = 1$ for all $i \in \mathbb{Z}$, or equivalently, the resultant $r = \operatorname{res}_x(f(x), g(x+y)) \in F[y]$ has no integer zeroes. When $f_1 = f_0 = 0$, for example, then $r = (y^2 + dy + g_0)^2$, and f and g are shift-coprime if and only if $g_0 \neq i(d - i)$ for all integers i.

Theorem 10.7. *Let F be a field of characteristic zero and $f = x^2 + f_1 x + f_0$, $g = x^2 + (f_1 + d)x + g_0$ in $F[x]$, with $f_1, f_0, g_0 \in F$ and $d \in \mathbb{N}_{\geq 2}$, and let $a, b, v \in F[x]$ be a Gosper-Petkovšek form of f/g. If f and g are shift-coprime, then we have $\deg v = 0$, and there is a unique polynomial $u \in F[x]$ of degree $d - 2$ solving (10.1).*

Proof. Substituting $x - f_1/2$ for x is an automorphism of $F(x)$ and does not change our arguments in the sequel, and we may assume that $f_1 = 0$. The fact that f and g are shift-coprime implies that (10.3) is already in Gosper-Petkovšek form, i.e., we have – up to multiplication by constants – $a = f$, $b = g$, and $v = 1$. Let $L = a\Delta + a - E^{-1}b$. Then $\delta_L = d - 2$ and Corollary 9.6 implies that $\deg u = \delta_L = d - 2 \geq 0$ if a solution u of the key equation (10.1) exists. (If $d < 2$ or $d \notin \mathbb{N}$, then Corollary 9.6 implies that (10.1) is unsolvable.)

We note that (10.1) reads $Lu = 1$. If $u = u_{d-2}x^{d-2} + \cdots + u_1x + u_0 \in F[x]$, then the coefficients of x^d and x^{d-1} in $Lu = a \cdot Eu - E^{-1}b \cdot u$ cancel, and hence L maps the $(d-1)$-dimensional vector space $P_{d-1} \subseteq F[x]$ of all polynomials of degree less than $d-1$ to itself. Since f and g are shift-coprime, so are a and $E^{-1}b$, whence (10.2) is unsolvable. By Lemma 9.1, it follows that L is injective, and hence the restriction of L to P_{d-1} is an automorphism of the vector space P_{d-1}. This implies that the key equation has a unique solution u, and Corollary 9.6 implies that $\deg u = \delta_L = d - 2$. $\square$

If we regard d as an indeterminate, then $\delta_L = d - 2 \notin \mathbb{N}$, we are in the first case of Corollary 9.6, and $\deg v - \deg L = -1$ implies that (10.1) is unsolvable.

We note that the proof of Theorem 10.7 hinges upon the fact that $\deg f = \deg g = 2$. If $f = x$ and $g = x + d$, then f and g are not shift-coprime, and if $f = x^k + f_{k-2}x^{k-2} + \cdots$ and $g = x^k + dx^{k-1} + g_{k-2}x^{k-2} + \cdots$ for some $k > 2$, then we can only show that the linear operator L maps polynomials of degree less than $d - 1$ to polynomials of degree less than $d + k - 3 > d - 1$.

Example 10.8. *Let $t \in \mathbb{N}_{\geq 2}$, and define the hypergeometric expression q by*

$$q = 2^{4x}\left(\frac{x+t}{x}\right)^{-2}\left(\frac{2x+2t}{x+t}\right)^{-2}.$$

Its term ratio is

$$\frac{Eq}{q} = \frac{(x+1)^2}{\left(x + \dfrac{2t+1}{2}\right)^2}.$$

If we let $d = 2t + 1$, then

$$\frac{f}{g} = \frac{(x+1)^2}{\left(x + \dfrac{d}{2}\right)^2} = \frac{x^2 + 2x + 1}{x^2 + dx + \dfrac{d^2}{4}},$$

so that f and g are shift-coprime since d is odd. Theorem 10.7 implies that there is a unique polynomial $u \in F[x]$ of degree $d - 4 = 2t - 3$ solving (10.1) for $a = f$, $b = g$, and $v = 1$, and thus the hypergeometric expression $p = u \cdot E^{-1}b \cdot q$ satisfies $\Delta p = q$ and $\sum_{0 \leq k < n} q(k) = p(n) - p(0)$ for all $n \in \mathbb{N}$.

For example, when $t = 6$ we obtain

$$u = \frac{4}{281302875}(4x+7)(8388608x^8 + 117440512x^7 + 658767872x^6$$

$$+1881800704x^5 + 2862755840x^4 + 2179846144x^3$$

$$+648167040x^2 + 504000x - 496125)\,,$$

$$\Delta p \;=\; \Delta\Big(\frac{(4x^2 + 44x + 121)(4x + 7)}{281302875}(8388608x^8 + 117440512x^7$$
$$+\; 658767872x^6 + 1881800704x^5 + 2862755840x^4$$
$$+\; 2179846144x^3 + 648167040x^2 + 504000x - 496125)q\Big) = q \;.$$

Example 10.9. *This is a variant of an example from Lisoněk, Paule & Strehl (1993). We let*

$$q = \frac{\big((-5/2)^{\overline{x+1}}\big)^2}{(-1/3)^{\overline{x+1}}(-2/3)^{\overline{x+1}}} \;,$$

where $c^{\overline{x+1}} = \Gamma(c + x + 1)/\Gamma(c)$ *for all* $c, x \in \mathbb{C}$. *In particular, we have* $c^{\overline{x+1}} = c(c + 1) \cdots (c + x)$ *for* $x \in \mathbb{N}$. *The term ratio is*

$$\frac{f}{g} = \frac{\big(x - \frac{5}{2}\big)^2}{\big(x - \frac{1}{3}\big)\big(x - \frac{2}{3}\big)} = \frac{x^2 - 5x + \frac{25}{4}}{x^2 - x + \frac{2}{9}} \;,$$

and f *and* g *are shift-coprime. Theorem 10.7 with* $d = 4$ *says that (10.1) has a unique polynomial solution* $u \in F[x]$ *of degree* $d - 2 = 2$ *for* $a = f$, $b = g$, *and* $v = 1$. *In fact, we have* $u = 4(2592x^2 - 12888x + 15985)/11025$ *and*

$$\Delta\Big(\frac{4}{11025}(2592x^2 - 12888x + 15985)\big(x - \frac{4}{3}\big)\big(x - \frac{5}{3}\big)q\Big) = q \;.$$

The following theorem shows that there are also cases where the denominator v in Gosper's algorithm 10.1 has high degree.

Theorem 10.10. *Let* F *be a field of characteristic zero,* $c \in F$ *and* $c \neq 1$, $d \in \mathbb{N}$, *and* $q = (x - 1)^{\underline{d}}c^x$. *Then* q *has term ratio*

$$\frac{Eq}{q} = \frac{cx}{x - d} \;,$$

$a = c$, $b = 1$, *and* $v = (x - 1)^{\underline{d}}$ *is a Gosper-Petkovšek form of* $(Eq)/q$, *and the key equation (10.1) has a unique solution* $u \in F[x]$ *of degree* d.

Proof. We have

$$\frac{cx}{x - d} = \frac{c}{1} \cdot \frac{x^{\underline{d}}}{(x - 1)^{\underline{d}}} \;,$$

and a, b, v is a Gosper-Petkovšek form as claimed. The key equation (10.1) reads

$$c \cdot Eu - u = (x - 1)^{\underline{d}} \;.$$

Since $c \neq 1$, the linear difference operator $L = cE - 1 = c\Delta + c - 1$ maps polynomials of degree k to polynomials of degree k, for all $k \in \mathbb{N}$. Thus L is bijective on $F[x]$, and the last claim follows. □

Example 10.11. *For $c = 2$ and $d = 10$, we obtain*

$$u = x^{10} - 75x^9 + 2580x^8 - 54270x^7 + 785253x^6 - 8316315x^5 + 66478670x^4$$
$$-401800380x^3 + 1770720696x^2 - 5140078560x + 7428153600 \,.$$

If we let $p = u \cdot E^{-1}b \cdot q/v = u2^x$, then

$$\Delta p = (2\,Eu - u)2^x = v2^x = q \,.$$

In the following, we give some high degree examples in the differential case. Here, we consider the key equation

$$Lu = b \cdot Du + (a + Db)u = v \,, \tag{10.4}$$

which we solve in step 2 of Algorithm 10.4, the continuous variant of Gosper's algorithm.

The analog of (10.3) is to take $f = -dx + f_0$ and $g = x^2 + g_1 x + g_0$ in $F[x]$, with $f_0, g_1, g_0 \in F$ and $d \in \mathbb{N}_{\geq 2}$.

Theorem 10.12. *Let F be a field of characteristic zero, $f = -dx + f_0$ and $g = x^2 + g_1 x + g_0$ in $F[x]$, with $f_0, g_1, g_0 \in F$ and $d \in \mathbb{N}_{\geq 2}$, and let $a, b, v \in F[x]$ be a GP'-form of f/g. If $\gcd(g, f - ig') = 1$ for all integers $i \in \mathbb{Z}$, then we have $a = f$, $b = g$, and $v = 1$, and (10.4) has a unique polynomial solution $u \in F[x]$ of degree $d - 2$.*

Proof. The proof parallels the proof of Theorem 10.7. After performing a suitable shift of variable if necessary, we may assume that $g_1 = 0$. The assumption that $\gcd(g, f - ig') = 1$ for all $i \in \mathbb{Z}$ implies that the resultant $\mathrm{res}_x(g, f - yg')$ has no integer roots, and hence Algorithm 8.28 returns the GP'-form $a = f$, $b = g$, and $v = 1$. Let $L = bD + a + b'$. Then $\delta_L = d - 2$, and Corollary 9.6 says that every solution $u \in F[x]$ of the key equation (10.4) has $\deg u = \delta_L = d - 2 \geq 0$. As in the proof of Theorem 10.7, we find that L maps the $(d - 1)$-dimensional vector space P_{d-1} of all polynomials of degree less than $d - 1$ to itself. The assumptions imply that $\gcd(b, -a - b' - ib') = 1$ for all $i \in \mathbb{Z}$, and Corollary 8.30 implies that the homogeneous equation $Lh = 0$ has no polynomial solution. It follows that L is injective on $F[x]$ and hence a linear F-automorphism of P_{d-1}. This in turn implies that the key equation has a unique solution u, and Corollary 9.6 says that $\deg u = \delta_L = d - 2$. $\square$

As in the difference case, the differential equation is unsolvable if d is an indeterminate.

Example 10.13. *Let $d \in \mathbb{N}_{\geq 3}$ be odd and $q = (x^2 + 1)^{-d/2}$. Then*

$$Dq = -\frac{d}{2}(x^2 + 1)^{-(d+2)/2} \cdot 2x = \frac{-dx}{x^2 + 1}q \,,$$

so that $g = x^2 + 1$ and $f = -dx$. Since d is odd, we have $\gcd(g, f - ig') = \gcd(x^2 + 1, -(d + 2i)x) = 1$ for all $i \in \mathbb{Z}$. Obviously q is not rational, and

Theorem 10.12 implies that the differential equation (10.4) has a unique solution $u \in F[x]$ of degree $d - 2$ for $b = g = x^2 + 1$, $a = f = -dx$, and $v = 1$, and hence $p = ubq = u \cdot (x^2 + 1)^{-(d-2)/2}$ satisfies $Dp = q$.

For example, if $d = 11$ then

$$u = \frac{x}{315}(128x^8 + 576x^6 + 1008x^4 + 840x^2 + 315)$$

and

$$Dp = D\left(\frac{x(128x^8 + 576x^6 + 1008x^4 + 840x^2 + 315)}{315(x^2 + 1)^{9/2}}\right) = \frac{1}{(x^2 + 1)^{11/2}} .$$

We may write the integrand as $q = (x^2 + 1)^{-(d+1)/2}\sqrt{x^2 + 1}$, and the degree of the "rational part" $(x^2 + 1)^{-(d+1)/2}$ seems to determine the degree of u. A similar remark applies to Example 10.8, where we may split off a factorial power.

The example illustrates that the "frontier" between rational and non-rational hyperexponential problems is thin: for even d, we obtain a rational integrand, which has no hyperexponential integral.

The next result, which is analogous to Theorem 10.10, is well-known (see, e.g, Kaltofen 1984). It implies that there are non-rational hyperexponential inputs for which the continuous analog of Gosper's algorithm returns a solution with a denominator of arbitrarily high degree.

Theorem 10.14. *Let F be a field of characteristic zero, $c \in F$ nonzero, $d \in \mathbb{N}$, and $q = x^d e^{cx}$. Then q has logarithmic derivative*

$$\frac{Dq}{q} = \frac{cx + d}{x} = \frac{c}{1} + \frac{d}{x} ,$$

$(a, b, v) = (c, 1, x^d)$ is a GP'-form of $(Dq)/q$, and the key equation (10.4) has a unique solution $u \in F[x]$ of degree d.

Proof. The key equation (10.4) reads $u' + cu = x^d$. The linear operator $L = D + c$ maps a polynomial of degree k to a polynomial of degree k, for all $k \in \mathbb{N}$, and hence there is a unique polynomial $u \in F[x]$ of degree d with $Lu = x^d$. □

Example 10.15. *For $c = 1$ and $d = 10$, we have*

$$u = x^{10} - 10x^9 + 90x^8 - 720x^7 + 5040x^6 - 30240x^5 + 151200x^4$$

$$-604800x^3 + 1814400x^2 - 3628800x + 3628800 .$$

Thus for $p = ubq/v = ue^x$, we have $Dp = (u' + u)e^x = ve^x = q$, or equivalently, $\int x^{10}e^x = ue^x$.

In the following theorem, we exhibit a subclass of the class of hyperexponential terms over $\mathbb{Q}(x)$ for which the values e and δ from Theorem 10.5 can be expressed in terms of degrees of the input polynomials. A slightly weaker result was proven by Kaltofen (1984). For a rational function $f/g \in \mathbb{Q}(x)$, where $f, g \in \mathbb{Z}[x]$ are nonzero and coprime, we let $\deg(f/g) = \deg f - \deg g$, $d(f/g) = \max\{\deg f, \deg g\}$, and $\mathrm{lc}(f/g) = \mathrm{lc}(f)/\mathrm{lc}(g)$.

Theorem 10.16. *Let* $c \in \mathbb{C}$ *be an arbitrary constant,* $m \in \mathbb{N}_{\geq 1}$, *and* $w, z \in \mathbb{Q}(x)$ *with* $\deg w \neq 0$.

(i) $q = c \cdot \exp(w) z^{1/m}$ *is hyperexponential over* $\mathbb{Q}(x)$, *with logarithmic derivative*

$$\frac{q'}{q} = w' + \frac{z'}{mz} = \frac{f}{g} \, ,$$

where $f, g \in \mathbb{Z}[x]$ *are coprime in* $\mathbb{Q}[x]$.

(ii) *All roots of the resultant* $r = \operatorname{res}_x(g, f - yg') \in \mathbb{Z}[y]$ *are in* $\frac{1}{m}\mathbb{Z}$ *and absolutely less than* $d(z)/m$.

(iii) *Let* $M = gD + f$. *If* $\deg w > 0$, *then* $\delta_M = \infty$. *Otherwise, if* $\deg w < 0$, *then* $\delta_M = -(\deg z)/m$.

The restriction that $\deg w$ be nonzero is not severe: If $\deg w = 0$, then there is a constant $v \in \mathbb{Q}$ and a rational function $w^* \in \mathbb{Q}(x)$ with $\deg w^* < 0$ such that $\exp(w) = \exp(v + w^*) = \exp(v)\exp(w^*)$, and we can incorporate the constant $\exp(v)$ into c.

Proof. Statement (i) follows from $q'/q = (\exp(w))'/\exp(w) + (z^{1/m})'/z^{1/m}$. For the proof of (ii), we note that the complex roots of r are the residues of f/g at its simple poles, by the remark following Fact 7.24. Since w' is the derivative of a rational function, its poles have order at least two, and hence the simple poles of f/g are the simple poles of z'/mz. Lemma 8.22 (viii) implies that

$$\operatorname{Res}_p\left(\frac{z'}{mz}\right) = \frac{1}{m}\operatorname{Res}_p\left(\frac{z'}{z}\right) = \frac{v_p(z)}{m}$$

for all irreducible polynomials p in $\mathbb{Q}[x]$, and (ii) follows from $|v_p(z)| \leq d(z)$.

We have $\deg z' \leq \deg z - 1$, and hence $\deg(z'/mz) \leq -1$, with equality if and only if $\deg z \neq 0$. The assumption $\deg w \neq 0$ implies that $\deg w' = \deg w - 1 \neq -1$. Thus $\deg f - \deg g = \deg(f/g) \leq \max\{\deg w - 1, \deg(z'/mz)\}$, with equality if $\deg w > 0$. In this case, we have $\deg f \geq \deg g$ and $\delta_M = \infty$. Now assume that $\deg w < 0$. Then $\deg f \leq \deg g - 1$, with equality if and only if $\deg z \neq 0$. If $\deg z = 0$, then $\deg f < \deg g - 1$ and $\delta_M = 0$. If $\deg z \neq 0$, then $\delta_M = -\operatorname{lc}(f/g)$. Now $\deg w' < -1 = \deg(z'/mz)$ implies that $\operatorname{lc}(f/g) = \operatorname{lc}(z'/mz) = \operatorname{lc}(z')/m\operatorname{lc}(z) = (\deg z)/m$, and the claim (iii) follows. □

If a hyperexponential term is given as in Theorem 10.16, then Algorithm 10.4 runs in polynomial time in the degrees and the coefficient lengths of the numerators and denominators of w and z, which, however, may be exponentially large in the degrees and the coefficients lengths of f and g, as Examples 10.14 and 10.13 show.

Koepf (1998) shows in Lemma 10.1 that any hyperexponential term over $\mathbb{C}(x)$ is of the form

$$q = \exp(w) \prod_{1 \leq j \leq k} (x - x_j)^{\alpha_j} \, ,$$

with $w \in \mathbb{C}(x)$ and $x_j, \alpha_j \in \mathbb{C}$ for all j. Thus any hyperexponential term over $\mathbb{Q}(x)$ is of the form

$$q = c \cdot \exp(w) \prod_{1 \le j \le k} (x - x_j)^{\alpha_j} \, ,$$

with $c \in \mathbb{C}$, $w \in \mathbb{Q}(x)$, and algebraic numbers x_j, α_j for all j such that the product on the right is invariant under the Galois group of $\overline{\mathbb{Q}}$ over $\mathbb{Q}$. The class from Theorem 10.16 is a proper subclass of all hyperexponential terms over $\mathbb{Q}(x)$. For example, $q = (x - \sqrt{2})^{\sqrt{2}}(x + \sqrt{2})^{-\sqrt{2}}$ is hyperexponential over $\mathbb{Q}(x)$, with logarithmic derivative $q'/q = 4/(x^2 - 2)$, but not of the form as stated in the theorem: the residues of q'/q at $\sqrt{2}$ and $-\sqrt{2}$ are $\sqrt{2}$ and $-\sqrt{2}$, respectively.

References

С. А. Абрамов (1971), О суммировании рациональных функций. Журнал вычислительной Математики и математической Физики **11**(4), 1071–1075. S. A. ABRAMOV, On the summation of rational functions, *U.S.S.R. Computational Mathematics and Mathematical Physics* **11**(4), 324–330.

С. А. Абрамов (1975), Рациональная компонента решения линейного рекуррентного соотношения первого порядка с рациональной правой частью. Журнал вычислительной Математики и математической Физики **15**(4), 1035–1039. S. A. ABRAMOV, The rational component of the solution of a first-order linear recurrence relation with rational right side, *U.S.S.R. Computational Mathematics and Mathematical Physics* **15**(4), 216–221.

С. А. Абрамов (1989a), Задачи компьютерной алгебры, связанные с поиском полиномиальных решений линейных дифференциальных и разностных уравнений. Вестник Московского Университета. Серия 15. Вычислительная Математика и Кибернетика **3**, 56–60. S. A. ABRAMOV, Problems of computer algebra involved in the search for polynomial solutions of linear differential and difference equations, *Moscow University Computational Mathematics and Cybernetics* **3**, 63–68.

С. А. Абрамов (1989b), Рациональные решения линейных дифференциальных и разностных уравнений с полиномиальными коэффициентами. Журнал вычислительной Математики и математической Физики **29**(11), 1611–1620. S. A. ABRAMOV, Rational solutions of linear differential and difference equations with polynomial coefficients, *U.S.S.R. Computational Mathematics and Mathematical Physics* **29**(6), 7–12.

S. A. ABRAMOV (1995a), Rational solutions of linear difference and *q*-difference equations with polynomial coefficients. In *Proceedings of the 1995 International Symposium on Symbolic and Algebraic Computation ISSAC '95,* Montreal, Canada, ed. A. H. M. LEVELT, ACM Press, 285–289.

S. A. ABRAMOV (1995b), Indefinite sums of rational functions. In *Proceedings of the 1995 International Symposium on Symbolic and Algebraic Computation ISSAC '95,* Montreal, Canada, ed. A. H. M. LEVELT, ACM Press, 303–308.

S. A. ABRAMOV (2002), Applicability of Zeilberger's Algorithm to Hypergeometric Terms. In *Proceedings of the 2002 International Symposium on Symbolic and Algebraic Computation ISSAC2002,* Lille, France, ed. TEO MORA, ACM Press, 1–7.

SERGEI A. ABRAMOV, MANUEL BRONSTEIN, and MARKO PETKOVŠEK (1995), On Polynomial Solutions of Linear Operator Equations. In *Proceedings of the 1995 International Symposium on Symbolic and Algebraic Computation ISSAC '95,* Montreal, Canada, ed. A. H. M. LEVELT, ACM Press, 290–296.

SERGEI A. ABRAMOV and MARK VAN HOEIJ (1999), Integration of solutions of linear functional equations. *Integral Transforms and Special Functions* **8**(1–2), 3–12.

S. A. ABRAMOV and K. YU. KVANSENKO [K. YU. KVASHENKO] (1991), Fast Algorithms to Search for the Rational Solutions of Linear Differential Equations with Polynomial

Coefficients. In *Proceedings of the 1991 International Symposium on Symbolic and Algebraic Computation ISSAC '91,* Bonn, Germany, ed. STEPHEN M. WATT, ACM Press, 267–270.

S. A. ABRAMOV and H. Q. LE (2002), A criterion for the applicability of Zeilberger's algorithm to rational functions. *Discrete Mathematics* **259**, 1–17.

S. A. ABRAMOV and M. PETKOVŠEK (2001), Canonical Representations of Hypergeometric Terms. In *Formal Power Series and Algebraic Combinatorics (FPSAC01),* Tempe AZ, eds. H. BARZELO and V. WELKER, 1–10.

S. A. ABRAMOV and M. PETKOVŠEK (2002a), Rational Normal Forms and Minimal Decompositions of Hypergeometric Terms. *Journal of Symbolic Computation* **33**(5), 521–543. Special Issue Computer Algebra: Selected Papers from ISSAC 2001, Guest Editor: G. Villard.

S. A. ABRAMOV and M. PETKOVŠEK (2002b), On the structure of multivariate hypergeometric terms. *Advances in Applied Mathematics* **29**(3), 386–411.

ALFRED V. AHO, JOHN E. HOPCROFT, and JEFFREY D. ULLMAN (1974), *The Design and Analysis of Computer Algorithms.* Addison-Wesley, Reading MA.

A. V. AHO, K. STEIGLITZ, and J. D. ULLMAN (1975), Evaluating polynomials at fixed sets of points. *SIAM Journal on Computing* **4**, 533–539.

GERT ALMKVIST and DORON ZEILBERGER (1990), The Method of Differentiating under the Integral Sign. *Journal of Symbolic Computation* **10**, 571–591.

E. BACH, J. DRISCOLL, and J. SHALLIT (1993), Factor refinement. *Journal of Algorithms* **15**, 199–222.

STANISŁAW BALCERZYK and TADEUSZ JÓZEFIAK (1989), *Commutative Rings: Dimension, Multiplicity and Homological Methods.* Mathematics and its applications, Ellis Horwood Limited, Chichester, UK.

M. A. BARKATOU (1999), Rational Solutions of Matrix Difference Equations. Problem of Equivalence and Factorization. In *Proceedings of the 1999 International Symposium on Symbolic and Algebraic Computation ISSAC '99,* Vancouver, Canada, ed. SAM DOOLEY, ACM Press.

ANDREJ BAUER and MARKO PETKOVŠEK (1999), Multibasic and Mixed Hypergeometric Gosper-Type Algorithms. *Journal of Symbolic Computation* **28**, 711–736.

LAURENT BERNARDIN (1999), *Factorization of Multivariate Polynomials over Finite Fields.* PhD thesis, ETH Zürich.

JOANNES BERNOULLIUS [JOHANN BERNOULLI] (1703), Problema exhibitum. *Acta eruditorum,* 26–31.

D. BINI and V. Y. PAN (1994), *Polynomial and matrix computations,* vol. 1. Birkhäuser Verlag.

LENORE BLUM, FELIPE CUCKER, MICHAEL SHUB, and STEVE SMALE (1998), *Complexity and Real Computation.* Springer-Verlag, New York.

LENORE BLUM, MIKE SHUB, and STEVE SMALE (1989), On a theory of computation and complexity over the real Numbers: NP-completeness, recursive functions, and universal machines. *Bulletin (New Series) of the American Mathematical Society* **21**(1), 1–46.

GEORGE BOOLE (1860), *Calculus of finite differences.* Chelsea Publishing Co., New York. 5th edition 1970.

A. BORODIN and R. MOENCK (1974), Fast Modular Transforms. *Journal of Computer and System Sciences* **8**(3), 366–386.

A. BORODIN and I. MUNRO (1975), *The Computational Complexity of Algebraic and Numeric Problems.* Theory of computation series **1**, American Elsevier Publishing Company, New York.

R. P. BRENT and H. T. KUNG (1978), Fast Algorithms for Manipulating Formal Power Series. *Journal of the ACM* **25**(4), 581–595.

MANUEL BRONSTEIN (1990), The Transcendental Risch Differential Equation. *Journal of Symbolic Computation* **9**, 49–60.

MANUEL BRONSTEIN (1991), The Risch Differential Equation on an Algebraic Curve. In *Proceedings of the 1991 International Symposium on Symbolic and Algebraic Computation ISSAC '91,* Bonn, Germany, ed. STEPHEN M. WATT, ACM Press, 241–246.

MANUEL BRONSTEIN (1992), On solutions of linear ordinary differential equations in their coefficient field. *Journal of Symbolic Computation* **13**, 413–439.

MANUEL BRONSTEIN (1997), *Symbolic Integration I—Transcendental Functions.* Algorithms and Computation in Mathematics **1**, Springer-Verlag, Berlin Heidelberg.

MANUEL BRONSTEIN (2000), On Solutions of Linear Ordinary Difference Equations in their Coefficient Field. *Journal of Symbolic Computation* **29**, 841–877.

MANUEL BRONSTEIN and ANNE FREDET (1999), Solving Linear Ordinary Differential Equations over $C(x, e^{\int f(x)dx})$. In *Proceedings of the 1999 International Symposium on Symbolic and Algebraic Computation ISSAC '99,* Vancouver, Canada, ed. SAM DOOLEY, ACM Press, 173–180.

MANUEL BRONSTEIN and MARKO PETKOVŠEK (1994), On Ore rings, linear operators, and factorization. *Programming and Computer Software* **20**(1), 14–26.

MANUEL BRONSTEIN and MARKO PETKOVŠEK (1996), An introduction to pseudo-linear algebra. *Theoretical Computer Science* **157**(1), 3–33.

W. S. BROWN (1971), On Euclid's Algorithm and the Computation of Polynomial Greatest Common Divisors. *Journal of the ACM* **18**(4), 478–504.

W. S. BROWN and J. F. TRAUB (1971), On Euclid's Algorithm and the Theory of Subresultants. *Journal of the ACM* **18**(4), 505–514.

V. CHVÁTAL (1979), The tail of the hypergeometric distribution. *Discrete Mathematics* **25**, 285–287.

FRÉDÉRIC CHYZAK (1998a), *Fonctions holonomes en calcul formel.* PhD thesis, École Polytechnique, Paris.

FRÉDÉRIC CHYZAK (1998b), Gröbner Bases, Symbolic Summation and Symbolic Integration. In *Gröbner Bases and Applications*, eds. BRUNO BUCHBERGER and FRANZ WINKLER. London Mathematical Society Lecture Note Series **251**, Cambridge University Press, Cambridge, UK, 32–60.

FRÉDÉRIC CHYZAK (2000), An extension of Zeilberger's fast algorithm to general holonomic functions. *Discrete Mathematics* **217**, 115–134.

FRÉDÉRIC CHYZAK and BRUNO SALVY (1998), Non-commutative Elimination in Ore Algebras Proves Multivariate Identities. *Journal of Symbolic Computation* **26**(2), 187–227.

RICHARD M. COHN (1965), *Difference algebra.* Interscience Tracts in Pure and Applied Mathematics **17**, Interscience Publishers, New York – London – Sydney.

G. E. COLLINS (1966), Polynomial remainder sequences and determinants. *The American Mathematical Monthly* **73**, 708–712.

GEORGE E. COLLINS (1967), Subresultants and Reduced Polynomial Remainder Sequences. *Journal of the ACM* **14**(1), 128–142.

GABRIEL CRAMER (1750), *Introduction a l'analyse des lignes courbes algébriques.* Frères Cramer & Cl. Philibert, Genève.

F. CUCKER, M. KARPINSKI, P. KOIRAN, T. LICKTEIG, and K. WERTHER (1995), On real Turing machines that toss coins. In *Proceedings of the Twenty-seventh Annual ACM Symposium on the Theory of Computing,* Las Vegas NV, ACM Press, 335–342.

J. H. DAVENPORT (1986), The Risch differential equation problem. *SIAM Journal on Computing* **15**(4), 903–918.

GEMA M. DIAZ-TOCA and LAUREANO GONZALES-VEGA (2001), Squarefree Decomposition of Univariate Polynomials Depending on a Parameter. Application to the Integration of Parametric Rational Functions. *Journal of Symbolic Computation* **32**(3), 191–209.

JOHN D. DIXON (1982), Exact Solution of Linear Equations Using P-Adic Expansions. *Numerische Mathematik* **40**, 137–141.

SHALOSH B. EKHAD (1990), A Very Short Proof of Dixon's Theorem. *Journal of Combinatorial Theory, Series A* **54**, 141–142.

SHALOSH B. EKHAD and SOL TRE (1990), A Purely Verification Proof of the First Rogers–Ramanujan Identity. *Journal of Combinatorial Theory, Series A* **54**, 309–311.

WINFRIED FAKLER (1999), *Algebraische Algorithmen zur Lösung von linearen Differential-gleichungen*. MuPAD Reports, B. G. Teubner, Stuttgart, Leipzig.

A. FRÖHLICH (1967), Local Fields. In *Algebraic Number Theory*, eds. J. W. S. CASSELS and A. FRÖHLICH, chapter I, 1–41. Academic Press, London and New York.

SHUHONG GAO (2001), On the Deterministic Complexity of Factoring Polynomials. *Journal of Symbolic Computation* **31**(1–2), 19–36.

JOACHIM VON ZUR GATHEN (1984), Hensel and Newton methods in valuation rings. *Mathematics of Computation* **42**(166), 637–661.

JOACHIM VON ZUR GATHEN (1990), Functional Decomposition of Polynomials: the Tame Case. *Journal of Symbolic Computation* **9**, 281–299.

JOACHIM VON ZUR GATHEN and JÜRGEN GERHARD (1997), Fast Algorithms for Taylor Shifts and Certain Difference Equations. In *Proceedings of the 1997 International Symposium on Symbolic and Algebraic Computation ISSAC '97*, Maui HI, ed. WOLFGANG W. KÜCHLIN, ACM Press, 40–47.

JOACHIM VON ZUR GATHEN and JÜRGEN GERHARD (1999), *Modern Computer Algebra*. Cambridge University Press, Cambridge, UK. Second edition 2003.

JOACHIM VON ZUR GATHEN and SILKE HARTLIEB (1998), Factoring Modular Polynomials. *Journal of Symbolic Computation* **26**(5), 583–606.

JOACHIM VON ZUR GATHEN and THOMAS LÜCKING (2002), Subresultants revisited. *Theoretical Computer Science* **297**, 199–239.

CARL FRIEDRICH GAUSS (1863), Disquisitiones generales de congruentiis. Analysis residuorum caput octavum. In *Werke* II, Handschriftlicher Nachlass, ed. R. DEDEKIND, 212–240. Königliche Gesellschaft der Wissenschaften, Göttingen. Reprinted by Georg Olms Verlag, Hildesheim New York, 1973.

KEITH GEDDES, HA LE, and ZIMING LI (2004), Differential Rational Normal Forms and a Reduction Algorithm for Hyperexponential Functions. In *Proceedings of the 2004 International Symposium on Symbolic and Algebraic Computation ISSAC2004*, Santander, Spain, ed. JAIME GUTIERREZ, ACM Press, 183–190.

JÜRGEN GERHARD (1998), High degree solutions of low degree equations. In *Proceedings of the 1998 International Symposium on Symbolic and Algebraic Computation ISSAC '98*, Rostock, Germany, ed. OLIVER GLOOR, ACM Press, 284–289.

JÜRGEN GERHARD (2000), Modular algorithms for polynomial basis conversion and greatest factorial factorization. In *Proceedings of the Seventh Rhine Workshop on Computer Algebra RWCA'00*, ed. T. MULDERS, 125–141.

JÜRGEN GERHARD (2001), Fast Modular Algorithms for Squarefree Factorization and Hermite Integration. *Applicable Algebra in Engineering, Communication and Computing* **11**(3), 203–226.

J. GERHARD, M. GIESBRECHT, A. STORJOHANN, and E. V. ZIMA (2003), Shiftless Decomposition and Polynomial-time Rational Summation. In *Proceedings of the 2003 International Symposium on Symbolic and Algebraic Computation ISSAC2003*, Philadelphia PA, ed. J. R. SENDRA, ACM Press, 119–126.

P. GIANNI and B. TRAGER (1996), Square-Free Algorithms in Positive Characteristic. *Applicable Algebra in Engineering, Communication and Computing* **7**, 1–14.

MARK WILLIAM GIESBRECHT (1993), *Nearly Optimal Algorithms for Canonical Matrix Forms*. PhD thesis, Department of Computer Science, University of Toronto. Technical Report 268/93.

R. WILLIAM GOSPER, JR. (1978), Decision procedure for indefinite hypergeometric summation. *Proceedings of the National Academy of Sciences of the USA* **75**(1), 40–42.

R. GÖTTFERT (1994), An acceleration of the Niederreiter factorization algorithm in characteristic 2. *Mathematics of Computation* **62**(206), 831–839.

R. L. GRAHAM, D. E. KNUTH, and O. PATASHNIK (1994), *Concrete Mathematics*. Addison-Wesley, Reading MA, 2nd edition. First edition 1989.

JOOS HEINTZ and JACQUES MORGENSTERN (1993), On the Intrinsic Complexity of Elimination Theory. *Journal of Complexity* **9**, 471–498.

PETER A. HENDRIKS and MICHAEL F. SINGER (1999), Solving Difference Equations in Finite Terms. *Journal of Symbolic Computation* **27**, 239–259.

C. HERMITE (1872), Sur l'intégration des fractions rationnelles. *Annales de Mathématiques,* 2ème *série* **11**, 145–148.

MARK VAN HOEIJ (1998), Rational Solutions of Linear Difference Equations. In *Proceedings of the 1998 International Symposium on Symbolic and Algebraic Computation ISSAC '98,* Rostock, Germany, ed. OLIVER GLOOR, ACM Press, 120–123.

MARK VAN HOEIJ (1999), Finite singularities and hypergeometric solutions of linear recurrence equations. *Journal of Pure and Applied Algebra* **139**, 109–131.

MARK VAN HOEIJ (2002), Factoring polynomials and the knapsack problem. *Journal of Number Theory* **95**(2), 167–189.

JORIS VAN DER HOEVEN (1997), Lazy Multiplication of Formal Power Series. In *Proceedings of the 1997 International Symposium on Symbolic and Algebraic Computation ISSAC '97,* Maui HI, ed. WOLFGANG W. KÜCHLIN, ACM Press, 17–20.

W. G. HORNER (1819), A new method of solving numerical equations of all orders by continuous approximation. *Philosophical Transactions of the Royal Society of London* **109**, 308–335.

ELLIS HOROWITZ (1969), *Algorithms for Symbolic Integration of Rational Functions.* Ph.D. Dissertation, University of Wisconsin, Madison WI.

ELLIS HOROWITZ (1971), Algorithms for partial fraction decomposition and rational function integration. In *Proceedings 2nd ACM Symposium on Symbolic and Algebraic Manipulation,* Los Angeles CA, ed. S. R. PETRICK, ACM Press, 441–457.

E. L. INCE (1926), *Ordinary differential equations.* Longmans, Green and Co. Reprinted 1956 by Dover Publications, Inc., New York.

N. JACOBSON (1937), Pseudo-linear transformations. *Annals of Mathematics* **28**(2), 484–507.

LIEUWE DE JONG and JAN VAN LEEUWEN (1975), An improved bound on the number of multiplications and divisions necessary to evaluate a polynomial and all its derivatives. *SIGACT News* **7**(3), 32–34.

CHARLES JORDAN (1939), *Calculus of finite differences.* Röttig and Romwalter, Sopron, Hungary. 3rd edition, Chelsea Publishing Company, New York, 1965.

ERICH KALTOFEN (1984), A Note on the Risch Differential Equation. In *Proceedings of EUROSAM '84,* Cambridge, UK, ed. JOHN FITCH. Lecture Notes in Computer Science **174**, Springer-Verlag, Berlin, 359–366.

E. KAMKE (1977), *Differentialgleichungen — Lösungsmethoden und Lösungen,* vol. I. Gewöhnliche Differentialgleichungen. B. G. Teubner, Stuttgart, 9th edition.

IRVING KAPLANSKY (1957), *An introduction to differential algebra.* Actualités scientifiques et industrielles **1251**, Hermann, Paris, second edition. Publication de l'Institut de Mathematique de l'Université de Nancago, V.

A. Карацуба и Ю. Офман (1962), Умножение многозначных чисел на автоматах. Доклады Академии Наук СССР **145**, 293–294. A. KARATSUBA and YU. OFMAN, Multiplication of multidigit numbers on automata, Soviet Physics–Doklady **7** (1963), 595–596.

MICHAEL KARR (1981), Summation in Finite Terms. *Journal of the ACM* **28**(2), 305–350.

MICHAEL KARR (1985), Theory of Summation in Finite Terms. *Journal of Symbolic Computation* **1**, 303–315.

DONALD E. KNUTH (1998), *The Art of Computer Programming, vol. 2, Seminumerical Algorithms.* Addison-Wesley, Reading MA, 3rd edition. First edition 1969.

WOLFRAM KOEPF (1995), Algorithms for m-fold Hypergeometric Summation. *Journal of Symbolic Computation* **20**, 399–417.

WOLFRAM KOEPF (1998), *Hypergeometric Summation.* Advanced Lectures in Mathematics, Friedrich Vieweg & Sohn, Braunschweig / Wiesbaden.

ELLIS R. KOLCHIN (1973), *Differential Algebra and Algebraic Groups.* Pure and Applied Mathematics **54**, Academic Press, New York.

L. KRONECKER (1882), Grundzüge einer arithmetischen Theorie der algebraischen Grössen. *Journal für die reine und angewandte Mathematik* **92**, 1–122. *Werke*, Zweiter Band, ed. K. HENSEL, Leipzig, 1897, 237–387. Reprint by Chelsea Publishing Co., New York, 1968.

E. LANDAU (1905), Sur quelques théorèmes de M. Petrovitch relatifs aux zéros des fonctions analytiques. *Bulletin de la Société Mathématique de France* **33**, 251–261.

DANIEL LAUER (2000), *Effiziente Algorithmen zur Berechnung von Resultanten und Subresultanten.* Berichte aus der Informatik, Shaker Verlag, Aachen. PhD thesis, University of Bonn, Germany.

D. LAZARD and R. RIOBOO (1990), Integration of Rational Functions: Rational Computation of the Logarithmic Part. *Journal of Symbolic Computation* **9**, 113–115.

HA LE (2002), Simplification of Definite Sums of Rational Functions by Creative Symmetrizing Method. In *Proceedings of the 2002 International Symposium on Symbolic and Algebraic Computation ISSAC2002,* Lille, France, ed. TEO MORA, ACM Press, 161–167.

HA QUANG LE (2003a), *Algorithms for the construction of minimal telescopers.* PhD thesis, University of Waterloo, Canada.

H. Q. LE (2003b), A direct algorithm to construct the minimal Z-pairs for rational functions. *Advances in Applied Mathematics* **30**, 137–159.

ARJEN K. LENSTRA, HENDRIK W. LENSTRA, JR., and L. LOVÁSZ (1982), Factoring Polynomials with Rational Coefficients. *Mathematische Annalen* **261**, 515–534.

ZIMING LI and ISTVÁN NEMES (1997), A Modular Algorithm for Computing Greatest Common Right Divisors of Ore Polynomials. In *Proceedings of the 1997 International Symposium on Symbolic and Algebraic Computation ISSAC '97,* Maui HI, ed. WOLFGANG W. KÜCHLIN, ACM Press, 282–289.

JOHN D. LIPSON (1981), *Elements of Algebra and Algebraic Computing.* Addison-Wesley, Reading MA.

PETR LISONĚK, PETER PAULE, and VOLKER STREHL (1993), Improvement of the degree setting in Gosper's algorithm. *Journal of Symbolic Computation* **16**, 243–258.

RÜDIGER LOOS (1983), Computing rational zeroes of integral polynomials by p-adic expansion. *SIAM Journal on Computing* **12**(2), 286–293.

JESPER LÜTZEN (1990), *Joseph Liouville 1809–1882: Master of Pure and Applied Mathematics.* Studies in the History of Mathematics and Physical Sciences **15**, Springer-Verlag, New York, Berlin, Heidelberg.

D. MACK (1975), *On rational integration.* Technical Report UCP-38, Department of Computer Science, University of Utah.

K. MAHLER (1960), An application of Jensen's formula to polynomials. *Mathematika* **7**, 98–100.

YIU-KWONG MAN (1993), On Computing Closed Forms for Indefinite Summations. *Journal of Symbolic Computation* **16**, 355–376.

YIU-KWONG MAN and FRANCIS J. WRIGHT (1994), Fast Polynomial Dispersion Computation and its Application to Indefinite Summation. In *Proceedings of the 1994 International Symposium on Symbolic and Algebraic Computation ISSAC '94,* Oxford, UK, eds. J. VON ZUR GATHEN and M. GIESBRECHT, ACM Press, 175–180.

MICHAEL T. MCCLELLAN (1973), The Exact Solution of Systems of Linear Equations With Polynomial Coefficients. *Journal of the Association for Computing Machinery* **20**(4), 563–588.

MAURICE MIGNOTTE (1989), *Mathématiques pour le calcul formel*. Presses Universitaires de France, Paris. English translation: *Mathematics for Computer Algebra*, Springer-Verlag, New York, 1992.

GARY L. MILLER (1976), Riemann's Hypothesis and Tests for Primality. *Journal of Computer and System Sciences* **13**, 300–317.

ROBERT MOENCK (1977), On computing closed forms for summation. In *Proceedings of the 1977 MACSYMA Users Conference*, Berkeley CA, NASA, Washington DC, 225–236.

ROBERT T. MOENCK and JOHN H. CARTER (1979), Approximate algorithms to derive exact solutions to systems of linear equations. In *Proceedings of EUROSAM '79*, Marseille, France, ed. EDWARD W. NG. Lecture Notes in Computer Science **72**, Springer-Verlag, Berlin Heidelberg New York, 65–73.

THOM MULDERS (1997), A note on subresultants and the Lazard/Rioboo/Trager formula in rational function integration. *Journal of Symbolic Computation* **24**(1), 45–50.

THOM MULDERS and ARNE STORJOHANN (1999), Diophantine Linear System Solving. In *Proceedings of the 1999 International Symposium on Symbolic and Algebraic Computation ISSAC '99*, Vancouver, Canada, ed. SAM DOOLEY, ACM Press, 181–188.

DAVID R. MUSSER (1971), *Algorithms for Polynomial Factorization*. PhD thesis, Computer Science Department, University of Wisconsin. Technical Report #134, 174 pages.

ISAAC NEWTON (1691/92), De quadratura Curvarum. The revised and augmented treatise. Unpublished manuscript. In: DEREK T. WHITESIDE, *The mathematical papers of Isaac Newton* vol. VII, Cambridge University Press, Cambridge, UK, 1976, pp. 48–128.

HARALD NIEDERREITER (1993a), A New Efficient Factorization Algorithm for Polynomials over Small Finite Fields. *Applicable Algebra in Engineering, Communication and Computing* **4**, 81–87.

H. NIEDERREITER (1993b), Factorization of Polynomials and Some Linear Algebra Problems over Finite Fields. *Linear Algebra and its Applications* **192**, 301–328.

HARALD NIEDERREITER (1994a), Factoring polynomials over finite fields using differential equations and normal bases. *Mathematics of Computation* **62**(206), 819–830.

HARALD NIEDERREITER (1994b), New deterministic factorization algorithms for polynomials over finite fields. In *Finite fields: theory, applications and algorithms*, eds. G. L. MULLEN and P. J.-S. SHIUE. Contemporary Mathematics **168**, American Mathematical Society, 251–268.

HARALD NIEDERREITER and RAINER GÖTTFERT (1993), Factorization of Polynomials over Finite Fields and Characteristic Sequences. *Journal of Symbolic Computation* **16**, 401–412.

HARALD NIEDERREITER and RAINER GÖTTFERT (1995), On a new factorization algorithm for polynomials over finite fields. *Mathematics of Computation* **64**(209), 347–353.

ÖYSTEIN ORE (1932a), Formale Theorie der linearen Differentialgleichungen. (Erster Teil). *Journal für die reine und angewandte Mathematik* **167**, 221–234.

OYSTEIN ORE (1932b), Formale Theorie der linearen Differentialgleichungen. (Zweiter Teil). *Journal für die reine und angewandte Mathematik* **168**, 233–252.

OYSTEIN ORE (1933), Theory of non-commutative polynomials. *Annals of Mathematics* **34**(22), 480–508.

M. OSTROGRADSKY (1845), De l'intégration des fractions rationnelles. *Bulletin de la classe physico-mathématique de l'Académie Impériale des Sciences de Saint-Pétersbourg* **4**(82/83), 145–167.

M. S. PATERSON and L. STOCKMEYER (1973), On the Number of nonscalar Multiplications necessary to evaluate Polynomials. *SIAM Journal on Computing* **2**, 60–66.

PETER PAULE (1994), Short and Easy Computer Proofs of the Rogers-Ramanujan Identities and of Identities of Similar Type. *The Electronic Journal of Combinatorics* **1**(# R10). 9 pages.

PETER PAULE (1995), Greatest Factorial Factorization and Symbolic Summation. *Journal of Symbolic Computation* **20**, 235–268.

PETER PAULE and VOLKER STREHL (1995), Symbolic summation — some recent developments. In *Computer Algebra in Science and Engineering,* Bielefeld, Germany, August 1994, eds. J. FLEISCHER, J. GRABMEIER, F. W. HEHL, and W. KÜCHLIN, World Scientific, Singapore, 138–162.

MARKO PETKOVŠEK (1992), Hypergeometric solutions of linear recurrences with polynomial coefficients. *Journal of Symbolic Computation* **14**, 243–264.

MARKO PETKOVŠEK (1994), A generalization of Gosper's algorithm. *Discrete Mathematics* **134**, 125–131.

MARKO PETKOVŠEK and BRUNO SALVY (1993), Finding All Hypergeometric Solutions of Linear Differential Equations. In *Proceedings of the 1993 International Symposium on Symbolic and Algebraic Computation ISSAC '93,* Kiev, ed. MANUEL BRONSTEIN, ACM Press, 27–33.

MARKO PETKOVŠEK, HERBERT S. WILF, and DORON ZEILBERGER (1996), *A=B.* A K Peters, Wellesley MA.

ECKHARD PFLÜGEL (1997), An Algorithm for Computing Exponential Solutions of First Order Linear Differential Systems. In *Proceedings of the 1997 International Symposium on Symbolic and Algebraic Computation ISSAC '97,* Maui HI, ed. WOLFGANG W. KÜCHLIN, ACM Press, 164–171.

ROBERTO PIRASTU (1996), *On Combinatorial Identities: Symbolic Summation and Umbral Calculus.* PhD thesis, Johannes Kepler Universität, Linz.

R. PIRASTU and V. STREHL (1995), Rational Summation and Gosper-Petkovšek Representation. *Journal of Symbolic Computation* **20**, 617–635.

J. M. POLLARD (1971), The Fast Fourier Transform in a Finite Field. *Mathematics of Computation* **25**(114), 365–374.

MARIUS VAN DER PUT and MICHAEL F. SINGER (1997), *Galois Theory of Difference Equations.* Lecture Notes in Mathematics **1666**, Springer-Verlag, Berlin, Heidelberg.

MARIUS VAN DER PUT and MICHAEL F. SINGER (2003), *Galois Theory of Linear Differential Equations.* Grundlehren der mathematischen Wissenschaften **238**, Springer-Verlag.

MICHAEL O. RABIN (1976), Probabilistic algorithms. In *Algorithms and Complexity,* ed. J. F. TRAUB, Academic Press, New York, 21–39.

MICHAEL O. RABIN (1980), Probabilistic Algorithms for Testing Primality. *Journal of Number Theory* **12**, 128–138.

DANIEL REISCHERT (1997), Asymptotically Fast Computation of Subresultants. In *Proceedings of the 1997 International Symposium on Symbolic and Algebraic Computation ISSAC '97,* Maui HI, ed. WOLFGANG W. KÜCHLIN, ACM Press, 233–240.

DANIEL RICHARDSON (1968), Some undecidable problems involving elementary functions of a real variable. *Journal of Symbolic Logic* **33**(4), 514–520.

ROBERT H. RISCH (1969), The problem of integration in finite terms. *Transactions of the American Mathematical Society* **139**, 167–189.

ROBERT H. RISCH (1970), The solution of the problem of integration in finite terms. *Bulletin of the American Mathematical Society* **76**(3), 605–608.

JOSEPH FELS RITT (1950), *Differential Algebra.* AMS Colloquium Publications **XXXIII**, American Mathematical Society, Providence RI. Reprint by Dover Publications, Inc., New York, 1966.

STEVEN ROMAN (1984), *The umbral calculus.* Pure and applied mathematics **111**, Academic Press, Orlando FL.

S. ROMAN and G.-C. ROTA (1978), The umbral calculus. *Advances in Mathematics* **27**, 95–188.

J. BARKLEY ROSSER and LOWELL SCHOENFELD (1962), Approximate formulas for some functions of prime numbers. *Illinois Journal of Mathematics* **6**, 64–94.

G.-C. ROTA (1975), *Finite Operator Calculus.* Academic Press, New York.

MICHAEL ROTHSTEIN (1976), *Aspects of symbolic integration and simplification of exponential and primitive functions.* PhD thesis, University of Wisconsin-Madison.

MICHAEL ROTHSTEIN (1977), A new algorithm for the integration of exponential and logarithmic functions. In *Proceedings of the 1977 MACSYMA Users Conference,* Berkeley CA, NASA, Washington DC, 263–274.

G. SCHEJA and U. STORCH (1980), *Lehrbuch der Algebra, Teil 1.* B. G. Teubner, Stuttgart.

CARSTEN SCHNEIDER (2001), *Symbolic summation in difference fields.* PhD thesis, Johannes Kepler Universität, Linz, Austria.

ARNOLD SCHÖNHAGE (1984), Factorization of univariate integer polynomials by Diophantine approximation and an improved basis reduction algorithm. In *Proceedings of the 11th International Colloquium on Automata, Languages and Programming ICALP 1984,* Antwerp, Belgium. Lecture Notes in Computer Science **172**, Springer-Verlag, 436–447.

ARNOLD SCHÖNHAGE (1998), Multiplicative Complexity of Taylor Shifts and a New Twist of the Substitution Method. In *Proceedings of the 39th Annual IEEE Symposium on Foundations of Computer Science,* Palo Alto CA, IEEE Computer Society Press, 212–215.

ARNOLD SCHÖNHAGE, ANDREAS F. W. GROTEFELD, and EKKEHART VETTER (1994), *Fast Algorithms – A Multitape Turing Machine Implementation.* BI Wissenschaftsverlag, Mannheim.

A. SCHÖNHAGE and V. STRASSEN (1971), Schnelle Multiplikation großer Zahlen. *Computing* **7**, 281–292.

M. SHAW and J. F. TRAUB (1974), On the Number of Multiplications for the evaluation of a Polynomial and some of its derivatives. *Journal of the ACM* **21**, 161–167.

VICTOR SHOUP (1995), A New Polynomial Factorization Algorithm and its Implementation. *Journal of Symbolic Computation* **20**, 363–397.

MICHAEL SHUB and STEVE SMALE (1995), On the intractability of Hilbert's Nullstellensatz and an algebraic version of "NP $\neq$ P?". *Duke Mathematical Journal* **81**, 47–54.

MICHAEL F. SINGER (1991), Liouvillian Solutions of Linear Differential Equations with Liouvillian Coefficients. *Journal of Symbolic Computation* **11**, 251–273.

R. SOLOVAY and V. STRASSEN (1977), A fast Monte-Carlo test for primality. *SIAM Journal on Computing* **6**(1), 84–85. Erratum in **7** (1978), p. 118.

VOLKER STRASSEN (1973), Die Berechnungskomplexität von elementarsymmetrischen Funktionen und von Interpolationskoeffizienten. *Numerische Mathematik* **20**, 238–251.

V. STRASSEN (1974), Some Results in Algebraic Complexity Theory. In *Proceedings of the International Congress of Mathematicians 1974,* Vancouver, 497–501.

VOLKER STRASSEN (1976), Einige Resultate über Berechnungskomplexität. *Jahresberichte der DMV* **78**, 1–8.

V. STRASSEN (1983), The computational complexity of continued fractions. *SIAM Journal on Computing* **12**(1), 1–27.

ROBERT GEORGE TOBEY (1967), *Algorithms for Antidifferentiation—Rational Functions.* PhD thesis, Harvard University, Cambridge MA.

BARRY M. TRAGER (1976), Algebraic Factoring and Rational Function Integration. In *Proceedings of the 1976 ACM Symposium on Symbolic and Algebraic Computation SYMSAC '76,* Yorktown Heights NY, ed. R. D. JENKS, ACM Press, 219–226.

XINMAO WANG and VICTOR Y. PAN (2003), Acceleration of Euclidean algorithm and rational number reconstruction. *SIAM Journal on Computing* **32**(2), 548–556.

HERBERT S. WILF and DORON ZEILBERGER (1990), Rational functions certify combinatorial identities. *Journal of the American Mathematical Society* **3**(1), 147–158.

HERBERT S. WILF and DORON ZEILBERGER (1992), An algorithmic proof theory for hypergeometric (ordinary and "*q*") multisum/integral identities. *Inventiones Mathematicae* **108**, 575–633.

DAVID Y. Y. YUN (1976), On Square-free Decomposition Algorithms. In *Proceedings of the 1976 ACM Symposium on Symbolic and Algebraic Computation SYMSAC '76,* Yorktown Heights NY, ed. R. D. JENKS, ACM Press, 26–35.

DAVID Y. Y. YUN (1977a), Fast algorithm for rational function integration. In *Information Processing 77—Proceedings of the IFIP Congress 77*, ed. B. GILCHRIST, North-Holland, Amsterdam, 493–498.

DAVID Y. Y. YUN (1977b), On the equivalence of polynomial gcd and squarefree factorization problems. In *Proceedings of the 1977 MACSYMA Users Conference,* Berkeley CA, NASA, Washington DC, 65–70.

HANS ZASSENHAUS (1969), On Hensel Factorization, I. *Journal of Number Theory* **1**, 291–311.

DORON ZEILBERGER (1990a), A holonomic systems approach to special function identities. *Journal of Computational and Applied Mathematics* **32**, 321–368.

DORON ZEILBERGER (1990b), A fast algorithm for proving terminating hypergeometric identities. *Discrete Mathematics* **80**, 207–211.

DORON ZEILBERGER (1991), The Method of Creative Telescoping. *Journal of Symbolic Computation* **11**, 195–204.

Index

Lecture Notes in Computer Science

For information about Vols. 1–3223

please contact your bookseller or Springer

Vol. 3274: R. Guerraoui (Ed.), Distributed Computing. XIII, 465 pages. 2004.

Vol. 3273: T. Baar, A. Strohmeier, A. Moreira, S.J. Mellor (Eds.), <<UML>> 2004 - The Unified Modelling Language. XIII, 454 pages. 2004.

Vol. 3271: J. Vicente, D. Hutchison (Eds.), Management of Multimedia Networks and Services. XIII, 335 pages. 2004.

Vol. 3270: M. Jeckle, R. Kowalczyk, P. Braun (Eds.), Grid Services Engineering and Management. X, 165 pages. 2004.

Vol. 3269: J. Lopez, S. Qing, E. Okamoto (Eds.), Information and Communications Security. XI, 564 pages. 2004.

Vol. 3266: J. Solé-Pareta, M. Smirnov, P.V. Mieghem, J. Domingo-Pascual, E. Monteiro, P. Reichl, B. Stiller, R.J. Gibbens (Eds.), Quality of Service in the Emerging Networking Panorama. XVI, 390 pages. 2004.

Vol. 3265: R.E. Frederking, K.B. Taylor (Eds.), Machine Translation: From Real Users to Research. XI, 392 pages. 2004. (Subseries LNAI).

Vol. 3264: G. Paliouras, Y. Sakakibara (Eds.), Grammatical Inference: Algorithms and Applications. XI, 291 pages. 2004. (Subseries LNAI).

Vol. 3263: M. Weske, P. Liggesmeyer (Eds.), Object-Oriented and Internet-Based Technologies. XII, 239 pages. 2004.

Vol. 3262: M.M. Freire, P. Chemouil, P. Lorenz, A. Gravey (Eds.), Universal Multiservice Networks. XIII, 556 pages. 2004.

Vol. 3261: T. Yakhno (Ed.), Advances in Information Systems. XIV, 617 pages. 2004.

Vol. 3260: I.G.M.M. Niemegeers, S.H. de Groot (Eds.), Personal Wireless Communications. XIV, 478 pages. 2004.

Vol. 3258: M. Wallace (Ed.), Principles and Practice of Constraint Programming – CP 2004. XVII, 822 pages. 2004.

Vol. 3257: E. Motta, N.R. Shadbolt, A. Stutt, N. Gibbins (Eds.), Engineering Knowledge in the Age of the Semantic Web. XVII, 517 pages. 2004. (Subseries LNAI).

Vol. 3256: H. Ehrig, G. Engels, F. Parisi-Presicce, G. Rozenberg (Eds.), Graph Transformations. XII, 451 pages. 2004.

Vol. 3255: A. Benczúr, J. Demetrovics, G. Gottlob (Eds.), Advances in Databases and Information Systems. XI, 423 pages. 2004.

Vol. 3254: E. Macii, V. Paliouras, O. Koufopavlou (Eds.), Integrated Circuit and System Design. XVI, 910 pages. 2004.

Vol. 3253: Y. Lakhnech, S. Yovine (Eds.), Formal Techniques, Modelling and Analysis of Timed and Fault-Tolerant Systems. X, 397 pages. 2004.

Vol. 3252: H. Jin, Y. Pan, N. Xiao, J. Sun (Eds.), Grid and Cooperative Computing - GCC 2004 Workshops. XVIII, 785 pages. 2004.

Vol. 3251: H. Jin, Y. Pan, N. Xiao, J. Sun (Eds.), Grid and Cooperative Computing - GCC 2004. XXII, 1025 pages. 2004.

Vol. 3250: L.-J. (LJ) Zhang, M. Jeckle (Eds.), Web Services. X, 301 pages. 2004.

Vol. 3249: B. Buchberger, J.A. Campbell (Eds.), Artificial Intelligence and Symbolic Computation. X, 285 pages. 2004. (Subseries LNAI).

Vol. 3246: A. Apostolico, M. Melucci (Eds.), String Processing and Information Retrieval. XIV, 332 pages. 2004.

Vol. 3245: E. Suzuki, S. Arikawa (Eds.), Discovery Science. XIV, 430 pages. 2004. (Subseries LNAI).

Vol. 3244: S. Ben-David, J. Case, A. Maruoka (Eds.), Algorithmic Learning Theory. XIV, 505 pages. 2004. (Subseries LNAI).

Vol. 3243: S. Leonardi (Ed.), Algorithms and Models for the Web-Graph. VIII, 189 pages. 2004.

Vol. 3242: X. Yao, E. Burke, J.A. Lozano, J. Smith, J.J. Merelo-Guervós, J.A. Bullinaria, J. Rowe, P. Tiño, A. Kabán, H.-P. Schwefel (Eds.), Parallel Problem Solving from Nature - PPSN VIII. XX, 1185 pages. 2004.

Vol. 3241: D. Kranzlmüller, P. Kacsuk, J.J. Dongarra (Eds.), Recent Advances in Parallel Virtual Machine and Message Passing Interface. XIII, 452 pages. 2004.

Vol. 3240: I. Jonassen, J. Kim (Eds.), Algorithms in Bioinformatics. IX, 476 pages. 2004. (Subseries LNBI).

Vol. 3239: G. Nicosia, V. Cutello, P.J. Bentley, J. Timmis (Eds.), Artificial Immune Systems. XII, 444 pages. 2004.

Vol. 3238: S. Biundo, T. Frühwirth, G. Palm (Eds.), KI 2004: Advances in Artificial Intelligence. XI, 467 pages. 2004. (Subseries LNAI).

Vol. 3237: C. Peters, J. Gonzalo, M. Braschler, M. Kluck (Eds.), Comparative Evaluation of Multilingual Information Access Systems. XIV, 702 pages. 2004.

Vol. 3236: M. Núñez, Z. Maamar, F.L. Pelayo, K. Pousttchi, F. Rubio (Eds.), Applying Formal Methods: Testing, Performance, and M/E-Commerce. XI, 381 pages. 2004.

Vol. 3235: D. de Frutos-Escrig, M. Nunez (Eds.), Formal Techniques for Networked and Distributed Systems – FORTE 2004. X, 377 pages. 2004.

Vol. 3234: M.J. Egenhofer, C. Freksa, H.J. Miller (Eds.), Geographic Information Science. VIII, 345 pages. 2004.

Vol. 3233: K. Futatsugi, F. Mizoguchi, N. Yonezaki (Eds.), Software Security - Theories and Systems. X, 345 pages. 2004.

Vol. 3232: R. Heery, L. Lyon (Eds.), Research and Advanced Technology for Digital Libraries. XV, 528 pages. 2004.

Vol. 3231: H.-A. Jacobsen (Ed.), Middleware 2004. XV, 514 pages. 2004.

Vol. 3230: J.L. Vicedo, P. Martínez-Barco, R. Muñoz, M. Saiz Noeda (Eds.), Advances in Natural Language Processing. XII, 488 pages. 2004. (Subseries LNAI).

Vol. 3229: J.J. Alferes, J. Leite (Eds.), Logics in Artificial Intelligence. XIV, 744 pages. 2004. (Subseries LNAI).

Vol. 3226: M. Bouzeghoub, C. Goble, V. Kashyap, S. Spaccapietra (Eds.), Semantics of a Networked World. XIII, 326 pages. 2004.

Vol. 3225: K. Zhang, Y. Zheng (Eds.), Information Security. XII, 442 pages. 2004.

Vol. 3224: E. Jonsson, A. Valdes, M. Almgren (Eds.), Recent Advances in Intrusion Detection. XII, 315 pages. 2004.